AF400707

LABOR ECONOMICS
FROM A FREE
MARKET PERSPECTIVE

EMPLOYING THE UNEMPLOYABLE

LABOR ECONOMICS

FROM A FREE

MARKET PERSPECTIVE

EMPLOYING THE UNEMPLOYABLE

WALTER BLOCK

Loyola University New Orleans, USA

NEW JERSEY • LONDON • SINGAPORE • BEIJING • SHANGHAI • HONG KONG • TAIPEI • CHENNAI

Published by

World Scientific Publishing Co. Pte. Ltd.

5 Toh Tuck Link, Singapore 596224

USA office: 27 Warren Street, Suite 401-402, Hackensack, NJ 07601

UK office: 57 Shelton Street, Covent Garden, London WC2H 9HE

Library of Congress Cataloging-in-Publication Data
Labor economics / edited by Walter Block.
    p. cm.
    With the exception of chapters 5, 17, and 20 this book consists of previously
published articles of [the author's].
    Includes bibliographical references and index.
    ISBN-13: 978-981-270-568-6 (alk. paper)
    ISBN-10: 981-270-568-6 (alk. paper)
    1. Labor economics. 2. Wages. 3. Industrial relations. I. Block, Walter, 1941–
HD4901.L1153 2007
331--dc22

                                          2007036227

British Library Cataloguing-in-Publication Data
A catalogue record for this book is available from the British Library.

Printed in Singapore by World Scientific Printers

Contents

Acknowledgments

The author would like to thank the publishers and publications who granted reprint permissions for the following chapters:

Chapter 1
Block, Walter and Robert Lawson (2006). Promotion, Turnover and Preemptive Wage Offers. *Humanomics*, 22(3), 133–138. Reprinted with kind permission of Emerald Group Publishing Limited.

Chapter 2
Block, Walter, Jerry Dauterive and John Levendis (2007). Globalization and the Concept of Subsistence Wages. *Journal of Income Distribution*, 16(1), 74–88. Reprinted with kind permission of Ad Libros Publications, Inc.

Chapter 3
Block, Walter (1996). Labor Market Disputes: A Comment on Albert Rees' "Fairness in Wage Distribution". *Journal of Interdisciplinary Economics*, 7(3), 217–230. Repeated efforts to contact the original publisher for permission to reprint this article were made but no response was received at the time of printing.

Chapter 4
Block, Walter (1990). The DMVP-MVP Controversy: A Note. *Review of Austrian Economics*, 4, 199–207. Reprinted with kind permission of The Ludwig von Mises Institute.

Chapter 6
Block, Walter (1991). Labor Relations, Unions and Collective Bargaining: A Political Economic Analysis. *Journal of Social Political and Economic Studies*, 16(4), 1991, 477–507. Reprinted with kind permission of the Council for Social and Economic Studies.

Chapter 7
Evans, Jason and Walter Block (2002). Labor Union Policies: Gains or Pains? *Cross Cultural Management*, 9(1), 71–79. Reprinted with kind permission of Emerald Group Publishing Limited.

Chapter 8
Block, Walter (September 5, 2005). The Yellow Dog Contract: Bring It Back! *LewRockwell.com* Reprinted with kind permission of LewRockwell.com.

Chapter 9
Block, Walter (November 10, 2006). Unions: Social Benefactors or Gangs of Thugs?: Debate with Boyd Blundell. *LewRockwell.com* Reprinted with kind permission of LewRockwell.com.

Chapter 10
Block, Walter (1996). Comment on Richard B. Freeman's "Labor Markets and Institutions in Economic Development". *International Journal of Social Economics*, 23(1), 6–16. Reprinted with kind permission of Emerald Group Publishing Limited.

Chapter 11
Block, Walter (March 9, 2004). A Primer on Jobs and the Jobless. *Mises.org Daily Articles* Reprinted with kind permission of The Ludwig von Mises Institute.

Chapter 12
McCormick, Paul and Walter Block (2000). The Minimum Wage: Does it Really Help Workers? *Southern Connecticut State University Business Journal*, 15(2), 77–80. Reprinted with kind permission of the Southern Connecticut State University.

Chapter 13
Block, Walter (January 28, 2002). Delusions of Rising Wages. *New Orleans City Business*, p. 28. Reprinted with kind permission of the New Orleans Publishing Group.

Chapter 14
Block, Walter (2001). The Minimum Wage Once Again. *Journal for Economic Educators* (formerly known as the *Journal of The Tennessee Economics Association*), 3(2). Reprinted with kind permission of Tennessee Economics Association.

Chapter 15
Block, Walter (October 2000). Heritage Stumbles on Minimum Wage. *The Free Market*, 18(10). Reprinted with kind permission of The Ludwig von Mises Institute.

Chapter 16
Block, Walter and William Barnett II (2002). The Living Wage: What's Wrong. *The Freeman: Ideas on Liberty*, 52(12), 23–24. Reprinted with kind permission of the publisher, Foundation for Economic Education, Irvington-on-Hudson, NY 10533, www.fee.org, all rights reserved.

Chapter 17
Economic Policy Institute (2006). Hundreds of Economists Say: Raise the Minimum wage. Washington DC: EPI. Available at http://www.epi.org/content.cfm/minwagestmt2006 Reproduced with kind permission of the Economic Policy Institute (www.wpi.org)

Chapter 18
Block, Walter (1998). A Libertarian Case for Free Immigration. *Journal of Libertarian Studies: An Interdisciplinary Review*, 13(2), 167–186. Reprinted with kind permission of The Ludwig von Mises Institute.

Chapter 19
Block, Walter and Gene Callahan (2003). Is There a Right to Immigration? A Libertarian Perspective. *Human Rights Review* 5(1), 46–71. Reprinted with kind permission of Springer Science and Business Media.

Chapter 20
Gregory, Anthony and Walter Block. On Immigration: Reply to Hoppe. *Journal of Libertarian Studies: An Interdisciplinary Review,* forthcoming. Reprinted with kind permission of The Ludwig von Mises Institute.

Chapter 21
Block, Walter (2002). On Reparations to Blacks for Slavery. *Human Rights Review*, 3(4), 53–73. Reprinted with kind permission of Springer Science and Business Media.

Chapter 22
Block, Walter and Guillermo Yeatts (1999–2000). The Economics and Ethics of Land Reform: A Critique of the Pontifical Council for Justice and Peace's "Toward a Better Distribution of Land: The Challenge of Agrarian Reform". *Journal of Natural Resources and Environmental Law*, 15(1), 37–69. Reprinted with kind permission of The University of Kentucky.

Chapter 23
Block, Walter (January 1993). Comments on Thomason and Burton, Bruce and Atkins, Anderson and Meyer, and Green and Riddell. *The Journal of Labor Economics*, 11(1) Part 2, S305–S326. Reprinted with kind permission of the University of Chicago Press.

Chapter 24
Anderson, Gary M. and Walter Block (January 1993). Comment on Hum and Simpson. *The Journal of Labor Economics*, 11(1), Part 2, S348–S363. Reprinted with kind permission of the University of Chicago Press.

Chapter 25

Block, Walter, Peter Klein and Per Henrik Hansen (April 2007). The Division of Labor under Homogeneity: A Critique of Mises and Rothbard. *The American Journal of Economics and Sociology,* 66(2), 457–464. Reprinted with kind permission by Blackwell Publishing.

Chapter 26

McGee, Robert W. and Walter Block (1991). Academic Tenure: A Law and Economics Analysis. *Harvard Journal of Law and Public Policy*, 14(2), 545–563. Reprinted with kind permission from Harvard Society for Law & Public Policy, Inc.

Chapter 27

Block, Walter (2001). Comment on Canice Prendergast's "A Theory of 'Yesmen'" *Quarterly Journal of Austrian Economics*, 4(2), 61–68. Reprinted with kind permission of The Ludwig von Mises Institute.

Chapter 28

Block, Walter (2001). Cyberslacking, Business Ethics and Managerial Economics. *Journal of Business Ethics*, 33(3), 225–231. Reprinted with kind permission of Springer Science and Business Media.

Chapter 29

Block, Walter (2000). Paternalism in Agricultural Labor Contracts in the U.S. South: Implications for the Growth of the Welfare State. *Unisinos: Perspectiva Economica*, 35(112), 81–93. Reprinted with kind permission of UNISINOS — Universidade do Vale do Rio dos Sinos.

The author wishes to thank his co authors; he also includes himself on this list, for identification purposes:

Gary M. Anderson, Ph.D.
Professor Emeritus
Economics Department
California State University Northridge
18111 Nordhoff Street
Northridge, CA 91330-8245
hceco015@email.csun.edu

William Barnett II, Ph.D., J.D.
Chase Distinguished Professor of International Business and Professor of Economics
Joseph A. Butt, S. J. College of Business
Loyola University New Orleans
6363 St. Charles Ave.
New Orleans, LA 70118-6143
(504) 864-7950
wbarnett@loyno.edu

Walter Block, Ph.D.
Harold E. Wirth Eminent Scholar Endowed Chair and Professor of Economics
Joseph A. Butt, S.J. College of Business Administration
Loyola University New Orleans
6363 St. Charles Avenue, Box 15, Miller Hall 318
New Orleans, LA 70118
Tel: (504) 864-7934
Fax: (504) 864-7970
wblock@loyno.edu

Jerry Dauterive, Ph.D.
Associate Professor of Economics and Interim Dean
Joseph A. Butt, S. J. College of Business Administration
Loyola University New Orleans
6363 St. Charles Ave.
New Orleans, LA 70118
(504) 864 7979
dauteriv@loyno.edu

Jason Evans
B.A. Economics
The University of Central Arkansas
Conway, AR 72035

Anthony Gregory
Research Assistant
The Independent Institute
100 Swan Way, Oakland, CA 94621-1428
http://www.independent.org
Office: (510) 632-1366
Fax: (510) 568-6040
AGregory@independent.org
Anthony.Gregory@gmail.com
http://www.anthonygregory.com

Per Henrik Hansen
Denmark
perhenrikhansen@hotmail.com

Peter G. Klein, Ph.D.
Agricultural Economics Department
Contracting and Organizations Research Institute
University of Missouri - Columbia
143 Mumford Hall
Columbia, MO 65211
573-882-7008
573-884-6572 (fax) 573-882-3958
pklein@missouri.edu
www.ssu.missouri.edu/faculty/pklein

Robert A. Lawson, Ph.D.
Professor of Economics and George H. Moor Chair
Capital University
1 College and Main
Columbus, OH 43209-2394
614-236-6138
614-236-6540 (fax)
rlawson@capital.edu

John Levendis, Ph.D.
College of Business Administration
Loyola University New Orleans
6363 St. Charles Avenue, Box 15, Miller 321
New Orleans, LA 70118
Office: (504) 864-79341
Dept: (504) 864-7944
Fax: (504) 864-7970
jlevendi@loyno.edu

Paul McCormick, Esq.
B.A. Economics
College of the Holy Cross
Worcester, MA 01610
195 Gould Ave., North Caldwell, NJ 07006
paul_m_mccormick@yahoo.com

Robert McGee, Ph.D., J.D., C.P.A.
Director, Center for Accounting, Auditing & Tax Studies
School of Accounting
College of Business Administration ACII 124
Florida International University
3000 NE 151st Street
N. Miami, Florida 33181
bob414@hotmail.com

Guillermo M. Yeatts, Chairman
Fundación Atlas 1853
Alicia Moreau de Justo 740
Piso 3 Oficina 1 (C1107AAR)
Buenos Aires, Argentina
Teléfono Atlas 54-11-5235-3988
gyeatts@atlas.org.ar
www.guillermoyeatts.com.ar

Foreword

Walter Block is one of the great economists of our time. Immensely productive, he has made contributions to virtually all fields of economic analysis and also excelled as a public intellectual. These qualities have gained him admirers from all over the world. They are also palpable in the present collection of his major papers pertaining to labor economics.

Block's prose is crystal clear and simple. While this is a patent blessing for his readers, it is also a token of the thorough command he has of his subject. In the 17th century, the French writer Nicolas Boileau wrote on the art of poetry: "Things that are rightly understood can be stated clearly and in words that flow easily." Boileau's countrymen have always held that his adage applies not only to poetry, but to just any sort of writing, and in particular to the sciences. Abstruse writing is a sign of immaturity, if not incapacity. Plain language and a lucid style are the marks of a master. They are certainly the marks that we find on the following pages. Here, Walter Block deals with some of the most subtle problems of economic analysis, as well as with basic principles that should be known to anyone — but which unfortunately need to be restated to confront ever-recurring errors, for example, the error that minimum-wage laws help to increase the living standards of the population.

There is no need for us to comment on any of these writings. However, there is one overall trait that deserves to be mentioned. Block usually sets forth his ideas through a discussion of the ideas of others. In fact, he is one of the most eminent polemicists in economics today, successor to distinguished polemicists of previous times such as Eugen von Böhm-Bawerk and Murray N. Rothbard.

Polemics is the art of learned contention. While poor polemics is not much more than the academic equivalent of bickering, good polemics is one of the most efficient means — and certainly the most entertaining one — of promoting progress in scientific analysis. Good polemics builds bridges between thinkers. It creates common ground. Today, this is more necessary than ever in economics, because the discipline has fallen into an unprecedented state of fragmentation. There is no longer an uncontested body of doctrine providing common ground for a fruitful division of labor among economists; whatever remnants might still exist of such a doctrine dwindle by the day. There are, today, only groups or loosely overlapping networks of scholars, each of which explores the implications of a different set of — often fictitious — hypotheses. Good polemics can bring these individuals and groups together, because there is after all a common ground called the real world, a world structured by scarcity and thus subject to economic laws. Good polemics can build bridges to this world, as the papers of this volume show. It is no accident that their author is a former student of Gary Becker and a disciple of Murray Rothbard. He is a most competent critic of both the neoclassical and the Austrian approaches, as well as of related theories in political philosophy.

One danger of the polemical form of presentation is to convey a twisted image of the author. Polemicists often appear more hard-nosed and aggressive than they really are. This certainly holds true in the present case. Walter Block is a passionate man and a vigorous champion of truth and liberty. But having known and befriended him for more than 10 years, what struck me most in his character are gentleness, indulgence, humility, and care for others. If I may be allowed to say so, he *would be* a perfect Christian if he had the grace of the Faith. He *is* a distinguished economist, not least of all because he relentlessly builds bridges toward his colleagues, saving them from the seclusion — or shall we say confinement? — that results from the way economic research is practised today.

May this volume reach many readers! I doubt not it will inspire them.

Guido Hülsmann
Angers, France,
June 2007

Introduction

Labor accounts for some 70–75% of the GDP. Thus, any book addressing labor issues is potentially an important one. This book does just that, and from an economic point of view. However, it is not a textbook on labor economics. If you are looking for an explication of the backward-bending supply curve of labor, or for an explanation of the pattern of labor force participation, or for why labor accounts for some 70-75% of the GDP, you will have to look elsewhere.

Instead, *Labor Economics from a Free Market Perspective* is an ideological book and thus might serve as a secondary reading in a course in labor economics, or as one of several texts for an instructor who values this particular perspective. The phrase "ideological" has a bad press in some quarters, but etymologically it means, merely, that this volume takes a position[1] on ideas. And the position it takes may be read directly from the title of the book: It will look at numerous labor market issues from a vantage point of free enterprise or libertarianism. Its contention throughout is that if we as a society want to cure unemployment, raise real wages, and in other such ways improve this sector of our economy, we will base public policy on private property rights, the non-aggression principle and the law of free association. In the free and prosperous society everyone may act precisely as he pleases, provided, only, that he does not initiate violence against non aggressors.

With the exception of Chapters 5, 17, and 20, this book consists of previously published articles of mine. Therefore, gentle reader, you can expect some repetition. Hopefully, some other aspects of this collection will more than compensate for this, if you find that problematic. But I make no excuses for this possible shortcoming. My philosophy on this matter is that as long as an evil exists, I am going to keep hacking away at it in my writings. I feel as did Heinlein (1966, p. 25) when the Mooniacs kept throwing rocks at that military installation in Colorado[2] until it was utterly pulverized:

---

[1] The opinions expressed in this book are solely those of the author and do not necessarily reflect the opinions of the publisher.

[2] "If three bombings on three rotations of Terra did not do it, we might still be throwing rocks in '77 — till they ran out of interceptors... or till they destroyed us (far more likely). For a century North American Space Defense Command had been buried in a mountain south of Colorado Springs, Colorado, a city of no other importance. During Wet Firecracker War the Cheyenne Mountain took a direct hit; space defense command post survived — but not sundry deer, trees, most of city and some of top of mountain. What we were about to do should not kill anybody unless they stayed outside on that mountain despite three days' steady warnings. But North American Space Defense Command was to receive full Lunar treatment: twelve rock missiles on

there are some legislative enactments that are so truly offensive (unions, and union-spawned minimum wage laws surely come under this rubric) that they deserve severe approbation. As long as these institutions are still part of our economic lives, I will continue to toss intellectual hand grenades at them, again and again, and, yes, once again. This book represents, among other things, the negative manner with which I view these phenomena that prey on the weak and helpless, and yet, paradoxically and horrendously, are enthusiastically supported by their very victims.

I am encouraged in this behavior of mine by Mises (1976, Chapter 3), who said: "There cannot be too much of a correct theory."[3] As long as unions and minimum wage laws exist, I am going to keep writing about their evils. This is like tossing mortars over to the bad guys, the more the merrier, as far as I am concerned. I also obtain solace in from Hayek (1991, pp. 35–36) who says: "[I]n economics you can never establish a truth once and for all but have always to convince every generation anew." And again (Hayek, 1960, p. 1) "If old truths are to retain their hold on men's minds, they must be restated in the language and concepts of successive generations."[4]

With this introduction to the philosophy underlying this undertaking, let me offer you an overview of its contents. Perhaps the most important thing for an economic theory to accomplish is to explain prices. Accordingly, we start off Section I of this book with material devoted to the determination of the price of labor, or wages. To give the story away, compensation for employees is based on how hard and smart they work, that is, their productivity (subject to appropriate qualifications that appear in the text). This depends, in turn, on the quantity and quality of capital goods at their disposal, which is based on savings and technological improvements, which emanates, ultimately, from how free is the economy, and how consistent is its legal system with private property rights.

Chapter 1 argues that information asymmetries in labor markets do not amount to a type of market failure where workers are not paid according to their marginal productivities. In Chapter 2 we cast doubt on the notion that "subsistence wages" exist. Chapter 3 takes a mainstream author to task for his charge that the market is "unfair" in its wage determination. Chapter 4 constitutes one of the exceptions to the rule that wages tend toward marginal revenue productivity: the time dimension must also be taken into account.

Section II is devoted to organized labor and starts off by establishing the case for regarding unions as incompatible with the libertarian philosophy in Chapters 5–7. Chapter 8 defends the "yellow dog contract" hated by all advocates of criminal unionism.

---

first pass, then all we could spare on second rotation, and on third — and so on, until we ran out of steel casings, or were put out of action… or North American Directorate hollered quits." I owe this citation to Juan Fernando Carpio.

[3]I owe this citation to BK Marcus and Roderick Long.

[4]I owe this citation to Sudha Shenoy.

Chapter 9 offers a debate on this issue. Chapter 10 is a critique of a defender of organized labor.

In Section III the minimum wage law is eviscerated, at least to the best of my ability. Herein are some of my most apoplectic writings. Don't ask, but I am filled with loathing and disgust for this legislation, and every paragraph reeks with such feelings. In my view, the claim that the minimum wage creates unemployment for everyone with a productivity level below that set by this pernicious law is not a matter to be settled by empirical "evidence," although this point is illustrated again and again in econometric findings. Rather, it is a matter of pure logic, or praxeology; its denial is a veritable self-contradiction. Mainstream economists of a market orientation would disagree: For them, it is presumably a matter of empirical studies. But it is my contention that if you scratch a good neoclassical economist deeply enough, you will find a praxeologist, at least on this issue. How else to explain their vituperative reaction to Card and Krueger, economists who purport to show the "benefits" of minimum wage legislation to the unskilled? This is why I regard opinions on this issue as a sort of litmus test: Those who fail it, e.g., think that this law benefits poor workers, are not really economists at all, their prizes and PhDs in the dismal science to the contrary notwithstanding. Chapter 11 sets the stage with some introductory material. In Chapter 12 through 17, I make the case against this pernicious law and deal with several supporters of it.

Chapter 17 is one of those written specifically for this publication. In it I take to task sentence by sentence and even word by word, a petition signed by hundreds of economists to the effect that the minimum wage promotes the welfare of unskilled labor. Why such focus on this one document? Why do I reprint that petition, accompanied by all 650 signatories to it? Because this is a highly problematic document, and deserves all the negative publicity it can garner. When and if justice ever prevails in the profession of economics, all of its signatories will have to deal with the negative repercussions of their words. We have recalls for everything from cars to tires to toasters; why not for economists?[5]

Immigration, the subject of Section IV, is a traditional part of labor economics; it constitutes a contribution to the supply curve of labor. Let me summarize my position on the ethics of this issue. People have the right to be fussy, very fussy, about who they associate with. The law of free association is, after all, an integral part of the libertarian philosophy. Indeed, they have the right to be as selective in this regard as were the condominium owners in the 1990 movie *The Green Card*[6] when they were making inquiries about the bone fides of the character played by Andie McDowell. If, in their

---

[5]When the latter misbehaved, that great philosopher Carmela Soprano said about her daughter Meadow: "There has to be consequences" (http://www.tv.com/the-sopranos/toodle-fucking-oo/episode/26463/summary.html). Just so in this case.

[6]http://www.imdb.com/title/tt0099699/

(unanimous) view, no Jews are "one too many" (Chapter 18), that is entirely within their rights.

However, this analogy between private individuals and their right to discriminate to their heart's content on the one hand, and a country's immigration policy on the other, only holds true given a crucially important proviso: Every inch of the nation's territory must be *privately owned*. That is, *every* square inch of it. There can be no such thing as a public park, a public roadway, a public library, a public recreation center, a public concert hall, and so on. Nor can there be any unowned deserts or mountains or lakes or rivers or molehills. For if there are, there is nothing in the libertarian philosophy that would prevent newcomers, *immigrants most certainly included*, from homesteading them. Nor do I interpret publicly owned property as "really owned" by the citizens of the country. No, this is *stolen* property, and if the citizenry does not have the gumption to seize it back from their rulers and slave masters, then it is fair game for outsiders to do so. Libertarianism, in my view, presents a very stark alternative to the electorate: Embrace laissez-faire free enterprise in its entirety, that is, anarcho-capitalism (Rothbard, 1973, 1998), or suffer the slings and arrows, whatever they are, of a completely open border policy. Chapters 18–20 argue that the effects of open borders will not be all that negative, and take issue with several critics of this perspective.

In Section V we ask how redistributive justice applies to the case of reparations for past slavery (in economic parlance, this constitutes a theft of labor). I take the position that possession is properly "nine-tenths of the law" and thus that the burden of proof should rest with those who seek compensation for past thefts of labor (coercive slavery). However, that it is at least conceivable that this burden can be met, and, as there should be no statute of limitations of justice in the free society, there is indeed a case to be made for reparations for past slavery.

The subject of Section VI is fringe benefits. We cover worker's compensation and unemployment insurance in Chapters 23 and 24, respectively.

Last but certainly not least is Section VII, devoted to other topics in labor economics. Herein some of the most interesting chapters in the book may be found. In the paper on homogeneity (Chapter 25), my coauthors and I make the case that a division of labor is so powerful it can exist even under the very stringent conditions whereby all human and non-human resources are exactly alike.[7] Is academic tenure justified? My coauthor and I argue (in Chapter 26) that this academic practice constitutes, merely, a long-run contract, and that all such matters are best left to the market, after full privatization, of course, of all colleges, schools and universities. In Chapter 27 I debunk Prendergast's claim of a labor market "failure"; that employees have too much of an incentive to butter up their bosses. Yet another supposed labor market failure is that workers will slack off too much

---

[7]My second coauthor agrees with me that most of the credit for this article should go to our third coauthor, who was only a graduate student when he made the point that eventuated in this paper at a Mises Institute seminar, over the strong objections of several faculty members with many years of seniority.

in the computer age; this one, too, is given the back of my hand in Chapter 28. And Chapter 29 defends paternalism in labor markets, when it emanates from voluntary agreements.

I would like to thank Jordan Schneider for serving as my research assistant/editor during the compilation of the material herein. I also acknowledge a debt of gratitude to Cheong Chean Chian, Sandhya, and Teng Poh Hoon of World Scientific Publishing for editorial assistance. I greatly appreciate the reprint permissions journals in which these chapters have first appeared as articles. A heartfelt thank you to my coauthors: Gary M Anderson, William Barnett II, Gene Callahan, Jerry Dauterive, Jason Evans, Anthony Gregory, Per Henrik Hansen, Peter Klein, Robert Lawson, John Levendis, Paul McCormick, Robert W McGee and Guillermo Yeatts. This book could not have been written without their contributions. I offer a heartfelt thank you to Guido Hülsmann for writing the splendid foreword to this book.

References

Hayek, FA (1991). *The Trend of Economic Thinking: Essays on Political Economists and Economic History.* The Collected Works of FA Hayek, Vol. 3, WW Bartley III and S Kresge (ed.). London: Routledge.

Hayek, FA (1960). *The Constitution of Liberty.* Chicago: University of Chicago Press.

Heinlein, R (1966). *The Moon is a Harsh Mistress.* New York: Berkley Medallion

Mises, LV (1976). *Epistemological Problems of Economics.* (G Reisman, Trans.). (Original work published 1933). New York: New York University Press.

Nozick, R (1974). *Anarchy, State and Utopia.* New York: Basic Books.

Rothbard, MN (1973). *For a New Liberty.* New York: Macmillan.

Rothbard, MN (1998). *The Ethics of Liberty.* New Jersey: Humanities Press.

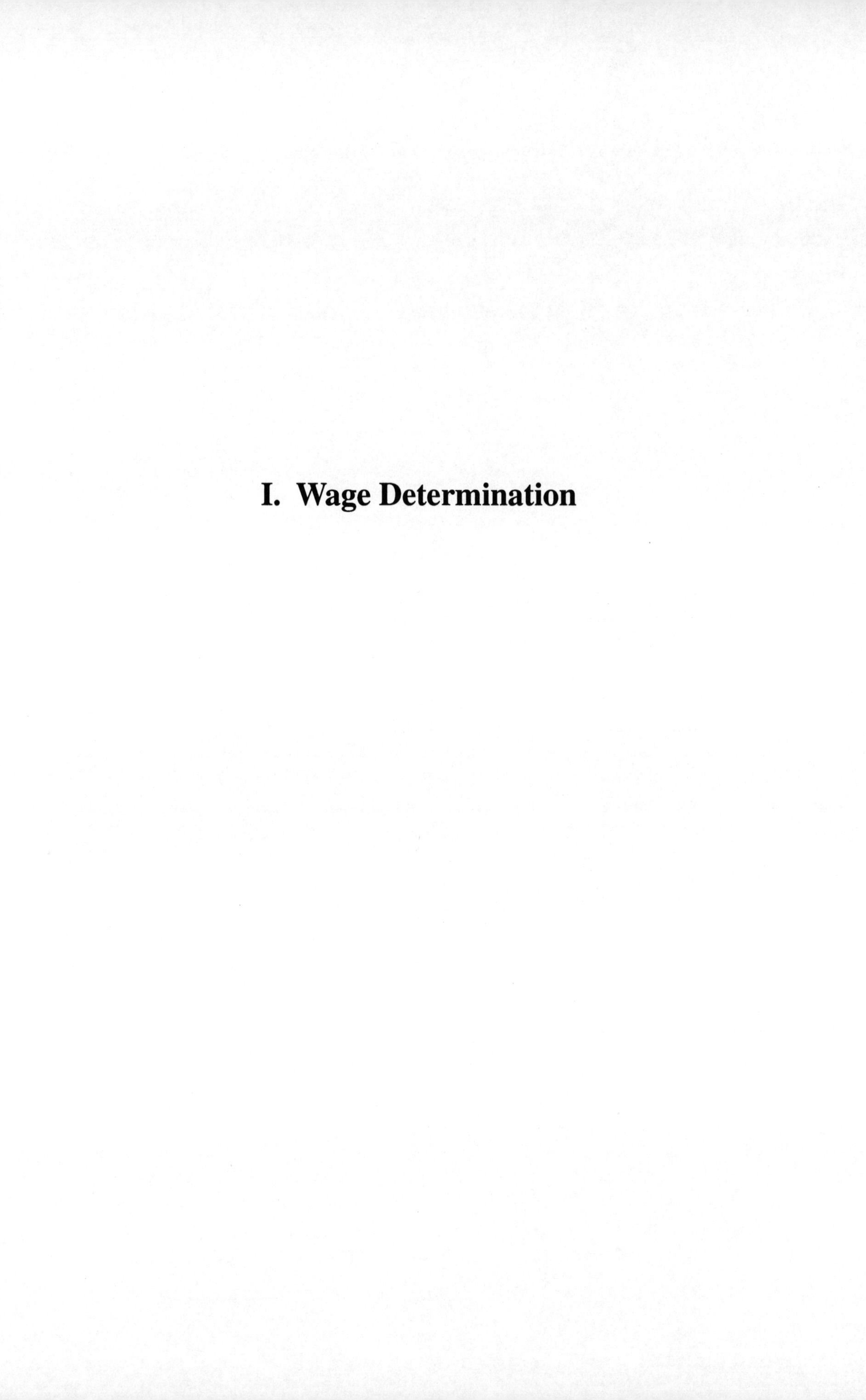

I. Wage Determination

The current issue and full text archive of this journal is available at
www.emeraldinsight.com/0828-8666.htm

Promotion, turnover and preemptive wage offers

Walter Block
*College of Business Administration, Loyola University,
New Orleans, Louisiana, USA, and*
Robert A. Lawson
*Department of Business Administration and Economics,
Capital University, Columbus, Ohio, USA*

Abstract

Purpose – The reason for writing the paper is to cast doubt on the claim that the informational asymmetries uncovered by Bernhardt and Scoones constitute a market failure.

Design/methodology/approach – The main method used is to quote these authors, and then critically comment upon their views. The theoretical scope of the paper is the premise that markets are efficient, effective and ethical.

Findings – It was found in the course of the work that Bernhardt and Scoones were in error in their contentions.

Research limitations/implications – Suggestions for future research; this paper is but the tip of the iceberg in terms of claims that markets perform badly, and are hence in need of governmental rectification. This entire literature cries out for more critical analysis.

Practical implications – The public policy and practical implications emanating from this paper is that laissez-faire capitalism is the best way to organize an economy.

Originality/value – What is new in the paper is that the work of Bernhardt and Scoones on informational asymmetries has not before been subject to critical analysis from the perspective of advocates of the free enterprise system. This is valuable, in that, when claims are subject to critique, they become more reliable. It is only through dialogue and debate of this sort that one can get that in closer to the truth.

Keywords Promotion, Information control, Employees, Jobs, Pay

Paper type Conceptual paper

Comment on "Promotion, turnover and preemptive wage offers" by Dan Bernhardt and David Scoones

Bernhardt and Scoones (1993 p. 771) examine the strategic promotion and wage policies of employers and claim to have discovered a new market failure: "... An employer has an incentive to exploit its private information about an able worker by not promoting the worker as quickly or often as is socially optimal" [1]. Why? Because there is an asymmetry of information[2]. Employer A, presumably, knows far more about his own employee, a, than alternative employer B, potentially competing for this employee. A cheap way for B to ascertain a's quality is to note whether or not a is promoted, given a raise, entrusted with greater responsibilities, etc.

In an ideal world, according to Bernhardt and Scoones, A would deal with a without considering B; that is, A would promote or give a raises solely in accordance with his productivity. A would not reward a more than a deserved – that way lies bankrupcty; nor

The authors would like to thank James T. Bennett for many helpful comments and suggestions which greatly improved this paper. All responsibility for remaining errors lies, of course, with the present authors. The authors would like to thank a referee of this journal, and its editor on a similar basis.

Humanomics
Vol. 22 No. 3, 2006
pp. 133-138
© Emerald Group Publishing Limited
0828-8666
DOI 10.1108/08288660610703302

H
22,3

134

would A unduly penalize a – otherwise A would risk losing his employees. But in the real world, assert Bernhardt and Scoones (1993, p. 771), A must always look over his shoulder to make sure that his treatment of a is not exploited by B: "... Employers balance the gain from placing able workers efficiently against the wage cost of revealing their abilities".

Ostensibly, Bernhardt and Scoones (1993, p. 771) examine this "market failure"[3] under conditions where a is more valuable to B than he is to A. Actually, however, this scenario is not as problematic as when a is equally valued by A and B, or when A values a more than B does. For in the former case it is in some sense efficient that a leave A and go to work for B. If this is done as the result of A promoting a or giving him a raise, and in this way making a more attractive to B, it would appear no cause for alarm, at least on grounds of welfare economics. In the present comment, then, we ignore the example where "employees may be more valuable to competing firm", and concentrate on the other two cases.

The problems with the Bernhardt and Scoones thesis are three[4]. First, they ignore several other strategies that could be adopted by employer A in order to make employer B's job of ascertaining the quality of employee a more difficult. Secondly, and somewhat paradoxically, their thesis is "too good." It can be applied to numerous markets other than that for labor; that is to say, it is vulnerable to the *reductio ad absurdum* argument. Third, their claim that "social optimality" is disrupted by this "demonstration" phenomenon is by no means as clear as they seem to think. Let us take up these three critiques in that order. We conclude in section 4.

1.

Which strategies are open to an employer A, who knows full well that if he promotes (well deserving employee) a, he risks losing him to competing employer B? At the very least, an employer A with an inordinate fear of losing a to B can borrow a leaf from stock market practice and engage in the "poison pill" strategy. That is, he can promote employees who are totally undeserving of this accolade, in the hope that B will snatch them up – and lose out thereby. He can do this without suffering too much loss of productivity in the interim period between the promotion of this "false" a promotion and B's raid of this "bait," by seeing to it that the newly promoted a is given a big office and an impressive sounding title – but no real responsibilities[5]. On the reasonable assumption that A knows far more about his own operation and personnel than does B, it is likely that B will be fooled[6]. If successful, this strategy[7] will impose great costs on B if he swallows the "bait." Even if he does not, in this one case, he will no longer be able to rely so heavily on Bernhardt and Scoones' demonstration effect[8].

Another strategy would be to offer big raises, but not ostentatious promotions (large offices, etc.), on the grounds that it is easier to hide the former than the latter[9].

A third possibility is to take the bull by the horns: even though there is this danger of "signaling" to other employers the higher productivity value of the newly promoted employee, there is also the risk of failing to take this action. For if A does not promote a, he risks alienating his entire labor force. If he does, on the other hand, his "promotion from within" strategy may well reduce his own quit rate. This might be the best way to promote employee loyalty, and to add to the inertia which typically tends to make one's present employer seem more safe than any alternative.

Another thing to remember is that a is not an innocent bystander. Workers can and often do move from job to job. If A does not promote a, then a can leave to work for B. In white-collar labor markets especially, there are well developed "head hunters" who facilitate exactly this kind of movement.

2.

The postulate of Bernhardt and Scoones is "too good" in that it can be made to apply to a whole host of phenomena besides the labor market. What their view comes down to, we shall show, is that there is something akin to market failure in all of these cases. To generalize the Bernhardt and Scoones theory, they claim that whenever anyone makes a decision choosing one thing and setting aside another[10], this constitutes a signal to his competitors that the thing he has chosen has more value than the competitor would have been aware of had this choice not been made. Therefore, the chooser necessarily undermines himself, by giving more power to his competitor than would otherwise be true.

The problem with this is that there are so many cases that seem to fit this bill[11]. For example, whenever a consumer buys peas instead of carrots, he thereby signals to other, competing consumers, that peas are "better" than carrots. This undermines the party of the first part, since if others act on this "signaling" of his, and copy him by also purchasing peas instead of carrots, the terms of trade will turn against the first consumer; that is, he will have to pay more for his beloved peas. Similarly, if Bank A lends money to borrower a, it "signals" to Bank B that a, who it might otherwise have turned down, might have more creditworthiness than it, B, had first supposed. Also, we can turn the Bernhardt and Scoones illustration on its ear. They talk in terms of employers signaling other competing employers about the productivity of employees; we can turn this around by considering the signaling of employees about employers. For example, if employee a agrees to work for employer A, he thereby "signals" to employee b, who might have contemplated working for B, that A actually constitutes a better option than he, b, had first supposed. Then of course there is the marriage market. When John asks Sue to marry him, he signals to George that Sue is actually a better marriage partner than he, George, had initially thought. John thereby "unleashes the whirlwind," provoking the very competition that he might be supposed to be most concerned to quell.

Again, as in our critique of Bernhardt and Scoones's own illustration, we can posit attempts to "hide" the true state of affairs, so as to obviate the unwelcome attention engendered by his original decision. For example, "secret" pea buying, "hidden" lending, "surreptitious" jobs.

There is, however, a silver lining. As far as the marriage market is concerned, this premise may well go part way toward explaining several sociological phenomena. For example, consider the following sociobiologically (Wilson, 1980) inspired scenario: a man and a woman meet; she is attracted to his Alfa Romeo, which indicates wealth and potential support for her and her future children; he is attracted to how she looks in a bikini. However, their marriage "signals" additional attractiveness to potential competitors for each. As a result, after the marriage, he tells his wife to dress more modestly, thus making her less attractive to other men, and she insists that he trade in his sports car for something more sedate, and thus become less appealing to other women. This may also, in part, explain the tendency of newly married couples to gain weight: each partner has an incentive to fatten up the other, so as to dampen just the sort of signals implied by Bernhardt and Scoones.

3.

As can be seen by the foregoing, there may be a kernel of truth in the Bernhardt and Scoones supposition. Yes, it is subject to a reductio, but it undoubtedly has a degree of explanatory power, nevertheless. However, even if it sheds light on some otherwise puzzling sociological phenomena, the claim that it constitutes a market failure of some sort is still problematic.

What sort of evidence for this latter claim is provided by these authors? Only the fact that all action is, to some extent, self-defeating, in that it sets up "signals" which tend to undermine the initial choice. One might as well label this a "market failure" as the fear of some environmentalists that all economic action uses up some resources, and hence is self-defeating in that there will be fewer assets left for further action. In like manner, every economic act has an alternative cost; it implies that there is an opportunity foregone; that, but for the given action, something else might have been undertaken. All of this is very regretful, to be sure. But to label any of this as not "socially optimal" as do Bernhardt and Scoones (1993, p. 771), would be at least highly misleading[12].

Indeed what they describe, if it exists at all, may not be socially inefficient in the least. Suppose workers reservations wages fall after they get a job. This means that an employer could slow down pay/promotion without losing employees, but it is not inefficient to do so. This merely alters the distribution of the gains from trade in favor of the employer, but does not alter the allocation of resources in any way.

There is a possibility as well that slower rates of promotion may be a correction for a "market failure" of sorts that can occur in the hiring process. If you envision the hiring of a worker as a type of auction, then it may be subject to the winner's curse (Thaler, 1994). When a worker's true worth is unknown, it is entirely likely that the winning bidder (i.e. the successful employer) will have overestimated his worth and paid too much. Slowing rates of promotion may be an *ex post facto* means to bring a worker's pay in line with this true marginal worth to the firm.

But more. These authors focus on the negative repercussions which boomerang against employer A in his decision to promote or otherwise reward his employee a. These external diseconomies, as it were, are perpetuated by the (possible) actions of B, the alternative employer. What Bernhardt and Scoones forget is that the negative implications for A give rise to positive ones for B. After all, the harm to A is no dead weight loss; on the contrary, it is of help to B, who now has further information about a furnished to him. In classifying this as non "socially optimal," Bernhardt and Scoones are implicitly claiming that the harm to A is greater than the benefit to B. But this they cannot do without resorting to unwarranted interpersonal comparisons of utility (Rothbard, 1977; Hoppe, 1993).

There is an implication here for the marriage market. The Bernhardt and Scoones theorem implies an over-optimal divorce rate – due solely to the demonstration or signaling effect[13] of this institution. Yes, each spouse who marries signals to all competitors that his or her mate has better characteristics than would have otherwise been apparent had they not chosen matrimony with this person. This is indeed a negative. But there is also a positive: new information has thereby been released. In order for their (implicit) claim of over-optimality of divorce to be sustained, it must be argued that the losses to the married couple swamp the informational gains to the rest of society. And this, again, cannot be done without utilization of invalid interpersonal comparisons of utility.

4.

The goal of Bernhardt and Scoones has been to demonstrate a new instance of "market failure" that at least implicitly calls for governmental action to "correct" it. Our objective has been to show that their case is logically and empirically flawed. To wit, we have established that their argument is problematic on three grounds. One, they fail to come to grips with inertia, poison pill strategies, and the fact that the extant employer almost necessarily knows more about his employee than does another firm which is now at best only a potential employer. Two, we have subjected the Bernhardt and Scoones thesis to devastating *reductios ad absurdum*. Third, we have dealt with their claim regarding "social optimality." To wit, that it would otherwise have been attained in the absence of this signaling phenomenon. What is this "social optimality" of Bernhardt and Scoones? It is equivalent to what is commonly called "economic welfare" (Rothbard, 1977). These authors maintain that "economic welfare" or "social optimality" is retarded, or reduced, by their signaling problems. Our conclusion is that they have failed to demonstrate this.

How does our paper address an aspect of ethics and social/economic theory? This should be clear. According to Bernhardt and Scoones, it is unfair to employees that employers will give them lower wages than they would otherwise have been paid for fear that competing firms will attempt to hire workers who get raises. In contrast, it is our view that there is nothing at all untoward about this necessary aspect of the competitive process. Every time, without exception, whether it is in the labor market, or in the marriage market, whenever any choice is made, a signal is sent out that what was chosen was more desirable to at least one person than would have been the case did he not make that decision. If this is "unfair," then life is unfair. Bernhardt and Scoones should get used to this phenomenon. It is here to stay.

Notes

1. Unless otherwise identified, all page notations refer to Bernhardt and Scoones (1993).

2. On the importance of information and knowledge in economics, see Kirzner (1973), Hayek (1948), Choudhury (2004), Sowell (1980).

3. This market failure amounts to plan discoordination. For further analysis of this phenomenon, see Anwar (1999), Bagwell and Ramey (1994), Boettke, (2001), Garrison (1994), O'Driscoll (1977a, b). For a critique of Bagwell and Ramey (1994), see Block (2003).

4. But there is a silver lining in the clouds of the second, as we acknowledge below.

5. We have here an entirely different explication of the "Peter Principle": promoting people until they achieve incompetence.

6. On the other hand, it may be more difficult to protect against lowered morale in firm A, when the other employees see the unworthy a being promoted ostensibly above them.

7. For the treatment of an analogous phenomenon, the principle agent problem, see McGarrity *et al.* (1999).

8. In army war games, typically, each side fires blanks at the other. In some cases, however, 1 in 500 or 1 in 1,000 bullets is "live," spread out equally amongst the two halves of the army, and unidentifiable by the soldiers. This is done to make sure the new recruits follow the sergeants' advice and "keep their heads down." Seeding false a's in and among the good a's promotion wise – in unknown proportions – can likewise be expected to have similar effects on B.

9. But again, this would have internal repercussions.

10. And of what else does economic action ever consist? (see Mises, 1966).

11. Ultimately, it might be claimed, all (of economic) action comes within the reach of the Bernhardt and Scoones thesis.

12. Happily, these authors do not take the further step, as is usual in such cases, of calling for government action to ameliorate the "market failure." They are wise in thus forbearing, for there is such a thing as "government failure" which is typically far more serious than any problem emanating from the marketplace.

H
22,3

13. So as to be clearly understood, all other claims about socially over-optimal divorce rates, e.g. no fault divorce laws, are strictly irrelevant to our present discussion.

138

References

Anwar, A.J. (1999), "Modelling (sic) altruistic behavior: a case of failure in coordination", *Humanomics*, Vol. 15 No. 4, pp. 94-122.

Bagwell, K. and Ramey, G. (1994), "Coordination economies, advertising, and search behavior in retail markets", *American Economic Review*, Vol. 84 No. 3, June, pp. 498-517.

Bernhardt, D. and Scoones, D. (1993), "Promotion, turnover and preemptive wage offers", *American Economic Review*, Vol. 83 No. 4, September, pp. 771-91.

Block, W. (2003), "Coordination economies, advertising and search behavior in retail markets by Bagwell and Ramey: a comment", *Cross Cultural Management*, Vol. 10 No. 1, pp. 80-6.

Boettke, P.J. (2001), *Calculation and Coordination: Essays on Socialism and Transitional Political Economy*, Routledge, London.

Choudhury, M. (2004), "The vantage point of the oppressed: a superior standpoint for understanding and interpreting the social world in terms of unique knowledge structures", *Humanomics*, Vol. 20 No. 3/4, pp. 58-65.

Garrison, R.W. (1994), "Hayekian triangles and beyond," in Jack, B. and Rudy van, Z. (Eds), *Hayek, Coordination and Evolution: His Legacy in Philosophy, Politics, Economics, and the History of Ideas*, Routledge, London, pp. 109-25.

Hayek, F. (1948), "The use of knowledge in society", in Hayek, F. (Ed.), *Individualism and Economic Order*, University of Chicago Press, Chicago, IL, pp. 77-91.

Hoppe, H.-H. (1993), *The Economics and Ethics of Private Property: Studies in Political Economy and Philosophy*, Kluwer, Boston, MA.

Kirzner, I.M. (1973), *Competition and Entrepreneurship*, University of Chicago Press, Chicago, IL.

McGarrity, J.P., Jim, B. and Jim, B. (1999), "Is there a principal – agent relationship between future employers and congressmen? An examination of the house vote on flag burning", *Humanomics*, Vol. 15 No. 4, pp. 44-7.

Mises, Ludwig von. (1966), *Human Action*, Regnery, Chicago, IL.

O'Driscoll, G.P. (1977a), *Economics as a Coordination Problem: The Contributions of Friedrich A. Hayek*, Sheed, Andrews and McMeel, Kansas City, KS.

O'Driscoll, G.P. Jr. (1977b), "Spontaneous order and the coordination of economic activities", *The Journal of Libertarian Studies*, Vol. 1 No. 2, Spring, pp. 137-51.

Rothbard, M.N. (1977), "Toward a reconstruction of utility and welfare economics", San Francisco, Center for Libertarian Studies, Occasional Paper #3.

Sowell, T. (1980), *Knowledge and Decisions*, Basic Books, New York, NY.

Thaler, R. (1994), *The Winner's Curse: Paradoxes and Anomalies of Economic Life*, Princeton University Press, Princeton, NJ.

Wilson, E.O. (1980), *Sociobiology*, Harvard University Press, Cambridge, MA.

Corresponding author
Walter Block can be contacted at: wblock@loyno.edu

Globalization and the Concept of Subsistence Wages

Walter Block
Jerry Dauterive
John Levendis
Loyola University New Orleans*

According to Malthus, there is an "Iron Law" for wages: they cannot stay above subsistence levels. When they do, increased population soon enough pushes them down to the previous level of immiseration. One might think that modern economics has long ago confined such views to the dustbin of history, however, belief in the "Iron Law" has made a comeback in this era of globalization. We argue that all versions of the Iron Law, new and old, are vulnerable to a knock-out critique. We argue that the Iron Law of Wages, and slavery for production and profit, are logically incompatible: if one ever existed, the other cannot.

Keywords: Malthus, subsistence wages, slavery
JEL Classifications: B120, B300, N300

Although most people, at least nowadays, reject large parts of Malthusianism, some continue to accept uncritically an important tenet of that doctrine: the specter of the subsistence-level wage rate. In fact, in recent years there has been increased discussion of the concept of subsistence wages due to the (alleged) impact of globalization and free trade. Have technology and globalization created the conditions in which the Iron Law of Wages[1] is at last free to function? In this paper, we review the Malthusian doctrine itself, the critiques that have been leveled against it by moderns, and the rise of the subsistence wage view tied to globalization. We then offer a fundamental argument against the "reality" of subsistence wages.

The Iron Law and Malthusian Population Doctrine

Adam Smith, in *The Wealth of Nations*, celebrated the capacity of the market economy to provide rising wages and improving standards of living for all classes in advancing, capital-accumulating societies like those of Europe, Britain, and North

* *Addresses for correspondence*: Walter Block, Harold E. Wirth Eminent Scholar Endowed Chair in Economics, College of Business Administration, Loyola University New Orleans, New Orleans, LA 70118, U.S.A. e-mail: wblock@loyno.edu; Jerry Dauterive, Associate Professor of Economics, College of Business Administration, Loyola University New Orleans, New Orleans, LA 70118, U.S.A. e-mail: dauteriv@loyno.edu; John Levendis, Assistant Professor of Economics, College of Business Administration, Loyola University New Orleans, New Orleans, LA 70118, U.S.A. e-mail: johndlevendis@yahoo.com

America. While aware of a linkage between wage rates and population growth rates, Smith expressed little concern about excessively rapid growth in population or labor supply; economic growth would automatically improve the lot of the poor (Smith [1776] 1991).

Malthus (2004) took a darker view of the capitalist economy's ability to provide prosperity throughout society. The basis of Malthusian pessimism lay in an assumption about the response of the lower classes to improvements in their income. He expected higher wage rates to lead to earlier and more prolific marriages, resulting in an increase, over time, in the size of the labor force and a subsequent decline in the wage rate to its original level.

Since the United States had, at the time, vast amounts of fertile and untouched land on which to farm, Malthus took from it the best-case scenario. Using sketchy data for the United States, he concluded "that population, when unchecked, goes on doubling itself every twenty-five years or increases in a geometrical ratio" (2004:21). The rate of fertility (the rate of actual births) will approach the rate of fecundity (the biological capacity to give birth). Thus, we follow Malthus' original definition[2] of the subsistence level of wages as that thin line between survival and starvation—a rate of wages where laborers are just able to sustain themselves and their progeny, with zero population growth.

The only problem with this obstetrician's delight is that sooner or later the food supply runs out. The food supply cannot also be increased in geometrical proportions. Diminishing returns set in.

> In the next twenty-five years, it is impossible to suppose that the produce could be quadrupled. It would be contrary to all our knowledge of the qualities of land. The very utmost that we can conceive is that the increase in the second twenty-five years might equal the present produce. Let us then take this for our rule, though certainly far beyond the truth, and allow that by great exertion, the whole produce of the Island might be increased every twenty-five years, by a quantity of subsistence equal to what it at present produces. The most enthusiastic speculator cannot suppose a greater increase than this. In a few centuries it would make every acre of land in the Island like a garden. Yet this ratio of increase is evidently arithmetical (Malthus [1798] 2004:22)

Every increment of the population is able to produce proportionately less and less food. Diminishing returns are due first to adding poorer and poorer land to cultivation, and second, when all available land is already under cultivation, to adding more and more people (and capital) to the fixed land factor. These diminishing returns make it impossible to increase the food supply in the same geometrical progression as the people. Instead, the food supply, in the Malthusian view, can only increase in arithmetic proportion.

> Taking the population of the world at any number, a thousand millions,
> for instance, the human species would increase in the ratio of – 1, 2, 4,
> 8, 16, 32, 64, 128, 256, 512, etc. and subsistence as – 1, 2, 3, 4, 5, 6, 7,
> 8, 9, 10, etc. In two centuries and a quarter, the population would be to
> the means of subsistence as 512 to 10: in three centuries as 4096 to 13,
> and in two thousand years the difference would be almost incalculable,
> though the produce in that time would have increased to an immense
> extent.

> No limits whatever are placed to the productions of the earth; they may
> increase for ever and be greater than any assignable quantity; yet still
> the power of population being a power of a superior order, the increase
> of the human species can only be kept commensurate to the increase
> of the means of subsistence by the constant operation of the strong law
> of necessity acting as a check upon the greater power (Malthus [1798]
> 2004:23)

An increase in population, particularly a geometric increase, cannot long continue without a similar increase in the food supply, and the food supply, we have
seen, cannot so increase. Therefore, the geometric increase in the population will
have to come to an end. The checks to population increase are brought to bear. The
positive checks include war, famine, and disease. The preventive check is moral
restraint: later marriages and fewer children. It was this pessimistic prognostica-
tion that led to economics being labeled the "dismal science". And no wonder: it
was the lot of mankind to be circumscribed by war, famine, disease, and sexual
restraint.

Malthus did maintain that wages would hover around the subsistence level,
both because lower-than-subsistence wages lead to starvation and to a decrease in
the supply of labor and because higher-than-subsistence wages lead to population
rises and increases in the supply of labor.

> The way in which, these effects are produced seems to be this. We
> will suppose the means of subsistence in any country just equal to the
> easy support of its inhabitants. The constant effort towards population,
> which is found to act even in the most vicious societies, increases the
> number of people before the means of subsistence are increased. The
> food therefore which before supported seven millions must now be di-
> vided among seven millions and a half or eight millions. The poor
> consequently must live much worse, and many of them be reduced to
> severe distress. The number of laborers also being above the propor-
> tion of the work in the market, the price of labor must tend toward a
> decrease, while the price of provisions would at the same time tend

to rise. The laborer therefore must work harder to earn the same as he did before. During this season of distress, the discouragements to marriage, and the difficulty of rearing a family are so great that population is at a stand. In the mean time the cheapness of labor, the plenty of laborers, and the necessity of an increased industry amongst them, encourage cultivators to employ more labor upon their land, to turn up fresh soil, and to manure and improve more completely what is already in tillage, till ultimately the means of subsistence become in the same proportion to the population as at the period from which we set out. The situation of the laborer being then again tolerably comfortable, the restraints to population are in some degree loosened, and the same retrograde and progressive movements with respect to happiness are repeated (Malthus [1798] 2004:24)

Malthus used the term "oscillation" to describe this pattern, but the relentless downward-to-subsistence pressure on wages has commonly been referred to as *the Iron Law of Wages*. Why were these oscillations not visible? Malthus believed the problem was a lack of data: historical research had focused, during Malthus' time, on the elites, not the working classes. He suggested that researchers examine

the comparative mortality among the children of the most distressed part of the community and those who lived rather more at their ease, what were the variations in the real price of labor, and what were the observable differences in the state of the lower classes of society with respect to ease and happiness, at different times during a certain period (Malthus [1798] 2004:25)

Malthus did recognize that improvements in agricultural technology could offset the falling wage level, since they would make it possible to feed more people with the given supply of land. However, he believed that such technological improvements would be unique historical events that would only put off for a limited time the inevitable stationary state of no growth and subsistence wages.

More recent Malthusians have pointed to medical breakthroughs that have enabled mankind to conquer many of the diseases that preyed unchecked upon him in the eighteenth century. The diminution of the death rate of adults, as well as the marked decrease in the rate of infant mortality, has taken much of the venom out of this "positive" check to population, but has only strengthened Malthus' argument. Modern medicine has made possible even greater increases in population

Modern Critiques against the Iron Law

This pessimistic world-view has not gone uncriticized. Modern economists with the vantage point of more history behind them have shown many weak points in the Malthusian population theory. Technological innovations and new techniques in farming have revolutionized the food-producing industries to a degree impossible for the contemporaries of Malthus even to envision. This has enabled food production to increase, if anything, at a greater than geometric progression.

The advent of birth-control techniques has diminished the relevance of the fecundity solution to population growth. The human capital approach (Becker, 1964) to population theory has instead stressed that the birth rate is itself a decision variable. Children are looked upon by this approach in a dual role: as a consumption good, and as a production or capital good. As a consumption good, children (the birth rate) are in competition with all other consumption goods; the final determination of the allocation of resources toward the production of this consumption good will depend upon income available, the costs of child rearing, and upon the alternatives foregone. As capital goods, the production of children will depend upon the rate of return, both pecuniary and non-pecuniary, upon the configuration of risk, and upon the present discounted values of the income streams available from alternative investments.

Von Mises (1966) notes that while life is an unceasing struggle against the fundamental problem of scarcity, those living and working in capitalistic societies will in fact "earn much more than is needed for bare sustenance". More specifically, Mises calls into question the concept of an Iron Law, which he states "is of no use for a catallactic theory of the determination of wage rates" (p. 604):

> One of the foundations upon which social cooperation rests is the fact that labor performed according to the principle of the division of labor is so much more productive than the efforts of isolated individuals that able-bodied people are not troubled by the fear of starvation which daily threatened their forebears. Within a capitalist commonwealth the minimum of subsistence plays no catallactic role... The 'iron law of wages' and the essentially identical Marxian doctrine of the determination of 'the value of labor power'... are the least tenable of all that has ever been taught in the field of catallactics......If one see in the wage earner merely a chattel and believes that he plays no other role in society, if one assumes that he aims at no other satisfaction than feeding and proliferation and does not know of any employment for his earnings other than the procurement of those animal satisfactions, one may consider the iron law as a theory of the determination of wage rates.

The most damning evidence against the Malthusian population theory is of course the sheer fact that real incomes *have* risen in the last century. It is true that population sizes have also increased, but according to the crude version of population growth we are considering, all the fruits of technological breakthroughs and capital accumulation should have resulted in population increases; none should be spread out among improved real wage. Pure rabbit-like behavior would convert all the fruits of technology and capital into more rabbits, conserving none of it at all for higher living standards. In fact, food is more plentiful, and many other kinds of goods and services are available that were not available two hundred years ago.

Per capita output has been steadily increasing for ages. But this leaves the data open to the standard Marxist criticism: the pie is getting bigger, but it is only the rich who eat more, while the worker still gets his subsistence wage. Thus we should not constrain ourselves to simple productivity statistics, but focus on workers' wages. Unfortunately for the Malthusians and the Marxists, these too have been rising. We report below an index of real wages of unskilled labor. Figure 1 below shows that, even for years as far back as 1800, there is no evidence of a declining real wage rate, even among the unskilled.

Figure 1

Source: Derived from data in McCusker (2001) and Williamson (2004).

Extending our view to the 20th century, we see in Figure 2 that, even one hundred years after Malthus made his dire prediction, there is no evidence of a declining real wage rate.

Technical progress in the production of food and in other fields has not been rare and accidental, but rather more or less continuous and cumulative. This improvement in technology has outrun population growth, leaving more and more people better off.

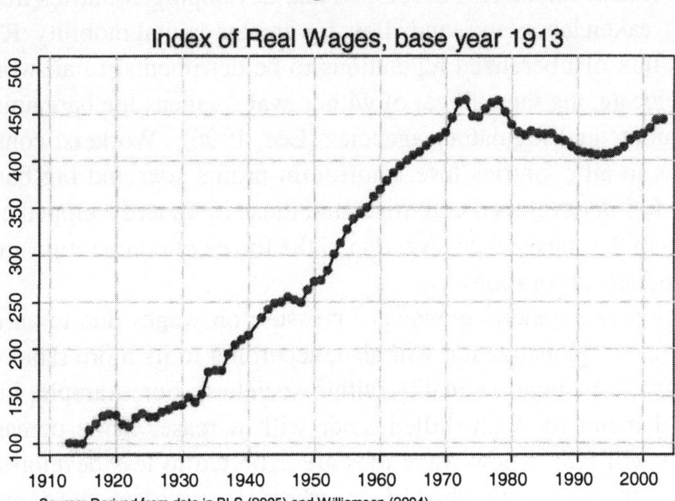

Globalization and the Return of the Iron Law

While modern economists have correctly criticized much of the Iron Law doctrine, we are today hearing more from those who claim that, thanks to globalization, the Iron Law is at last free to function (Brown, Deardorff and Stern, 1996; Hanson, 2001; Klevorick, 1996; Lee, 1996; Mehmet and Tavakoli, 2003; Nader, 1993; Rodrik, 1997; Wilson, 1996, and Wood, 1995 are all examples of the latter position). According to this view, contrary to the assumption of Malthus, population growth was curbed in the societies experiencing the industrial revolution and thus workers have been able to "protect" their interests. Today, however, through free trade, the free circulation of capital and resulting cut-throat competition, employees are in a globally competitive environment, which is creating what these critics call a "race to the bottom" (RTB) for workers.

Among the most damaging of the many negative effects of free trade, according to critics, is that the forces of globalization have created a world safe for capitalist investment in (or, in their terms, exploitation of) the poorest countries. Rather than this investment's being a positive factor, it instead creates a world in which capital moves inevitably to countries with the lowest wages. In those countries, workers will be paid subsistence (if not below-subsistence) wages. Corporations are able to pit country against country in a race to see who can offer the lowest wage levels (along with the lowest environmental standards and the poorest workplace safety standards). The result is a downward spiral in wages, an RTB, to a subsistence wage level.

At the heart of RTB theories is the belief that foreign direct investment increases the elasticity of demand for labor. It is argued that, in their attempt to attack foreign investment funds, less developed and developing countries liberalize their trade laws, weaken labor laws, and allow for greater capital mobility. RTB theorists believe this mix of liberalized regulations to be detrimental to all workers. Once firms can relocate, the mere threat of withdrawal weakens the bargaining power of workers, unions, and regulatory agencies (Lee, 1996). Workers, consumers, and communities in all countries lose; short-term profits soar and big business wins. Real wages fall in developed countries, and those in underdeveloped countries are driven down to the subsistence level due to the forces of competition and the power of multi-national corporations.

In addition to a general downward pressure on wages due to an elasticity of demand for labor, globalization will also, according to its more rabid critics, contribute to greater income inequality within societies. For example, in the United States, the demand for high-skilled labor will increase, while demand for less-skilled labor will fall, as low-wage jobs are exported to less developed countries. This will raise the wages of high-skilled workers along with profits of multi-national corporations, while lowering those of less-skilled laborers, in the United States. Some economists claim that expansion of trade with underdeveloped countries is the main cause of the "deteriorating situation of unskilled workers in developed countries" (Wood, 1995; Greenaway and Nelson, 2001). As international trade wipes out manufacturing jobs in developed countries, the displaced workers seek jobs in the service sector, exerting downward pressure on the wages of maintenance and custodial workers, taxi drivers, and fast-food cooks. To make matters worse, globalized trade and investment make companies more likely to jettison workers, when wages rise even modestly, since producers can easily switch to lower-wage options abroad (Hanson, 2001).

The list of evils that spring from free trade is, according to such critics, extensive and pervasive. In addition to the outcomes listed above, the list may include environmental degradation, greater exploitation of women and children, more unsafe products, less investment in education, and a reduction in workplace safety (Nader, 1993). It is, however, the impact on wages that is of interest here. In the market economy, according to Malthus, misery and poverty were the long-run fates of the masses, due to the response of the lower classes (increased population growth) to any short-term improvements in their income. Two hundred years later, despite all of the advances in technology and real incomes, many believe that the prognosis for the lower classes is equally grim, but for a different reason: increased competition due to globalization will ensure the wages of the poor are driven down to the subsistence level. Thus, we enter the twenty-first century with a renewed focus on the same dismal outcome of subsistence wages that Malthus predicted at the beginning of the eighteenth century.

A New Critique

Arguments against free trade go way back, but it is the renewed attention they have received in the past decade from the forces of anti-globalization and the link they present to the concept of a subsistence wage level which we address here. Very solid, theoretically sound arguments for free trade have been presented by economists with a remarkable degree of agreement for years (Dreuil et al., 2003; Block, Horton and Walker, 1998). As Mises summarizes,

> The theoretical demonstration of the consequences of the protective tariff and of free trade is the keystone of classical economics. It is so clear, so obvious, so indisputable, that its opponents were unable to advance any arguments against it that could not be immediately refuted as completely mistaken and absurd (von Mises, 1985:130)

It is not our purpose to confront (once more) the arguments of those who claim that free trade will lead to a variety of ills for workers in the twenty-first century. Rather, we consider here what is once again being forecast, that more fundamental fallacy predicted since the time of Malthus, the concept of the subsistence wage.

What we shall attempt to show is not that no one ever dies of starvation, not that there are not some people (the blind and deaf, quadruple amputees, severely mentally handicapped, etc.) whose productivity and hence wage rate might be less than subsistence level, nor yet that at times in history, and even in the present, there are not isolated cases of people in primitive conditions who have starved to death due to the harsh conditions of nature, or what have you. These cases are all true, unfortunately. What we shall attempt to show, however, is that such *widespread* starvation or near starvation on a nation-wide level due to a subsistence level of income as is alleged to have occurred *could not* have existed throughout all of recorded history. In a nutshell, the reason why a subsistence level could not have taken place is that it is inconsistent with the institution of slavery. To put it another way: if slavery for production and profit *ever* existed, then Malthus' Iron Law of Wages never did or will. The two, we will argue, are mutually inconsistent. Since we have always had slavery, throughout all of recorded history,[3] we could not have had a widespread subsistence level economy also. Let us explore this in some detail.

A widespread productivity level equal to subsistence implies a tendency toward a subsistence level wage rate. We know this because of the marginal revenue theory of wage determination. A productivity level of subsistence implies, in turn, that the worker can produce no more than is necessary for his own survival, plus the survival of enough other people (his children), such that the population level does not fall. The subsistence-level worker can just barely stay alive himself and can produce only just enough to keep up his numbers.

Th subsistence level of productivity simply cannot, however, suffice if the institution of slavery as an input into production is to exist, for the slave owner cannot possibly earn any return from the ownership of slaves whose productivity is equal to their subsistence level. If all of the slave's product must go to feed and care for the slave and his progeny, there will remain nothing for the master. And if the slave and his progeny are not given this full product, slavery will soon die out when the progeny of the slave are no longer alive to reproduce. If it will not pay the slave owner even *to keep* his slaves, it can readily be seen that it certainly will not pay to invest in their capture and guarding, to say nothing of the risks of slave uprisings and rebellions.

We have, unfortunately, had slavery throughout all of recorded history – in the sixteenth through nineteenth centuries, in medieval times, in the Greek and Roman eras, and even in the biblical epoch. It must therefore be true that the productivity of such slaves (who most assuredly *were* enslaved) was *greater* than the level that would ensure mere subsistence.

For slavery to continue to exist (and it did for centuries), slave productivity had to be greater than subsistence. But this is not to say that slave *wages* were higher than subsistence: wages are not synonymous with productivity. Slaves received paltry "wages", and the difference between a slave's output and the costs of maintaining a slave was the profit of the owner.

There is, however, a simple way to illustrate whether slaves' "wages" were below or above subsistence levels. In the United States, slave importation was illegal after 1808. If wages were below subsistence, the number of slaves should have decreased after 1808. It did not. Slave-owners did not mind the 1808 law much, since the local slave population was, by this time, self-sustaining. It is estimated that by 1680, "native-born blacks made up a majority of the slave population in the U.S. colonies... [and] by the end of the American Revolution, the African-born component of the black population had shrunk to 20 percent... By 1860 all but one percent of U.S. slaves were native-born" (Fogel and Engerman, 1974:23–24). "While the imports of Africans certainly contributed to the growth in the slave population of the U.S. colonies, they were of secondary importance in explaining that growth after 1720" (Fogel and Engerman, 1974:25). Put crudely, the antebellum slave population was home-grown and home-fed.

In their book *Time on the Cross* (1974), Robert Fogel and Stanley Engerman inaugurated a new era of historical research into the economics of slavery. They employed current economic theory and econometric techniques to examine the economics of slavery. Among their groundbreaking conclusions was that slaves were well fed and well cared for. (For his work in economic history Robert Fogel won the 1993 Nobel Prize in Economics, which he shared with economic historian Douglass North.) This is not to say that they thought slavery was morally acceptable, but they point out that there is much empirical data to support the assertion that slaves were better fed, clothed, and cared for than many of their contemporaries. For example, they found that the average life expectancy at birth for US slaves was much higher than many free Europeans at the time. The life expectancy for US black slaves was 36 years, exceeding the life expectancies of Italians in 1885, Austrians in 1875, and even Chileans as late as 1921 (Fogel and Engerman, 1974:125).

There is a rich biometric literature supporting Fogel and Engerman's conclusion, and by extension, our conclusion as well. Biometrics describes people's growth patterns, and the deviations from these patterns, across racial lines (Tanner, 1962; Eveleth and Tanner, 1976; Heald, Remmel and Mayer, 1969; Frisch and Revelle, 1969). Malnutrition, disease, family size, even stress (such as abuse or divorce) are known to affect physical development in measurable and predictable ways.

Drawing on this biometric tradition, Richard Sutch (1975; 1976) arrived at the conclusion that, though not as varied as Fogel and Engerman had first estimated, slave diet had a caloric intake even higher than Fogel and Engerman believed.

In the same tradition, Steckel (1979) compared the age-height profiles of slaves with those of contemporary white North Americans and Europeans. He found that American slaves were taller than contemporary Europeans, and slightly shorter than northern whites. Genetic differences were not relevant, since black Americans, white Americans, and white Europeans currently have identical heights at maturity. Considering that the disease environment in the South was harsher than in Europe or the American North, Steckel concluded that slaves were relatively well fed, with their medical care on a par with that of Europeans.

Margo and Steckel (1982) re-examined the heights of American slaves using more recently uncovered data. Their results confirm Steckel's earlier work. Margo and Steckel compared the heights of Virginian slaves with those of free blacks in Virginia and found that "at the very least, free blacks in Virginia were as well-fed as Virginia slaves" (p. 533). Taken in aggregate, this supports our thesis: slaves are fed a diet in excess of subsistence levels (since they were better fed than free Europeans at the time). Free people in the United States were even better fed, making it even less likely that they were paid subsistence wages.

We still have not fully answered the allegations of subsistence-level productivity (and wage rates), for we have not yet shown that the productivity of the free workers in, for example, eighteenth-century England and Europe as well as in the present-day underdeveloped countries (the subject of much concern from anti-globalists), was/is greater than subsistence level. (We have so far only shown that the productivity of the slave must be greater than subsistence.) Perhaps the easiest way to show that the productivity of free men is also greater than subsistence is to show that it is larger than that of the slave. Since we know that the slave's productivity is greater than subsistence, we can then deduce that the free worker's productivity must also be greater than subsistence if we can show that it is superior to the slave's productivity.

This task seems easy enough, however, since the productivity of free workers must be greater than the productivity of slaves, if for no other reason than that forcing slaves to work entails the extra costs of capture and guarding them, which do not arise in the employment of free workers. Thus, if we start with the assumption that "slaves" and free workers were originally equally productive (before the "slaves", once captured, became slaves), we must conclude that net of capturing and guarding, free workers must be more productive than slaves and hence more productive than the subsistence level.[4] We can also take into account the fact that the type of employment to which slaves may be put will have to be much more restricted than that which applies to free people. Slaves will pretty much be limited to types of employment where the costs of overseeing and guarding will be minimal. This means that the slaves will all have to be employed in close proximity to each other. Losses from slaves' running away would also have to be minimized. (It is probably no accident that a popular employment for slaves was rowing on galley ships.)

It is of course true that slaves may be allowed some latitude (which will increase their productivity) if they become psychologically dependent on their masters (e.g., "Uncle Tom") or if they have loved ones under the control of the slave master to ensure their "good behavior." Even with this extra latitude slaves are still most unlikely to be as economically productive in the master's behalf as free people for themselves, whether because of loss of initiative in the case of the subservient, or because of resentment in the case of the extorted. Threat is also not as powerful as it might seem, because if it is carried out and the loved ones of the slave with some degree of latitude are killed or hurt, the slave master will suffer economic losses. This is not to say that extortion is valueless; it is not. It is just not all-powerful. In any case, it is only a small percentage of slaves who are likely to fall into the category of victims of either pure extortion[5] or of emotional dependency. As for the majority of slaves, as well as for the members of these two categories, we must state that when account is taken of the costs of capture, guarding, and overseeing, of the incentives and abilities to work while under enslavement, and of the constant dangers of rebellion, it can only be concluded that the productivity of the free worker must be greater than that of the slave.

If the productivity of the slave *must* be greater than that necessary for subsistence level, and if the productivity of the free worker *must* be greater than that of the slave, then it necessarily follows that the productivity level and wage of the free worker must be *higher* than that necessary for subsistence, all allegations to the contrary not withstanding.

Let us conclude by dealing with a claim made by certain "radicals" in the realm of economic history, sometimes made with only a slight sense of exaggeration. The claim is that as bad as slavery is, it was better than the lot of the exploited worker during the industrial revolution, for at least the slave owner would keep his slaves alive, if only out of self interest, while the hard-hearted capitalist-pig employer had no such incentive. If the workers of the latter die off, he can easily replace them with other members of the "reserve army of the unemployed".

In order to answer this assertion, we must assume either that either a below-subsistence wage level is claimed for the time period of the industrial revolution, or that it is not. If this is assumed, then we can conclude on the basis of our previous analysis that this "radical" claim is false on the ground that the slave owner would have had no such incentive to keep his slaves alive. He would have had no such incentive because the institution of slavery *could not* have existed under our present assumption of a below-subsistence wage level. It is of course true that under this assumption neither would the exploiting capitalist employer have had any incentive to keep his workers alive. This does not, however, prove that an employer must necessarily be an "exploiter". It is merely the result of our assumption of a below subsistence average wage level. While things are grim with this assumption, our analysis tells us they are inapplicable to reality.

If we make the more realistic assumption that there is no below-subsistence wage level, this "radical" claim is still false for there is now no reason why workers should be dying off. The presumption and the tendency are for them to earn in accordance with their marginal productivity, and if this is greater than the subsistence level, there should be no question of their dying off in droves. Since this claim can be shown to be false whichever assumption we make, and since these two assumptions are mutually exclusive and exhaustive, we will have to reject this "radical" claim.

Conclusions

The concept of an "Iron Law of Wages," in which the forces of capitalism drive incomes down to subsistence, has long been criticized, if not dismissed, by most economists. Two hundred years after Malthus put forth this grim forecast, there is, however, a resurgence of concern about the inevitability of a subsistence wage level, and economics, to many, is once again deserving of its "dismal science" label. Although the reasons behind the Iron Law have changed, for some, from population growth to globalization, the results are the same: the institution of capitalism will force the poor throughout the world to live at subsistence. Despite these dire (and incorrect) projections, the arguments for the positive consequences of free trade are, as Mises stated, "obvious and indisputable". We have not attempted to use these indisputable arguments to counter this renewed support for an Iron Law, though that would be easy to do. Instead, we call into question the entire concept of a subsistence wage level as a possible outcome of economic activity in capitalist societies. We do not mean to say that there is no such thing as a level of nutrition, a subsistence level, which results in zero population growth. Rather, we have argued that in any capitalist society that has seen the institution of slavery—and every one has—the prevailing wage rate would always have been above subsistence. Slavery and the Iron Law of Wages are mutually incompatible.

Notes

[1] The "Iron Law of Wages" (supposedly) drives wages down to subsistence levels.

[2] In his second edition of *An Essay on the Principle of Population*, Malthus changed his definition of a subsistence wage to include non-nutritional remuneration (Malthus [1803] 2004:127). The subsistence wage became defined as any wage level where the population does not grow. This could include wages where the wage might buy a color TV, but that parents choose not to have more kids if they can't raise them without Cable. This re-definition, under the label "moral restraint", deprived the subsistence concept of scientific validity.

[3] It should be noted here that we assume slaves were primarily inputs into the production process. Slaves might have been consumer goods, but we doubt seriously whether this was the case to any large degree.

[4] What might be considered a counter argument goes as follows: Even net of the monitoring and capture costs, slavery was very profitable. How can this be, unless slave labor was more productive or efficient than free labor? Free labor was available. Why did plantation owners choose to buy rather than rent (i.e. purchase a slave rather than hire a worker)? Only because it was more efficient to buy. That is, there was a greater gap between the cost of the slave and the present discounted value of his services, than between the discounted marginal revenue product of the free worker and his wage. Of course, in equilibrium, which we never arrive at but are always approaching, these two gaps must be identical, and zero.

[5] For the distinction between extortion and blackmail, see Block (2000); Block, Kinsella and Hoppe (2000)

References

Becker, Gary. 1964. *Human Capital*. New York: The National Bureau of Economic Research.

Block, Walter. 2000. "Threats, Blackmail, Extortion and Robbery And Other Bad Things." *University of Tulsa Law Journal* 35:333–351.

Block, Walter, Joseph Horton and Debbie Walker. 1998. "The Necessity of Free Trade." *Journal of Markets and Morality* 1:192–200.

Block, Walter, Stephan Kinsella and Hans-Hermann Hoppe. 2000. "The Second Paradox of Blackmail." *Business Ethics Quarterly* 10:593–622.

Brown, D.K., A.V. Deardorff and R.M. Stern. 1996. International labor standards and trade: a theoretical analysis. In *Fair Trade and Harmonization, Prerequisites for Free Trade? Vol. 1: Economic Analysis*, ed. J. Bhawati and R.E. Hudec. Cambridge: MIT Press pp. 227–280.

Dreuil, Emile, James Anderson, Walter Block and Michael Saliba. 2003. "The Trade Gap: The Fallacy of Anti World-Trade Sentiment." *Journal of Business Ethics* 45:269–281.

Eveleth, P.B. and J.M. Tanner. 1976. *Worldwide Variation in Human Growth*. Cambridge: Cambridge Univ. Press.

Fogel, Robert William and Stanley L. Engerman. 1974. *Time on the Cross: The Economics of American Negro Slavery*. Boston: Little, Brown.

Frisch, R. and R. Revelle. 1969. "Variation in Body Weights and the Age of the Adolescent Growth Spurt among Latin American and Asian Population, in Relation to Calorie Supplies." *Human Biology* 41:185–212.

Greenaway, D. and D. Nelson. 2001. "The assessment: globalization and labour-market adjustment." *Oxford Review of Economic Policy* 16:1–11.

Hanson, Gordon H. 2001. "The Globalization of Production." *NBER Reporter* .

   URL: *http://www.nber.org/reporter/spring01;hanson.html*

Heald, F.P., P.S. Remmel and J. Mayer. 1969. Caloric Protein and Fat Intakes in Children and Adolescents. In *Adolescent Nutrition and Growth*, ed. F.P. Heald. New York: Appleton-Century-Crofts pp. 17–35.

Klevorick, A.K. 1996. Reflections on the race to the bottom. In *Fair Trade and Harmonization, Prerequisites for Free Trade? Vol. 1: Economic Analysis*, ed. R. He. Hudec. Cambridge: MIT Press pp. 459–467.

Lee, E. 1996. "Globalization and employment: is anxiety justified?" *International Labour Review* 135:5:485–497.

Malthus, T.R. 2004. An Essay on the Principle of Population. In *An Essay on the Principle of Population: Text, Sources, and Background Criticism*, ed. Philip Appleman. New York: Norton pp. 13–124.

Mehmet, Ozay and Akbar Tavakoli. 2003. "Does Foreign Direct Investment Cause A Race to the Bottom?" *Journal of the Asia Pacific Economy* 8:2:133–156.

Nader, Ralph, ed. 1993. *The Case Against Free Trade: GATT, NAFTA, and the Globalization of Corporate Power.* San Francisco: Earth Island Press.

Rodrik, D. 1997. *Has Globalization Gone Too Far?* Washington: Institute for International Economics.

Smith, Adam. 1991. *The Wealth of Nations.* New York: Prometheus.

Sutch, R. 1975. "The Treatment Received by American Slaves: A Critical Review of the Evidence Presented in Time on the Cross." *Explorations in Economic History* 12:335–438.

Sutch, R. 1976. The Care and Feeding of Slaves. In *Reckoning with Slavery*, ed. P.A. David. New York: Oxford University Press pp. 231–301.

Tanner, J.M. 1962. *Growth at Adolescence.* Springfield: Charles C. Thomas.

von Mises, Ludwig. 1966. *Human Action: A Treatise on Economics.* Third ed. San Francisco: Fox & Wilkes.

von Mises, Ludwig. 1985. *Liberalism.* Irvington: The Foundation for Economic Education.

Wilson, J.D. 1996. Capital mobility and environmental standards: is there a theoretical basis for the race to the bottom? In *Fair Trade and Harmonization, Prerequisites for Free Trade? Vol. 1: Economic Analysis*, ed. J. Bhawati and R.E. Hudec. Cambridge: MIT Press pp. 393–427.

Wood, Adrian. 1995. "How Trade Hurt Unskilled Workers." *Journal of Economic Perspectives* 9:3:57–80.

The Journal of Interdisciplinary Economics, 1996, Vol. 7, pp. 217–230
0260-1079/96 $10

Labor Market Disputes: A Comment on Albert Rees' "Fairness in Wage Distribution"

Walter Block

Economics Department, College of the Holy Cross, Worcester, MA 01610, USA

Abstract

Albert Rees criticizes neo classical labor economics for paying too little attention to fairness and utility function interdependence. As a result, he claims, this theory cannot fully come to grips with wage determination as it actually exists in the real world. The present author takes issue with Rees, and attempts to defend traditional labor economics against his criticisms.

1. INTRODUCTION

Albert Rees, author of the canonical textbook in labor economics (1973), looks back on a long career in this field, and doesn't much like all that he surveys. On the contrary, in his present view, the neoclassical wage theory with which he has long been associated is deficient in at least one respect: It fails to take into account the effect of how others are doing, payment wise, on the general determination of wages. That is, traditional theory overemphasizes considerations such as labor productivity and insufficiently takes into account how interdependence affects wage determination. Although Rees explicitly warns against being interpreted as "recanting" his distinguished contributions to neoclassical labor economics (1993, p. 251),[1] one may perhaps be forgiven for interpreting him in precisely this way. This is so particularly when he states that he doesn't "think that neoclassical theory is wrong . . . but it is incomplete" (p. 251). The lacunae? It fails to incorporate into the analysis "fairness," and "interdependence of utility functions" (p. 247).

What are the specifics of Rees' arguments? He acknowledges that in neoclassical analysis, wage determination is explained by the interaction of a supply curve based on worker utility considerations and a demand schedule based on marginal revenue productivity. But then he relates a

218

series of anecdotal experiences, not from the academic world, but rather from his career as a member of wage stabilization boards, as a corporate director, university provost, and college trustee: "In none of those roles did I find the theory I had been teaching for so long to be the slightest help" (p. 243).

II. FACES OF FAIRNESS

Why not? For one thing, because in the real world workers have a passion for "fairness," something not contemplated in textbook labor economics. And of what does "fairness" consist? It "involves the concept that if workers in one union or group receive a certain wage increase, the workers in another union or group are entitled to the same increase" (p. 244). So strong is this passion for justice on the part of organized labor that "a union in the retail food industry allow(ed) an employer to close a statewide group of stores rather than agree to a wage increase smaller than that just received by its traditional comparison group" (245).

Now this is more than curious. One problem is that Rees speaks only in terms of changes in wages, not of their initial levels. This runs counter to the usual assumption that the former is the dog, the latter only the tail. After all, unless there is a theory of the equilibrium level of wages, we shall never know whether a given change is a move toward or away from the point at which supply and demand intersect.

Another difficulty arises in using organized labor as a paradigm case of fairness. Unions are the only nongovernmental institution in society to be vested with the "right" to ban others ("scabs") from competing with them (Block, 1991). Suppose that there were two gangs of armed robbers, and one of them managed to mulct an extra $100,000 out of a hapless victim. Thereupon the other, in response, wrested the same amount from other innocent prey. Would we be tempted to describe the actions of the latter as "just" or "fair"? Hardly. All that could be said, at best, of the actions of the second group of criminals was that they had "kept pace" with those of the first.

John L. Lewis's coal miner's union shut down pretty much an entire state. After his depredations, West Virginia was an economic basket case for decades. From this we are to deduce fairness? And this applies as well to the San Francisco plumbers local which insisted upon "me too plus a dollar" (p. 245). If another union received a pay boost, this one, which sought parity with it, would insist on the same increase, plus $1. It must be nice to wield so much coercive power, but what this has to do with fairness is rather elusive.

Speaking of gangsters, it must be recognized, as Rees does not, that there is more than a superficial resemblance between these organ-

219

izations and labor union leaders: each of them is guilty of using violence, or the threat of violence, to gain their ends. In the case of the latter, although many commentators have interpreted their actions as being aimed at employers, more sophisticated ones (Reynolds, 1984; Petro, 1957; Block, 1991; Williams, 1982) have pointed out that they are more often directed at competing workers.[2] To characterize the desires of these entities as "fair" is thus almost a travesty of justice. Yes, "unions may regard themselves as entitled to the same absolute (or percentage, it matters not one whit) increase previously won by another union" (p. 245), but their mere subjective desires have very little relationship to fairness.

It is important to keep the distinction between normative and positive economics in mind. This is necessary at any time in economic public policy analysis, but it is crucial when matters of "fairness" are raised. The definition employed by Rees is something akin to equality. If one employee receives a raise, the other should get it too. But why is this necessarily "fair?" Consider the biblical parable, "The Workers in the Vineyard" (*Matthew*, 20). Here, the employer paid the same amount to a person who worked a full day as to the one who started only in the late afternoon. Was this "fair?" It was, at least according to one common sense definition of that word, that which holds as fair any and all "capitalistic acts between consenting adults" (Nozick, 1973). Certainly, this is the view of the Bible, which gives the definitive reply of the employer to the disgruntled workers who think they are hard done by:

> " 'Listen, friend,' the owner answered one of them, 'I have
> not cheated you. After all, you agreed to do a day's work for
> one silver coin. Now take your pay and go home. I want to
> give this man who was hired last as much as I gave you.
> Don't I have the right to do as I wish with my own money?
> Or are you jealous because I am generous?' " (Matthew, 20).

Rees informs us that concern with comparisons are by no means a monopoly of the employee side of the equation. Employers, too, look at such matters in this way. For example, consider the firm which refused to "give the red-blooded (nonunion) Americans who built this company less than those radicals in the union" (p. 245). But why is it unfair to pay off moral debts? This employer obviously felt positively toward the nonorganized sector of his work force for past support. Perhaps, too, he looked to them for backing in the future. Rees' implicit view that it was improper for him to pay both types of employees equally, when he was not forced to do so, would appear to imply that gifts and/or inducements to good future behavior are per se unfair.

It is the same with regard to executive salaries. According to Rees, "when executives or directors set salaries they refer to surveys of

220

salaries in comparable companies or institutions" (p. 245). This may well be so. But purely as a matter of positive economics, there is always that little matter of marginal revenue productivity lurking somewhere in the background. If executive's salaries (or those of anyone else for that matter) deviate from MRP, there are strong market forces which tend to bring these two figures back toward equality with one another. It seems strange to have to mention this in connection with an article written by Albert Rees, but if wages fall below MRP,[3] the quit rate will rise; failing that, other firms will scoop up these labor bargains. This market force is powerful enough to explain the fact that growers in the U.S. and Canada, thousands of miles away from Mexico, go down there every year to take advantage of the cheap labor there. They "exploit" low cost workers in that country by bidding their below market wages closer to productivity levels. If wages are above MRP on the other hand, a firm is courting bankruptcy. It is only when they are equal that there are no market forces unleashed to change labor market conditions. Top management can refer all it wants to other, supposedly comparable salaries, but if this process leads to pay which deviates from productivity contributions, it is untenable in the long run.

Of course "most organizations state that they would like to be in the upper half of what they regard as the relevant salary distribution" (p. 245). People will say anything on such surveys; if such appraisals were accurate, we would all be driving a Mercedes. The only "disturbing . . . macroeconomic implications of these preferences" (p. 245) would appear to be cost push inflation. But as the monetarists have shown (Friedman and Schwartz, 1963; Rothbard, 1983; Mises, 1971), price rises do not result from wage push; rather, they are caused by excessive monetary creation.

This analysis applies, as well, to those who whine that their wages are not as high as their long seniority would suggest. States Rees of such a person " 'Why do I make less than Y when I have given this outfit devoted service for twenty years, while he was only hired last year?' " (p. 246) The answer, which seems to have escaped Rees, is that sunk costs are sunk. Past contribution usually matters little, except insofar as it accurately predicts future usefulness. But the key is future expectations; *that* is why the new kid on the block is paid more. Merit, of course, is an entirely different matter, despite Rees' conflation of the two. If merit is interpreted as a proxy for productivity, then the market will tend to reward it.

Consider in this regard Rees' numerical example. A full professor with 20 years of satisfactory service earning $52,000 is appalled to learn that a newly-minted Ph.D. was offered $38,000. In his view, the proper differential between full and assistant professors would entitle him to at least $60,000. States Rees:

"This is an example of the general proposition that wage inequities must be remedied by raising the wage that is too low, and not by lowering the one that is felt by others to be too high" (p. 246).

A more confusing analysis would be hard to imagine. First, this appears to be a non *sequitur*. From what principle did we derive the conclusion that the way to promote equity is to raise a "too-low" wage, not to lower one that is too high? Even on his own grounds of equity as fairness, this by no means follows. Maybe it is better, more "fair," to do the very opposite; or to lower the one that is too high, while at the same time raising the one that is too low, until a midpoint is reached for both. This, at least, seems to have the advantage of symmetry, but in the absence of any criterion vouchsafed us we are flying in the dark.

Second, a university setting is perhaps amongst the most unfortunate of choices to illustrate the workings of an economic system, in that there are few if any profit and loss incentives which can impose discipline on the wage relationship. In a public university, if salaries bear no relationship to productivity, the threat of bankruptcy hardly looms. Subsidies courtesy of the taxpayer can paper over any number of economic sins. And even in the private sector, large endowments are likely to shield decision makers from paying for the error of their ways.

Third, this analysis bears an improper level of subjectivity. "Lowering the one that is felt by others to be too high," indeed. Are mere "feelings" as to others' wages supposed to play any meaningful role in economic analysis? If so, the sky is the limit. Anyone can announce himself as vexed at any economic phenomenon, thus casting doubt on its "fairness." Fourth, but not least, just how was it determined that there was an inequity in this particular case? What is the criterion? How can the proposition be tested? Rees, unhappily, is silent on all these issues.

III. FAIRNESS AND WAGES

Undaunted, however, Rees leaps to his next point: "Employers do not insist on fairness—workers and their unions do" (p. 246). Now this is rather one sided. It is truly rare, in any dispute, that one party is totally on the side of the angels. Even worse, this claim stands contradicted out of Rees' own word processor, for it was he who referred to the employer who refused to "give the red-blooded (nonunion) Americans who built this company less than those radicals in the union" (p. 254). And if this isn't an attempt to be fair, nothing is.

Upon reading Rees, one comes to the conclusion that he really isn't referring to "fairness," at least not as this word is used in normal speech. Instead, he seems almost to be alluding to reducing controversy,

222

or unhappiness, or, perhaps even better, dispute creation. This is seen in his contrast between the new worker and the experienced, long-serving one, who are both offered a wage deemed unfair. In the former case, "he will simply decline the offer," but in the latter there will be created a "sense of grievance" (p. 246). Interpreting Rees in this manner —he is interested in the positive economic explanation of labor conflict, not in the normative discussion of equity—at least has the advantage of providing another set of lenses though which his contribution can be understood.

The next thicket to ensnare Rees has to do with the interdependence of utility functions. According to his analysis, neoclassical economics must be modified in order to incorporate this insight. But this is hardly accurate. On the contrary, economists have been aware of the fact that we tend to be affected, positively or negatively, by the well-being of others since at least the time of Smith (1817) and Canard (1801). Stigler (1965, pp. 100–101)[4] refers to:

> "The inclusion of the quantities consumed by other people in the utility function of the individual. Thus one's pleasure from diamonds is reduced if many other people have them (or if none do!), and one's pleasure from a given income is reduced if others' incomes rise. This line of thought is very old, but it was first introduced explicitly into utility analysis (by Fisher) in 1892."

Apart from this aspect of the history of thought, Rees is somehow led from his focus on interdependence to the view that "envy (is) a local phenomenon while compassion is often a more distant one." That is, we are usually envious of the person "at the next work station or in the next town, while the beneficiary of charity may be half a world away" (p. 247). Yes, true enough, this sometimes occurs. But it cannot be denied that there is also such a thing as local charity, collected in the same town as it is disbursed, and that many Americans are envious of the Japanese (and vice versa), each of whom live thousands of miles away from the other. The point is, interdependence of utility functions is a theoretical construct; it does not logically imply the identity of the persons who will enter into a given individual's utility function, nor whether the welfare of these others will impact this utility in a positive or negative direction.

Next consider Rees' example of secretarial pay:

> "On Monday, you inform your secretary that she has received a pay increase of $20 per week. On Tuesday, she discovers that all the other secretaries in the organization have received increases of $30 per week. Clearly, she will be

223

less happy on Tuesday than she was on Monday. Moreover, it is highly likely that she will be less happy on Tuesday than she was the previous Friday—that is, she would prefer that no one get an increase to getting a smaller one than her fellow workers" (p. 247).

There is little doubt that the secretary will be less happy on Tuesday than she was on Monday. She would likely prefer both absolute and relative wage hikes, unless, of course, the pay received by her co-workers enters very heavily indeed into her own utility function. On the other hand, if she was the only one to receive a heftier pay envelope last month, and feels that everyone else in the office resents her, or if she feels her head near the chopping block due to her "excessive" salary, then on the contrary she may be delighted with the scenario depicted by Rees.

Moreover, even if Rees' is a true analysis of the secretary's reaction, it doesn't show a lacunae in traditional labor economics. As this author himself concedes, "the general idea of interdependence of utility functions" has been used reasonably widely in the economics literature (p. 247).[5] How, then, will incorporating it into neoclassical labor economics make good any serious oversights?

Further, how does the much vaunted "unfairness" enter into the picture? If fairness requires absolute (or proportional) equality, then the secretarial pay example may indeed depict an inequitable situation. But under other assumptions, as we have seen, this need not be true. Consider the case where all the other secretaries are nieces of the boss. Is nepotism necessarily unjust? If so, wouldn't this mean that Christmas or birthday gifts are also unfair, given that the boss will likely bestow them on his relatives, not strangers? That is, how are we to even conceptually distinguish Christmas gifts to nieces from pay raises to them. Yes, one is a once-and-for-all occurrence (at least for this year) and the other takes place every week, but surely there is some rate of discount that can equate them.

IV. EXECUTIVE COMPENSATION

From secretarial salaries we move to remuneration in the executive suite. As "the contribution of an executive to the output of the firm is extremely hard to measure, equity considerations . . . will loom large." These, in turn are driven by comparisons "with other executives in the same firm and with executives in the same function at other firms in the industry." In other words, all corporate officers rise or fall (mainly the former) together, thanks, in part, to the fact that "compensation for these positions is public information." (p. 248). This is why salaries

224

above the glass ceiling have begun to catapult. The "conscience" of the people in charge of setting these astronomical levels "is almost the only demand-side constraint that exists for executive pay" (p. 249).

Sensible as this may sound upon first reading, there are difficulties. This analysis, first off, appears to contradict Rees' explanation of union wage setting. There, he referred to a "linkage." He maintained that ". . . wages in different occupations in an enterprise are tied together in an accepted structure that is costly to disturb . . . probably because in a union situation information on occupational rates is available to everyone" (p. 248). But he just got finished proclaiming that such knowledge, "public information," was also available in the executive suite. Yet in the latter context, there are, seemingly, no barriers at all to continuing salary inflation, except, perhaps, for "conscience." How can two very different results be explained by the very same (informational availability) causal conditions? Rees does not suggest that union wages are on an uncontrolled upward roller coaster, as in the case of corporate officers. How can this be if the same explanatory factor is in operation in both cases?

No. We shall have to seek elsewhere for our explication. The reason executive salaries are on an uncontrolled flight path is because our legal system functions so as to overpower the market forces that could otherwise have nipped this in the bud. It is almost as if we are being overrun by rabbits, after having engaged in a protracted and thorough wolf kill program, and now are wondering at the spread of these long eared furry creatures.

Well, who is the wolf of the piece, the person who could have called a halt to escalating top managerial salaries, the one who as a matter of fact *was* engaged in just this activity, when he was imprisoned? His name is Michael Milken (Manne, 1966; McGee and Block, 1989). This entrepreneur was involved in raising vast sums to purchase poorly managed business firms. The corporations with the worst excesses with regard to top management salaries were precisely the ones whose stock prices most seriously underestimated their true values. Wall Street undervalued them expressly because of these uncontrolled raids on the corporate till, e.g., outrageous managerial compensation levels. In other words, the firms guilty of the behavior Rees so correctly denigrates were the ones subjected to "attack" by Milken, in a series of "unfriendly" corporate raids.

These takeovers were of course only "hostile" to the members of the bloated executive suites, rich beyond the dreams of avarice. For long-suffering stock and bond holders, these takeovers were extremely "friendly." Nevertheless, the corporate mandarins were able to convince the legal powers that be that Milken's fund raising was based on "junk" bonds; that he was guilty of "insider trading" and "fraud." As a result, this "wolf" was incarcerated, the "rabbits" in the executive suite are gorging themselves on their illgotten gains, and people like Rees (see also Berle and Means, 1932) are running around (to mix our metaphors even further) like chickens without a head in their attempt to explain this phenomenon.

Nor is Rees unaware of the Milken debacle. He even goes so far as to mention that sainted company Drexel Burnham Lambert, not as an example of the *cure* for the problem, but rather as a case in point of "conspicuous cases of large executive bonuses in firms losing market share, or even . . . about to go bankrupt" (p. 249). Now the person with the largest "executive bonus" in all of business history was none other than Michael Milken, employee of Drexel Burnham Lambert. But the *reason* for its financial difficulties have nothing to do with Milken's very high level of compensation. On the contrary, he was worth every penny to them. The explanation, rather, is the illicit acts of the various attorneys general, who both prosecuted and persecuted Milken. In terms of our analogy from the animal kingdom, then, it is as if Rees is blaming the wolf for *being* a rabbit.

We need not fear unduly, however. The system of profit and loss is still grinding away. If the career of one "wolf" is terminated with semi-extreme prejudice, others will arise to take his place. The economic axiom that wages tend to equal marginal revenue product is still operational. If top managers are given salaries in excess of their contribution to the production process, this necessarily sets up forces which will tend to obviate such practices. Governments can place obstacles in the path of such a process, but they cannot block it entirely. The share prices of such corporations will be pushed below what their potential earnings (based on more accurate executive salaries) would otherwise have entailed. This will set up incentives for others to purchase these companies at bargain prices. Stopping Milken, and thereby frightening all other would be Milkens will make this process more difficult, to be sure. But just as the "War Against Drugs" will always fail because with every "success," profits in this enterprise are driven up, thus encouraging even more illegal entrepreneurial activity (Boaz, 1990; Judson, 1974; Thornton, 1991; Block, 1993), so too will the machinations of these internal corporate raiders (e.g., the unconstrained looting executives) be brought back down to earth.

Nor should there be any mystery about why salaries at state colleges, National Institutes of Health, and Congress (p. 249) should be pegged at nonequilibrium levels. The market forces which tend to equate wages and MRP can only operate in the private sector, where penalties are paid when this economic axiom is violated. As this simply does not take place in the governmental sector, we cannot expect wage rationality to occur there.

226

V. MARKET FORCES AND SALARY STRUCTURES

An appreciation of this fact is what is missing from Rees' analysis. Consider now the case where one professor receives an offer from a comparable outside institution and threatens to leave if it is not matched by his present employer. Yes, this sets up all sorts of problems:

> "The other members of the department are now likely to feel unfairly treated whichever option is chosen. If the offer is matched, the salary differentials within the department will be viewed as inequitably large. It is not matched and the faculty member 'leaves, the remaining members will feel that they are being paid less than they could earn elsewhere" (p. 250).

The problem is that few, if any, universities function under the sort of bottom line thinking familiar in business. There are no profit and loss sanctions, or at least they are greatly attenuated. As a result, the deans and college presidents who make salary offers, operate outside of the usual market constraints. Why should we expect rationality from such a system, any more than we could from the planning authorities of the Soviet Union or Cuba? (Boettke, 1990; Hoppe, 1989)

In the event, however, even a modicum of common sense might be expected to come into play. Take Rees' two cases. First, if the outside offer is matched, those who still labor under the old lower salaries will indeed view the new differentials as "inequitably large." If so, they have an alternative: attain another legitimate offer, and use it to bargain for their own raises. If they cannot, then all talk of inequity is just so much blather. Similarly, for the second case, if the outside offer is not matched, and the faculty member leaves. Again the less attractive (in the sense of lower alternative cost) members of the department may feel "they are being paid less than they could earn elsewhere" but in the absence of evidence for this contention, namely, a bone fide outside offer, their claim is at best unproven.

Thus it is simply untrue to assert, as does Rees, that "pay equity is a goal that is constantly being pursued but is never reached. In this respect, it is not unlike market equilibrium" (p. 250). On the contrary, in the case of market equilibrium, there are forces pushing the economy toward this point. True, we may never reach it, because other exogenous changes requiring a new equilibrium may well arise long before the old one is attained, but at least these forces are always in operation. In the case of "equity," in contrast, there are no such analogous economic forces. Even more problematic, no one, Rees specifically, has ever managed to unambiguously define "equity," at least not in a way that does not imply voluntary accord. That is to say, we already have a perfectly good notion of justice in wage settlements: the salary to which both employer and employee agree. It is incumbent on those such as Rees who challenge this view to come up with a coherent alternative; so far, this has not been done.[6]

This analysis applies as well to Rees' two other cases in point: when "a new member is hired into a department at a salary substantially above those already at the same rank because this salary was required to recruit him" and the question of whether a university should "maintain the same top salaries across different disciplines even when the outside market pays different salaries" (p. 250). If universities are forced to function in a market the ordinary factors of supply and demand will solve these "problems." If not, there is no nonarbitrary way in which this can be done. This is something that the societies of Russia and Eastern Europe are now in the process of learning. Unfortunately, it has been forgotten by some western academics.

VI. TWO-TIER WAGE STRUCTURES

The next candidate for inequitability is the "two-tier wage structure." Yes, it is indubitable that these phenomena "afford a test of the importance of the concept of fairness in wage determination" (p. 251). It is indeed difficult to imagine anything more inequitable than the employer paying identical workers the same salary. But matters become more clear when we reflect on the fact that the employer is only likely to acquiesce in any such system *against his will*.[7] This is because the firm will always be tempted to fire higher-cost workers and replace them "lower tier" employees. But this is no challenge to neoclassical economic analysis or to ordinary morality. The reason for this patently unfair structure is easy to discern. It comes about as the result of the operation of an inequitable institution, namely, unionism.

Actually, two-tier wages make explicit the unfortunately implicit illegitimate power of unions. If otherwise homogeneous workers are compensated at two very different levels, to what other institution besides coercive unionism can we point as an explanation?

A strong analogue to rent control (Tucker, 1990) may be found in the two-tier wage system. It would make perfect sense, given that rent control is a given, to allow tenants "protected" by this legislation to sublet their apartments. This would reduce if not eliminate most of the negative effects of this law: the excess demand, the failure to maintain and upkeep, and the low vacancy rates.[8] The only problem with this plan is that it would make all too explicit the forced transfer of wealth from landlord to the tenant who is allowed to sublet. That is why this proposal has proven unsatisfactory to both sides: to tenants, because the issue of totally repealing rent control would then "naturally" arise, and

228

to the landlords, since the original theft they were forced to suffer would not be ameliorated.

The surprise, then, is not that the two-tiered system has not already disappeared; Rees correctly envisions this occurring "when the lower paid workers become a majority in the bargaining unit" (p. 251). Given that this is so problematic, the surprise is that it was ever allowed to occur in the first place.

VII. HOMOGENEOUS LABOR

The simplest neoclassical assumption in this regard is of course that all labor is equally productive. But Rees does not tax traditional economists with so simple a model. He realizes that it is "perfectly consistent with neoclassical theory to have two or more classes of labor, each homogeneous within itself, and treated as a separate factor of production" (p. 251). However, he charges, that even this level of sophistication will not suffice for the conditions

> "actually facing the large firm . . . At this extreme, the assumption of homogeneous labor breaks down utterly, and salary determination will necessarily involve an element of arbitrary decision making or of bargaining" (251).

My view, in contrast, is that neoclassical theory is by no means so feeble. At a purely formalistic level, "two or *more* classes of labor" can cover as many employees as could possibly *be* employed by the modern firm, no matter how large. Of course, if there were a separate class of homogeneous labor for each individual person on the payroll, there wouldn't be too much of a point of talking about classes in the first place. But, as a practical matter, it is unlikely in the extreme that amongst the literally tens of thousands of employees of any large corporation there wouldn't be found *some* homogeneities. Yes, each human being is unique, and beloved of his friends and family in a special way. Surely however there would be *some* jobs where the heterogeneities would be so small as to disappear for all intents and purposes.

But, even in the extreme scenario of no homogeneity whatsoever implicitly pictured by Rees, it would not still not follow that salary determination would be arbitrary. To be sure it would be much more complicated in a world of total and complete heterogeneity; certainly it would be more expensive to tailor salaries to individual workers under such extreme assumptions. Arbitrariness, however, is an attack on the economic axiom that wages tend to equal MRP. This tendency would still hold, as shown by assuming it not to hold. As stated before,

229

whether wages deviate from MRP on one side or the other, there are incentives brought to bear which tend to bring the two into equality, at least in equilibrium.

VIII. CONCLUSION

There may well be flaws in neoclassical labor economics; to doubt this even as a possibility is to leave the realm of science and to enter that of faith and creed. But Rees has not succeeded in showing any basic defects. However, he is to be congratulated at least for calling what might otherwise pass for verities into question. Surely, his skeptical attitude is one that bears emulation.

NOTES

1. All otherwise unidentified citations refer to Rees (1993).
2. True, unions do "good works." They contribute to charity, run bowling leagues, help in the political and educational process, are active in promoting pensions. However, the Mafia, too, is well-known for its support of religious and other public spirited concerns. It is also undeniable that unions, not gangs, are legal institutions in the U.S. and elsewhere. But to equate this with fairness or justice is to commit the sin of legal positivism: whatever the law states is right. To see this error, reflect on the fact that murderous "white citizens councils" were legally recognized entities in the post civil war confederate states, and that the Nazi Party was a legal entity in Germany from 1933 to 1945.
3. More technically, below the alternative cost of MRP, namely the MRP that would obtain in the next best alternative to present employment. See Block (1990).
4. I owe this citation to Frank Petrella.
5. Given this, it is difficult to understand his complaint that neoclassical economics must be modified in order to incorporate the interdependence of utility functions.
6. Rawls (1971) and members of the public choice school (e.g., Buchanan and Tullock, 1971; Buchanan, 1979, p. 196; 1990, p. 9) have attempted to define equity by resort to the "veil of ignorance." Equity is defined as that allocation of whatever (wealth, income, salaries, beauty, etc.) that would have been chosen had no one known what levels of these traits or goods would have been assigned to them if we were to start off again, de novo. But why should this be equitable? Surely risk preferrers, neutrals and avoiders will make very different choices in this regard. By definition, they cannot all be "equitable," if for no other reason than that they will all be different. (This holds if equity is defined in terms of static income distributions, as is typically done. If it is defined in terms of a just process, then of course very different outcomes can all be seen as equitable.) But unless all but one of these three perspectives on risk can itself be shown to be "inequitable," there is no hope that this mental experiment will be able to pinpoint a proper distribution. See Nozick (1974) for a devastating critique of the "veil of ignorance" ploy.
7. We have already seen that if the employer *agrees* to a two-tiered wage system, or even more so, *initiates* it, as in the case of voluntary nepotism, or the parable of the vineyard, this is not necessarily "unfair."
8. About the only thing it would not ameliorate would be the reduced new construction rates, but this is because old investors-owners would still remain expropriated even with this amendment to the law. Similarly, would be newcomers to this industry might reasonably fear that the same treatment would be imposed upon them, in the future.

230

REES BIBLIOGRAPHY

Berle, A.A., Jr., and Means, Gardner C., *The Modern Corporation and Private Property* New York: Commerce, 1932.

Block, Walter, "Drug Prohibition: A Legal and Economic Analysis," *Journal of Business Ethics*, 1993, Vol. 12, pp. 107–118.

Block, Walter, "Labor Relations, Unions and Collective Bargaining: *Journal of Social, Political and Economic Studies*, Vol. 16, No. 4, Winter 1991, p. 477–507.

Block, Walter, "The Discounted Marginal Value Product—Marginal Value Product Controversy: A Note," *Review of Austrian Economics*, Vol. IV, 1990, pp. 199–207.

Boaz, David, ed., *The Crisis in Drug Prohibition*, Washington D.C.: The Cato Institute, 1990.

Boettke, Peter J., *The Political Economy of Soviet Socialism*, Boston: Kluwer, 1990.

Buchanan, James M., "Public Choice and Public Finance," in *What Should Economists Do?*, Indianapolis: Liberty Press, 1979.

Buchanan, James M., "The Contractarian Logic of Classical Liberalism," Ellen Frankel Paul and Howard Dickman, eds., *Liberty, Property, and the Future of Constitutional Development*, Albany: State University of New York Press, 1990.

Buchanan, James M., and Gordon Tullock, *The Calculus of Consent: Logical Foundations of Constitutional Democracy*, Ann Arbor: University of Michigan, 1971.

Canard, N.F., *Principles d'economie politique*, chap. v, Paris: Buisson (1801).

Fisher, Irving, *Mathematical Investigations in the Theory of Value and Prices*, New Haven: Yale University Press, 1892 (1937).

Friedman, Milton, and Anna J. Schwartz, *A Monetary History of the U.S., 1867–1960*, New York: National Bureau of Economic Research, 1963.

Hoppe, Hans, *A Theory of Socialism and Capitalism*, Boston: Dordrecht, 1989.

Judson, Horace Freeland, *Heroin Addiction in Britain*, New York: Harcourt, Brace, Jovanovish, 1974.

Manne, Henry A., *Insider Trading and the Stock Market*, New York: The Free Press, 1966a.

McGee, Robert, and Block, Walter, "Information, Privilege, Opportunity and Insider Trading," *Northern Illinois University Law Review*, Vol. 10, No. 1, 1989.

Mises, Ludwig von, *The Theory of Money and Credit*, New York: The Foundation for Economic Education, 1971.

New Testament, "The Workers in the Vineyard" (Matthew, 20).

Nozick, Robert, *Anarchy, State, and Utopia*, New York: Basic Books Inc., 1974.

Petro, Sylvester, *The Labor Policy of the Free Society*, New York, Ronald Press, 1957.

Rawls, John, *A Theory of Justice*, Cambridge: Harvard University Press, 1971.

Rees, Albert, *The Economics of Work and Pay*, New York: Harper and Row, 1973.

Rees, Albert, "The Role of Fairness in Wage Determination," *Journal of Labor Economics*, 1993, Vol. 11, No. 1, Part 1, pp. 243–252.

Reynolds, Morgan O., *Power and Privilege: Labor Unions in America*, New York: Manhattan Institute for Policy Research, 1984.

Rothbard, Murray N., *The Mystery of Banking*, NY: Richardson and Snyder, 1983.

Smith, Adam, *The Theory of Moral Sentiments*, Boston: Wells and Lilly, (1817).

Stigler, George, "The Development of Utility Theory," *The Journal of Political Economy*, Vol. LVIII, August and October, 1950; reprinted in *Essays in the History of Economics*, Chicago: University of Chicago Press, 1965.

Stigler, George, "The Development of Utility Theory," *The Journal of Political Economy*, Vol. LVIII, August and October, 1950.

Thornton, Mark, *The Economics of Prohibition*, Salt Lake City: University of Utah Press, 1991.

Tucker, William, *The Excluded Americans*, Chicago: Regnery-Gateway, 1990.

Williams, Walter, E., *The State Against Blacks*, New York, McGraw-Hill, 1982.

The DMVP-MVP Controversy: A Note

Walter Block*

We are all familiar with the process of discounting the future. From the earliest courses in economics we are taught that money receivable right now is not the equivalent of money receivable one year hence; that money receivable one year from now is not equivalent to money which will fall in to our clutches after a period of two years. And not just because inflation may erode part of the value, or because of the risk of never seeing the money. Even in a perfectly certain world of no inflation, where all accounts receivable were fully guaranteed, we would still value money more, the sooner we were to receive it.

If this were not so, we could never act in the present,[1] for every action done now *could* have been done in the future. The fact that we choose to act in the present, when we could have waited, shows that we prefer the present; that we enjoy goods, the sooner, the better. But the future will present the same alternatives: action and non action. *Future* action will thus *also* imply time preference for the present, paradoxically. By acting in the immediate future, instead of waiting for the even more distant future, we also show ourselves as present oriented. The only way to illustrate a lack of preference for the present is never to act at all—a manifest impossibility for human beings.

One implication of the foregoing is that we discount money receivable in the future. This is done in accordance with the rate of interest. Simply put, we prefer a dollar today to a dollar tomorrow because we can always put our present dollar in the bank, collect the interest payment, and have more than a dollar. Given a non-inflationary world and a guarantee that the bank will not renege, we are sure to have more in the next period. If the rate of interest is 10 percent, then $1.00 today will be worth $1.10 at the end of one year.

Alternatively, we can say that payments receivable in the future

*Walter Block is senior research fellow at the Fraser Institute and director of its Center for the Study of Religion and Economics; he is also co-editor of *The Review of Austrian Economics*.

[1]Ludwig von Mises, *Human Action*, 3rd ed. (Chicago: Henry Regnery, 1966), p. 484.

The Review of Austrian Economics, Vol. 4, 1990, pp. 199-207
ISSN 0889-3047

are *discounted* to obtain present discounted values. Thus $1.00 due at the end of one year is worth $.90 today, for $.90 is the amount of money that has to be put in the bank today for it to turn into $1.00 at the end of the year (ignoring rounding errors and compound interest). We can say, then, that $.90 is the present discounted value of $1.00 receivable in one year.

All of this is elementary, and accepted by the entire economics profession. It would not be worth mentioning, but for the fact *that* virtually all economists refuse to *apply* the doctrine of discounting future income streams to the case of marginal productivity. Specifically, in the view of most economists, there is a tendency, on the market, for factor payments to equal the Marginal Value Products (MVP) of the factors. Abstracting from questions of perfect or imperfect competition, this means, for example, that in the view of the profession, wages will come to equal the value of the marginal product of labor (the marginal physical product of labor multiplied by the price at which the product can be sold).

In contrast, the Austrian school[2] insists that what tends toward equality with wages is not MVP, but *discounted* MVP, or DMVP. There is no real point at issue when work on immediate consumption goods is considered. For example, the wage of the grocer's clerk, it is admitted by both sides, will tend to equal his MVP, because there is virtually no time that elapses between the labor and the consumption of the final good. Since there is no time under which the discounting process can work, DMVP reduces to MVP.

The divergence between the Austrian and orthodox schools is reached in the cases where labor is added to the value of intermediate or higher order goods. Consider a year's labor on a process that will not reach the consumption stage for a number of years. Here, the Austrians insist that cognizance be taken of the time element; that just as we all commonly discount values receivable only in the future, we not falter when it comes to applying this insight to discounting the value of labor imputed to products which will not be usable until some years have passed. The Austrians argue, in other words, that *all* values receivable in the future be discounted by the rate of interest, even the values of the marginal product of labor, or any other factor, when such value cannot be used in consumption until an elapse of time has taken place.

Why do the non-Austrian economists refuse to follow the Austrians

---

[2]Murray N. Rothbard, *Man, Economy, and State* (New York: Van Nostrand, 1962), pp. 406-09 and 431-33; and Eugen von Böhm-Bawerk, *Capital and Interest*, vol. 2 (South Holland, Ill.: Libertarian Press, 1959), pp. 302-12.

on this seemingly straightforward application of the principle of discounting held by all? This is difficult to answer since most economists completely ignore DMVP, concentrating on MVP instead. Therefore the few orthodox economists who even *mention* DMVP (rejecting it in favor of MVP) are of great interest.

In the view of Sir John Hicks,[3] DMVP and MVP are consistent with each other; they are, in effect, alternatives, and either can be reasonably chosen. In Professor Hicks's words: "This conception [DMVP] is intermediate between 'net productivity' and 'marginal productivity,' as we have defined them; just as they are consistent with each other, since they describe the same phenomenon under slightly different assumptions, so 'discounted marginal productivity' is consistent with them."[4] And what are these "slightly different assumptions" that distinguish "net" and "marginal" productivity? Hicks answers: "'Net productivity' assumes the methods of production to be fixed; marginal productivity assumes them to be variable."[5] But this is puzzling, for it is nonsense to suppose that the methods of production are fixed. What makes these proceedings mysterious indeed is that no one knows this better than Professor Hicks himself, for in his very next sentence he tells us: "In fact, there can be very little doubt that [the methods of production] nearly always are variable to some extent; and consequently the marginal productivity theory has a deeper significance than the [net productivity theory]."[6] If this is so, it seems hard to conclude that "net" and "marginal" productivity theories are equivalent.

But what of our main point: Are DMVP and MVP theories equivalent? What reason does Professor Hicks give in support of his view that these latter two are consistent with each other? In point of fact, he gives *no* reason to support this conclusion. What he does say is that if we make the highly artificial assumption that the period of production ("the length of time elapsing between the payment of labor and the sale of the product") is fixed, then, "in order to maintain the condition of equality of selling price and cost of production, the cost of [any] additional circulating capital [equal to the wage paid multiplied by the period of production] must be deducted from the marginal product, i.e.[,] the marginal product (estimated in this manner) must be 'discounted.'"[7]

But this statement poses more problems than it answers. First

---

[3]John R. Hicks, *The Theory of Wages*, 2nd ed. (New York: St. Martin's Press, 1963).
[4]Ibid., pp. 17ff.
[5]Ibid., p. 14.
[6]Ibid.
[7]Ibid., pp. 17ff.

there is the question of exactly *what* is to be deducted from the MVP. In the Austrian view, the deduction is equivalent to discounting the MVP by the rate of interest. In Hicks's view, what is to be deducted from the MVP is nothing based on the interest rate, but rather, "the cost of additional circulating capital ... [which comes about] ... when the amount of labour employed slightly increases."[8] Circulating capital, it will be remembered, is equal to "the wages paid, multiplied by the length of time elapsing between the payment of labor and sale of the product." Why this amount is selected, rather than any other, is never explained. Nor are we given any reason to believe that a discount, so constructed, is equivalent to the discount based on the market rate of interest.

On the contrary, there is every reason to suppose that the two methods will give *different* results. In the Austrian view, the discounting period is between the time of the payment of labor and the *final* sale to the consumer. In the Hicksian vision, the relevant time, the period of production, is measured from payment of labor to the sale of the product. For Hicks, then, *any* sale will do, whether or not it is to the final consumer of the good.

For Austrians this matter is not at all arbitrary. The reason final consumption is insisted upon is that this alone is consistent with the essence of the whole process of production. The end, the goal, the final aim of production is *consumption*. It is not until the process has reached the consumption phase that it can be said to be completed in any meaningful sense. A worker's efforts have no value whatsoever if they are not eventually carried through to the consumption level. These efforts, then, must be discounted back to the present from the time that they come to fruition, that is, from the time that they become embodied in an item of final consumption. If this were not so, then the concept of DMVP would make no sense. For if every time a change in vertical integration of industry occurred, and there were greater or fewer stages of production between the worker's efforts and the final consumption stage, this would mean an increase or decrease in the number of sales that the good had to go through before it reached the consumer. But if this is so, it would necessarily imply a change in the "length of time elapsing between the payment of labor and the sale of the product." Thus, every time vertical integration increased, and more stages of production were created, this "period of production" would decrease; if the period of production decreases, then, for Hicks, the circulating capital must fall, since circulating capital is the wage multiplied by the period of production. And if

---

[8]Ibid.

circulating capital falls, then the DMVP must rise, since DMVP equals MVP minus a decreasing circulating capital, and MVP stays the same. Alternatively, vertical disintegration would imply a decrease in DMVP. Thus, a purely *legal* phenomenon, the ownership and organization of business enterprise, would intimately affect a purely *economic* phenomenon, the DMVP, which is defined in terms of productivity and the interest rate, and not at all in terms of mere legalistic ownership and sale.

Hicks gives no reason for wanting to "maintain the condition of equality of selling price and cost of production." Indeed, the Austrian view would be the diametric opposite. Here, there is no assumption that merely because businessmen invested in a product, and undertook certain expenses and costs, that *therefore* the consumer will spend an amount of money necessary to make the process profitable. This could only occur if we assumed perfect knowledge and hence an evenly rotating economy, an experience denied to man on this side of the Garden of Eden.

Finally, and most importantly, this scenario of Hicks's is *not* an indication that DMVP and MVP theories are consistent with each other, as Hicks supposedly sets out to show. Rather, it is a *denial* of that claim. If we accept all the assumptions made, it is an *acceptance* of the DMVP view ("the marginal product must be 'discounted'") and *hence* a *rejection* of the MVP theory, which denies that any such deduction must be made.

We need not, of course, accept the fixity of the period of production; we can, with Hicks, in his very next paragraph, "assume that the period of production is variable."[9] If we do, we will learn that "the additional product created by additional labour under the circumstances (of variability of the period of production) is a true marginal product, which in equilibrium must equal the wage, without any discounting."[10] So we see Hicks in his true colors: a complete reversal of field, where the MVP theory is now to be accepted, fully, and the DMVP theory to be rejected; again, far from his stated view that they are equivalent.

Undaunted by this, in his most recent conclusion, Professor Hicks completely reverses field once again and concludes: "Such a modernized wage-fund [the DMVP theory, with the realistic assumption of a variable period of production] is perfectly consistent with marginal productivity [MVP]; and I have often been tempted to use it on a considerable scale in this book. But I have concluded that the advantages

---

[9]Ibid., p. 14.
[10]Ibid., pp. 17ff.

of such a treatment would not compensate for the obstacles it would probably place in the way of readers brought up on the English tradition."[11] In other words, DMVP and MVP theory are once again fully compatible, but MVP theory is preferable on aesthetic grounds! What is to be done? I think we can conclude that MVP and DMVP theory are logically inconsistent, one denying the need for any discounting of MVP and the other insisting upon it.

I turn next to Professor Earl Rolph,[12] who also sees a possible reconciliation of the DMVP and the MVP theories. Defining the former as the view that "[factors] receive the discounted value of their marginal products," Professor Rolph sees the dispute as merely a verbal one: "An examination of the context in which these two propositions appear in economic discussions reveals that the term 'product' does not mean the same thing."[13] In the MVP view, "'product' refers to the *immediate* results of present valuable activities" while "in contrast, the term 'product' in the phrase 'discounted value of marginal product' refers to some *remote* product" (emphasis is mine).[14]

Now this "remote product," to the Austrian, is *consumption*, the be-all and end-all of production. True, if one is prepared to admit that any immediate results of an industrial process, such as a hole in the ground, in preparation for a new dwelling, that will not result in consumption goods for years to come, are *equivalent* to a final product, then one can agree with Professor Rolph that "the only apparent difference between the two views is a choice of words to say virtually the same thing."[15]

The Austrians, however, are not willing to make such a facile equation. It is only in the evenly rotating economy, where full and perfect information of all future events is given to all market participants, that each and every immediate result of an industrial process in the higher orders of capital goods will be guaranteed to come to fruition, eventually, as a consumption good. In the real world, *not all* "immediate results" of production will be so blessed. Many holes in the ground will remain just that—holes in the ground. Be the intentions of the entrepreneurs ever so well motivated, they will not all be filled up with houses.

[11]Ibid., pp. 17-18ff.

[12]Earl Rolph, "The Discounted Marginal Productivity Doctrine," in *Readings in the Theory of Income Distribution* (Homewood, Ill.: Richard D. Irwin, 1951), pp. 278-93.

[13]Ibid.,p. 279.

[14]Ibid., pp. 279-80.

[15]Ibid., p. 282.

Moreover, even if all intermediate efforts are crowned, eventually, with final consumption results, the equation of DMVP and MVP is still invalid. Even in this case there would be a *time element* differential to distinguish between them. The higher the order of production, the further removed, in time, from consumption.

As Professor Rothbard states:

> Every activity may have its immediate "results," but they are not results that would command any monetary income from anyone if the owners of the factors themselves were joint owners of all they produced until the final consumption stage. In that case, it would be obvious that they do not get paid immediately; hence, their product is not immediate. The only reason that they *are* paid immediately (and even here there is not strict immediacy) on the market is that capitalists *advance* present goods in exchange for those *future* goods for which they expect a premium, or interest return. Thus, the owners of the factors are paid the *discounted* value of their marginal product.[16]

It must be concluded, then, that an immediate result of a higher order production process is *not* equivalent to consumption; and that factors do *not* receive the undiscounted value of their immediate marginal products. Rather, factors tend, in the unhampered market, to receive the *discounted* value of what their marginal products are thought to be worth as potential, future consumption goods.

In the remainder of this paper I shall construct another objection to DMVP theory, and then try to show that it too fails to disprove the validity of DMVP.

According to this objection, DMVP theory is satisfactory for the intertemporal level, but not on the intratemporal. Intertemporally, it makes sense for the value of a factor to be determined, in part, by how many years away from final consumption it lies. If factor A is to be used *now*, and factor B one year from now, then the price of B must be adjusted downward accordingly; B must sell for less then A. But suppose A and B are identical! If intratemporal equilibrium is to be attained, then identical factors must receive the same remuneration. B's price cannot then be adjusted downward by the discount, as DMVP theory would have it.

First, suppose that there are two equally skilled carpenters: Ike and Mike. They are exactly alike insofar as carpentry abilities are concerned. They each, therefore, have the same MVP. An entrepreneur, employing several other carpenters, will benefit (lose) by the exact same amount whether he hires (fires) Ike or Mike. His revenues will change by the same amount regardless of which carpenter he

[16]Rothbard, *Man, Economy, and State*, p. 432.

                    The Review of Austrian Economics, Volume 4

deals with. Under such assumptions, intratemporal equilibrium must require that Ike and Mike receive equal wages. If they do not, the familiar market forces will be set up in motion to make sure they do.

But suppose Ike takes a job in a consumption industry, where his work is practically simultaneous with consumption, and Mike finds employment in a higher-order production process, whose fruits will not be available for consumption for 10 years. It would seem, according to DMVP theory, that Mike's wages would have to be heavily discounted, and hence much lower than Ike's. But if this is so, it is in violation of the intratemporal equilibrium that must exist, since we are dealing with equally productive workers, by assumption.

Consider, also, two identical 100 pound bags of coal. Intratemporal equilibrium demands that they receive the exact same price. But if one of them is used for heating a home right now, and the other used in the beginning step of a process which will not be completed for one year, then it would seem that this latter bag of coal will have to sell at a lower price, low enough to reflect the discount called for by the DMVP theory.

The examples could be multiplied without limit.[17] Fish is used for immediate consumption—and also for salting and curing. Some wine is allowed to ferment for one year. But other wine, identical to the first, at the outset, is allowed to ferment for longer periods of time. DMVP theory, it is contended, cannot be correct if it calls for different prices for the same identical good, service, or factor. And yet if this is not what would satisfy DMVP, it is hard to see what would.

The way to solve this paradox is to take this objection "by the horns" and show it to be without merit. Accordingly, for the sake of argument, assume its analysis is correct: if the MVP of the bag of coal to be used up for consumption is $100, and the rate of interest is five percent, then it follows ineluctably that the equilibrium DMVP of an identical bag of coal, to be used in a one year long process, is $95, ignoring compounding complications. So the intertemporal or time market may be in equilibrium, but the spot coal market certainly cannot, for one bag of coal sells for $100, while another, identical to the first in every way, sells for $95. The only problem is, entrepreneurs at the higher level of production will not be able to buy any coal! Why should they be able to if they are only willing to pay $95, for something that coal owners are able to charge $100 for?

What must then happen? The entrepreneurs at the higher stage

---

[17]See Böhm-Bawerk, *Capital and Interest*, for an enumeration, as well as for an eloquent and fully complete analysis.

of production will have to abstain from all projects using coal that cannot attain a DMVP of at least $100, the alternative cost of coal. But at a five percent interest rate, in order to reach a DMVP of $100, the MVP must be $105.

In the words of Professor Rothbard:

> The more remote the time of operation is from the time when the final product is completed, the greater must be the difference allowed for the annual interest income earned by the capitalists who advance present goods and thereby make possible the entire length of the production process. The *amount* of the discount from the MVP is greater here because the higher stage is more remote than the others from final consumption. Therefore, in order for investment to take place in the higher stages, their MVP has to be far higher than the MVP in the shorter processes.[18]

Thus we see that this objection is without merit. The DMVP's must be equated, in the evenly rotating economy, in all areas of production, not the MVP's. Coal will have the same price (assuming equal quality) wherever it is used in the structure of production: for consumption goods, or in long-term heavy industry. But the further away, in time, from consumption a process is, the higher will its MVP have to be to make its employment there profitable, and to result in a DMVP equivalent to the lower orders of production, and in consumption.

[18]Rothbard, *Man, Economy, and State*, p. 409.

II. Unions

Is it Possible to Reconcile Unions with the Libertarian Legal Code?

Walter Block

The present paper subjects unions to a libertarian analysis and finds this organizational structure highly problematic from the perspective of the criminal law. Libertarianism is defined as that philosophy which opposes the initiation, or the threat thereof, of violence against non-aggressive people. Unions are characterized as groups which although need not in principle act contrary to this stricture, as a matter of fact always and ever do so. Hence, organized labor, as presently constituted, cannot be reconciled with libertarian principles of non-aggression. However, it is clear that although they cannot be reconciled with libertarian principles, unions are not breaking any current laws, and therefore cannot be considered to be criminal or illegitimate from a legal perspective. They can only be looked upon as criminal or illegitimate from a libertarian point of view.

Keywords: Unions; threats; aggression; labor; criminal law.

I. Introduction

This paper is an attempt to analysis unionism from a libertarian point of view. In Section II this perspective is defined, and applied to organized labor, to the detriment of the latter. In Section III we address the special problems created by public-sector unions and in Section IV we ask if it can ever be legitimate for a libertarian to join this form of organized labor. Section V is devoted to possible criticisms of the thesis adumbrated herein.

II. Libertarianism and Unionism

Libertarianism is the political economic philosophy that has at its core the "non-aggression axiom": That it shall be illicit and impermissible for anyone to initiate either the threat of, or actual violence against, a person who is not himself an aggressor (Nozick, 1974; Rothbard, 1998/1982; Hoppe, 2001). At this level of non-specificity, there is scarcely a non-comatose commentator who will disagree. Who, after all, favors invasions? Even invaders invariably invoke self-defense, or retaliation for previous wrongs, or some such. The difference between supporters of this view, and the median political philosophy, then, is not in its basic premise; it is found, rather, in the degree to which libertarians are willing to carry this perspective.

For on the basis of it, unions are incompatible with the libertarian legal code, a position that will strike the average theorist as rather idiosyncratic,[1] at best. Nevertheless, for the libertarian, the notion must be resisted that we have a "right to unionize" or that unionization is akin, or worse, an implication of the right to freely associate. Yes, theoretically, labor organizations *could* limit themselves to organizing mass quits unless they got what they wanted, and hence would be acting licitly. That would indeed be an implication of the law of free association.

But every extant union, with no exceptions, reserves the right to employ violence (that is, to initiate violence) against competing workers, e.g., scabs,[2] whether in a "blue-collar way" by utilizing physical aggression, e.g., beating them up, or in a "white-collar way" by getting laws passed compelling employers to deal with themselves,[3] and not with potential employees competing against them, that is, scabs.[4]

But what of the fact that there are many counter examples: unions that have not actually engaged in the initiation of violence? Moreover, there are even people associated for many years with organized labor who have never witnessed the outbreak of actual violence.

Let me clarify my position. My opposition is not merely to violence, but, rather, to "violence, *or* the threat of violence." My position is that, often, no actual violence is needed, if the threat is serious enough, which, I contend, always obtains under unionism, at least as practiced in the US, Canada and other such Western countries.

Probably, the Internal Revenue Service never once engaged in the actual use of physical violence in its entire history.[5] This is because it relies on the courts and police of the US government that have overwhelming power. But it would be superficial to contend that the IRS does not engage in "violence, or the threat of violence."[6] This holds true also for the state trooper who stops you and gives you a ticket. They are, and are

---

[1] Why, then, should it even be of interest to the typical non-libertarian? Because he has "liberty envy." He is loath to admit that his own viewpoint can be inconsistent with this basic premise (non-aggression) of libertarianism.

[2] In Hayek's (1959, p. 47) view, unions "are the one institution where government has signally failed in its first task, that of preventing coercion of men by other men and by coercion I do not mean primarily the coercion of employers but the coercion of workers by their fellow workers."

[3] Or taking advantage of them when they are passed with the instigation of others.

[4] I know of no counter example to this claim. I would be happy to be shown one such. I once thought I had found a candidate: The Christian Labor Association of Canada. But based on a telephone interview with them I can say that while they do indeed eschew "blue-collar" aggression, they unfortunately support the "white-collar" version.

[5] It is mostly composed of nerds, not physically aggressive people.

[6] For the argument that "taxes are theft," see Sabrin (1995), Gordon (1994), Sechrest (1999).

trained to be, exceedingly polite. Yet, "violence, or the threat of violence" permeates their entire relationship with you.[7]

It cannot be denied, moreover, that sometimes, management also engages in "violence, or the threat of violence." My only contention is that it is possible to point to numerous cases where they do *not*, while the same is impossible for organized labor, at least in the countries under discussion.

The threat emanating from unions is objective, not subjective. It is the threat, in the old blue-collar days, that any competing worker, a "scab,"[8] would be beaten up if he tried to cross a picket line, and, in the modern white-collar days, that any employer who fires a striking employee union member and substitutes for him a replacement worker as a permanent hire, will be found in violation of various labor laws.[9]

Suppose a small scrawny hold-up man confronts a big burly football-player type and demands his money, threatening that if the big guy does not give it up, the little one will kick his butt. This is an objective threat, and it does not matter at all if the victim laughs himself silly in reaction. Second scenario: Same as the first, only this time the little guy whips out a pistol, and threatens to shoot the big guy unless he hands over his money.

Now, there are two kinds of big guys. One will feel threatened, and hand over his money. The second will attack the little guy.[10] Perhaps he is feeling omnipotent. Perhaps he is wearing a bullet-proof vest. It does not matter. The threat is a threat is a threat, regardless of the reaction of the big guy, regardless of his inner psychological response.

Now let us return to labor management relations. The union objectively threatens scabs, and employers who hire them. This, nowadays, is purely a matter of law, not psychological feelings on anyone's part. In contrast, while it cannot be denied that sometimes employers initiate violence against workers, they need not *necessarily* do it, qua employer. (Often, however, such violence is in self-defense.)

This is similar to the point I made about the pimp in Block (1991/1976): For this purpose, I don't care if each and every pimp has in fact initiated violence. Nor does it matter if they do it every hour on the hour. This is not a *necessary* characteristic of being a pimp. Even if there are no non-violent pimps in existence, we can still imagine one such. Even if all employers always initiated violence against employees, still, we can *imagine* employers who do not. In very sharp contrast indeed, because of the labor

---

[7]But could it be counter-argued that these police are only defending the private property rights of their employers? That the motorist who enters into their "turf" is morally obligated to follow the rules of the road? For the denial of these arguments, see Block (1996).

[8]Why is it not "discriminatory" and "hateful" to describe workers willing to take less pay, and to compete with unionized labor, as "scabs?" Should not this be consider on a par with using the "N" word for blacks, or the "K" word for Jews?

[9]For example, the Wagner Act. See http://concise.britannica.com/ebc/article?eu=407516; also http://www.mackinac.org/article.asp?ID=4020

[10]In self-defense, of course.

legislation they all support, we cannot even imagine unionized labor that does not threaten the initiation of violence.

Murray N. Rothbard, the father of modern libertarianism, was bitterly opposed to unions. This emanated from two sources. First, as a libertarian theoretician, because organized labor necessarily threatens violence (Rothbard, 1962, pp. 620–632). Second, based on personal harm suffered at their hands by his family (Raimondo, 2000, pp. 59–61).

III. Public Sector Unions

In my view, public-sector unions present theoretical libertarianism with a very complex challenge (albeit in a slightly different manner than do private-sector unions: In this case, they are not necessarily incompatible with the free society, but, as it happens, there are no actual cases in existence of such employee organizations that *are* consistent with economic freedom.)

The complexity presented by public-sector unions is that, on the one hand, from a libertarian perspective they can be seen as a counterweight to governments that act improperly, while on the other hand they constitute an attack on innocent citizens. Each of these different roles calls for a somewhat different libertarian response.

Let us take the first case first. For the limited government libertarian, or minarchist, the state is illegitimate if, and to the extent it exceeds its proper bounds. These, typically, include armies (for *defense* against foreign powers, not *offense* against them), police to keep local criminals in check (that is, rapists and murderers and so on, not victimless "criminals" such as drug dealers and prostitutes), and courts to determine guilt or innocence. Some more moderate advocates of laissez-faire add to this list roads, communicable disease inoculations, fire protection and mosquito control. For the anarcho-libertarian, of course, there is no such thing as a licit government.

What, then, are libertarians to say about a public-sector teachers' union, on strike against a state school? (A similar analysis holds for public-sector unions in garbage collection, post office, buses, or in any other industry where government involvement is improper in the first place). It is my contention that the correct analysis of this situation is: "A plague on both your houses." For not one, but *both* of these organizations act incompatibly with the libertarian legal code that forbids invasions. There is no libertarian who can favor government schools, whether anarchist or minarchist (Milton Friedman, who championed public schools, as long as they operate under a voucher system, thus falls outside the realm of libertarianism on this question). So, on one side of this dispute, there is behavior that falls afoul of libertarian law. But the same applies to the other, the union side, as we have demonstrated above. Thus, there are here two contending forces, both of them in the wrong. From a *strategic* point of view, we may well even support the union vis-à-vis the government, since they are the weaker of our two opponents. But from a principled perspective, my main interest here, we must look upon the two of them as would all men of good will witness a battle between the Blood and the Crips, or between Nazi Germany and Communist USSR. Root for both of them!

Now, let us consider the second case. Here, we note that the public-sector union does much more than attack illegitimate government. It also vastly inconveniences practically the entire populace. When schools are closed, garbage is not collected, the buses do not run — because public sector unions utilize violence and the threat thereof to these ends — then the libertarian response is clear: opposition, root and branch.

Let us take one last crack at public-sector unions, which brings about a further complication. Consider the ABC *20/20* news story on how public-employee unions are fighting against people who volunteer in a way that supplants them.[11]

The general issue is that citizens have been volunteering to do things, like helping public-sector unionists collect trash in parks, aiding them in planting flowers, helping them stack books in public libraries, and the unions have reacted viciously, as is their wont.

Before we can shed libertarian light on this contentious issue, let us first ask, what is the libertarian analysis of ordinary people volunteering to help the government do jobs it should not be doing in the first place? To put it in this way is almost to answer the question.

There is no difference in principle between volunteering to help the state perform reprehensible acts (of course, these are not illicit, per se, as are concentration camps; rather, it is improper, in libertarian theory, for *governments* to take on such responsibilities) such as regarding libraries, schools, parks, and so on, and sending them monetary donations for such purposes. In either case, one is aiding and abetting evil, and risks being found guilty of crimes against humanity by a future libertarian Nuremberg trial court.

Repeat after me: free enterprise, good, (excessive) government, bad. Once again from the top: *free enterprise, good, (excessive) government, bad!* The appellation "libertarian" is an honorific. It is too precious to be bestowed on all those who claim it. It is my contention that people who support (excessive) government are simply not entitled to its use. (At least in the specific context in which they violate the non-aggression axiom. John Stossel is indeed a libertarian on many other issues, but certainly not on this one).

Here is a lesson for libertarians. If you want to be worthy of this designation, and desire to contribute money to a good cause, do not give to a government that goes beyond its legitimate authority. There are many worthy causes that *oppose* statist depredations, not *support* them. If you want to be worthy of this honorific, and wish to donate time to a good cause, e.g., by collecting garbage, planting flowers, or filing books, etc., then

---

[11]See the "Give Me a Break" segment hosted by John Stossel: No Good Deed Goes Unpunished: Are Volunteers Taking Workers' Jobs? ABCNEWS.com at
http://abcnews.go.com/sections/2020/GiveMeABreak/GMAB_Volunteers_031226-1.html.

do so for the relevant *private* groups, whether charitable or profit-seeking, it matters not one whit.

IV. May a Libertarian Join a Union?

Is it proper, if it is even logically possible, for a libertarian to join a coercive union? Much as I hate to be controversial (OK, OK, I don't mind it a bit) my answer is yes. There are many issues upon which I disagree with William F. Buckley, but his decision to join American Federation of Television and Radio Artists is not one of them. (This was the requirement imposed upon him for being allowed to air his television show, *Firing Line*.)

Why would I take such a seemingly perverted stance? Let me answer by indirection. Given that it is unacceptable for the government to run schools and universities, is it proper for a libertarian to join them, whether as a student or a professor? Given that it is improper for the government to organize a post office, is it acceptable for a libertarian to mail a letter? Given that it is inappropriate for the government to build and manage roads, streets and sidewalks, is it illegitimate for a libertarian to utilize these amenities?

True confession time. I have been a student of public schools: grade school, high school and college. I have even been a professor at several public colleges and universities. I regularly purchase stamps from the evil government post office, and mail letters. I walk on public sidewalks, and avail myself of streets and highways. Mea culpa? Not a bit of it.

If Ayn Rand's heroic character Ragnar Danneskjold has taught us anything, it is that the government is not the legitimate owner of what it claims. Why, then, should we respect its "private property rights" when there is no practical reason to do so? If this means that libertarians can partake of services for which they favor privatization, then so be it.

Similarly, with coercive unions. If a hold-up man demands your money at the point of a gun, giving it up is *not* incompatible with libertarianism, even though it amounts to acquiescing in theft. If organized labor threatens you with bodily harm unless you join with it and pay dues to it, I cannot think that agreeing to do so per se removes the victim from the ranks of libertarianism. Buckley, to give him credit, never ceased inveighing against the injustice done to him in this way. If he had reversed field, and starting *defending* unions, then what little claim he has to be a libertarian would have vanished. In this regard, there is all the world of difference between a Marxist professor at a public university who promotes interventionism, and a libertarian who opposes it.

But suppose that there were two libertarians, pure as the driven snow, who wanted to start a workers' organization, or, a "legitimate" union.[12] Would it even be logically possible for them to do so, given what we have said about functioning in a milieu where government laws giving them unfair advantages are omnipresent?[13]

At first glance, it would appear not to be possible. For, given this threat, the boss might be inclined to grant them terms he otherwise would not have agreed to. Complying with the requirements to become a union is almost equivalent to "firing a shot across the firm's bows," to borrow an expression from naval warfare. Nor would it be legal for such a labor organization to forswear the advantages accorded to it by union legislation. They could forswear all they wanted to, but woe betide the employer who took advantage of any such declarations. He would be subject to the full penalty of the law; and if not, would always be open to later blackmail on the part of the "legitimate union."

The analogy with a libertarian occupying a rent-controlled apartment[14] is a strong one. Here, too, there is an overarching law, rent-control legislation, which violates landlord rights. So, could a libertarian lease such a dwelling? At the outset, again, the answer would appear to be in the negative. For he would still be taking advantage of the landlord, paying him less than the market value he could otherwise have obtained. Never in a million years would a building owner approach his tenants, to see which of them might be susceptible to a request for a higher, illegal rent. That way lies a jail cell. But suppose such a tenant were to approach the landlord, mention he was a libertarian and therefore opposed to such legislation, and would pay him full value henceforth. If the owner of the building agreed, he could be imprisoned for making such a contractual agreement. At the very least, if he gave a receipt for the full amount, he would open himself up for later blackmail on the part of the tenant. Other risks include anyone seeing the tenant paying more than the controlled rent. An even greater problem would be met were the libertarian philosophy to require a tenant to make such a declaration, moving to the side for the moment that it could not in the nature of things be efficacious. For this would imply placing a positive obligation[15] on the libertarian.

---

[12]I am greatly indebted to my friend Michael Edelstein for raising this issue with me, and then discussing it with me in depth.

[13]That is, suppose that CLAC took a different position than the one I attribute to them. They forswear all unfair and improper advantages afforded them by labor legislation.

[14]*Anarchy, State and Utopia* (1986) tells of libertarian Robert Nozick's attempt to utilize the rent-control law against his landlord, the non-libertarian author of "Love Story" and fellow Harvard professor Erich Segal. In my view, the former will have difficulty reconciling his philosophical position with his actions as a tenant. The only justification for such behavior, in my view, is if Nozick were to have had some legitimate other grievance against Segal that could only be assuaged by use of this type of government coercion. There is no evidence that this was, in fact, the case.

[15]I have argued elsewhere (Block, 2004; Block and Whitehead, 2005) that libertarian strictures against preempting constitute an exception to the general rule that the imposition of (seemingly) positive obligations and libertarianism are incompatible.

On the other hand, as far as the landlord is concerned, he does not at all mind having an advocate of this philosophy occupying his apartment. True, he cannot receive a market rent from him, but, then, he could not do so with anyone else either. In a sense, the libertarian tenant is not stealing money from him, at least no more than any other occupant. Were the libertarian to cut off his nose to spite his face and renounce the residential bargain[16] he is being offered, the landlord would be no better off.

My claim is that he would be as justified in keeping this apartment as he is in walking or driving on the public street, mailing a letter in the socialist post office, or joining a union. The only problem he will face is justifying this choice when and if there comes to be created a libertarian Nuremberg Court. How could a tenant do this? By starting, or joining, a group called Tenants Against Rent Control.[17] Nor is such exotic activity required; any ordinary garden-variety libertarian activity will qualify as a defense against acting the part of a quasi-thief: contributing money to the freedom cause, working for the Libertarian Party, writing and speaking in promotion of these ideas, and so on.

It is similar with unions. Whenever a worker organization, composed entirely of purist libertarians it matters not one whit, qualifies for union status, certain realities kick into gear. These redound to the benefit[18] of the employees, and to the detriment of employers and competing scabs. This will occur regardless of the intentions of these libertarian unionists, in creating a group that would otherwise be entirely licit. Should they be forced to give up their rights to organize, just because of unjust laws? Not at all. It is my contention that just as libertarians may properly occupy rent-controlled apartments, so may they create workers' associations. The problem, as before, is to distinguish oneself from all other union members (tenants), and this can be done as outlined before: by speaking out, contributing money to groups that fight for liberty, and so on.

V. Possible Criticisms of the Thesis

1. *Unionism is Legitimate*

Many readers of the present article would insist that theoretically, unions are compatible with the free society. I agree, I agree. Nothing said above should be taken to be inconsistent with this view. All such a union would have to do is to eschew both white- and blue-collar crime. I only argue that it has never happened in fact,[19] not that it would be impossible for it to occur.

---

[16]For arguments that in the long run rent control does not so much reduce rents as decrease the quality of apartments, see Baird, 1980; Block, 2002; Grampp, 1950; Hayek, 1981; Johnson, 1982; Salins, 1980; Tucker, 1990). We are now abstracting from this issue.

[17]"Tenants Against Rent Control" at http://www.wclf.org/articles/2.2.1.html

[18]Well, the short-run benefit, in any case.

[19]See Hutt (1973, 1989) and Petro (1957).

However, I am something of a stickler about language.[20] Surely, a workers' association that totally eschews the initiation of violence, or even the threat thereof, deserves different nomenclature from organizations it only superficially resembles, e.g., unions. My suggestion is that we not characterize as a union any labor organization that strictly limits itself to the threat of quitting en masse.

What, then, should we call a group of workers who eschew both beating up scabs and laws compelling employers to bargain with them? Here are some possibilities: workers' associations, employees' groups, organizations of staff members.

Thus, are workers' associations as defined above compatible with free enterprise? You bet your boots they are. Do unions, or organized labor qualify in this regard? No, a thousand times no.

2. *Not Aware of Violence*

Let us now consider the objection that numerous members of the rank and file are not aware of any violence in their own unions. But, many employees of the IRS are probably not aware that what they are doing amounts to the threat of the initiation of violence. I don't see why all union members should necessarily be aware of this for my thesis (this is hardly original with me) to be correct. My understanding is that after the British left India, the government of the latter began polling people in far-removed rural villages as to their thoughts on this matter; they had to stop when they learned that the villagers were not aware that the British had even arrived there in the first place. Heck, there are probably some people out there who still think the earth is flat, or that socialism is an ethical and efficacious system! That does not make it so.

3. *Self-defense*

Union violence does indeed exist, but is justified on the ground that this was only in self-defense, against employers, scabs, or foreigners. Let us consider each of these in turn.

Yes, employers are violent too. The Pinkertons spring immediately to mind in this regard. Some of these cases were justified in self-defense, against prior union aggression; some were not. In the former case, there is certainly no warrant for invasive behavior on the part of organized labor. But even the latter cases cannot serve as justification for pervasive union aggression, even against non-invading employers.[21] At best, this can validate self-defense on the part of the rank and file in those cases of employer aggression only.

---

[20]See Block (14 February 2000, 1 April 2000, 20 April 2000).

[21]It is only a Marxist who would claim that employers are necessarily offensive; for an exposure of this fallacy, see Bohm-Bawerk, Eugen (1959/1884); see particularly Part I, Chapter XII, "Exploitation Theory of Socialism-Communism."

And what of scabs? The claim, here, is that scabs are stealing, or, better yet, attempting to steal, union jobs. But the scab can only "steal" a job if it is *owned*, like a coat or a car. However, a job is very different. It is *not* something anyone can own. Rather, a job is an *agreement* between two parties, employer and employee. But when an employer is trying to hire a scab and fire the unionist, this shows he no longer *agrees*. Do not be fooled by the expression "my job." It does not denote ownership, any more than "my wife," "my husband," "my friend," "my customer," "my tailor" indicates possession in any of those contexts. Rather, all of these phrases are indicative of voluntary interaction, and end (apart from marriage laws which may prohibit this) when the agreement ceases.

Then, there is the supposed "threat" imposed by Mexican workers (or Indian or Japanese workers, whoever is the economic Hitler of the day). Remember that "giant sucking sound?" The best remedy for this bit of economic illiteracy is to read up on free trade. Henry Hazlitt's book, *Economics in One Lesson*, would be a great place to start.

4. *But they Signed a Contract*

What of the argument that since the employer signed a labor contract, he should be forced to abide by its provisions? My response is that the employer should not have to honor a contract that was signed under duress. Suppose I held a gun on you, threatened to shoot you unless you signed a "contract" with me, promising to give me $100 per week. Later on, when you were safe, you reneged on this "contract." Certainly, you'd be within your rights.

5. *Maximize Income*

Let us consider the following question: "How else is a man who sells the only product most of us have, labor and time, going to maximize the return of his investment, other than by joining a union?"

First of all, even if this were true, any criminal could say no less. A hold-up man, too, wants to "maximize the return of his investment" and does so by committing aggression against non-aggressors. How is the unionist any different than the hold-up man in this regard?

Secondly, it is by no means clear that organized labor is the last best chance for economic well-being on the part of the working man. Anyone ever hear of the Rust Belt? Unions from Illinois to Massachusetts demanded wages and fringe benefits in excess of productivity levels, and employers were powerless to resist. The result was "runaway shops." Either they ran into bankruptcy, or they located to places like Alabama, Mississippi and Louisiana, where unionism was seen more for the economic and moral scourge that it is, than in Taxachussetts. If organizing workers into unions is the be-all and end-all of prosperity, how is it that wages and working conditions are very good in computers, insurance, banking, and a plethora of other non-unionized industries?

6. *Hierarchy is the Real Problem*

"Hierarchy is the real problem. Why pick on unions, which are only one hierarchical structure among many. Surely, firms are even more hierarchical than is organized labor." Conceded by this objection is the premise that unions are incompatible with libertarianism; I certainly agree with this, but not for the reason given. Their real problem is not, however, that they are hierarchical. There is no reason whatsoever to oppose *all* hierarchical organizations, which would certainly include employers.

Libertarians oppose the initiation of coercion or the threat thereof, not hierarchy. Yes, all groups that violate the non-aggression axiom of libertarians are hierarchical. Governments, gangs, rapists, impose their will by force on their victims. They give orders. And yes, in all hierarchies, people at the top of the food chain give orders to those below them. But the difference, and this is crucial, is that the recipients of orders in the latter case have *agreed* to accept them, but this does not at all apply in the former case.

When the rapist orders the victim to carry out his commands, this is *illegitimate* hierarchy. When the conductor orders the cellist to do so, this is an aspect of *legitimate* hierarchy. I oppose unions not because they are hierarchical, but because the scabs have never agreed to carry out their orders.[22]

7. *Management Bullies*

Unions protect their members from bullies in management. This, if unions are an evil, they are a necessary one.

I would not wish to deny that it is often the case that organized labor acts as a counterweight to tyrannical managers. However, there are other better and more moral remedies; unionism is neither "necessary" nor ethical as a response to these bads.

It is not morally defensible since unions act not so much directly against firms, but indirectly, by attacking scabs who are innocent of any wrongdoing in this context. It is one thing to hit back at your attacker. It is quite another to lash out at a guiltless passer by.

Nor are these groups necessary. When managers exceed their proper role, they lose profits for their firms. The owners, then, have a financial incentive to root out managers who hurt a company by engendering a higher-than-optimal quit rate.

---

[22]But are corporations gangs? No. When they engage in uninvited border crossings, they necessarily go to jail if caught. Not so for unions, the only institution in the country that can legally use force, apart from government itself. Let us put this more moderately; yes, sometimes corporations can indeed be composed of gangsters, and thus may properly be indicted in the same manner in which we have dealt with organized labor. For example, claims in this regard would include illicit use of the Pinkertons, and, more recently, several of the scandals for which executives were imprisoned. The difference, however, is that unions are *necessarily* criminal organizations, while this does not at all apply to corporations. Only *some* of them can properly be considered gangsters. For a defense of the corporation, see Hessen (1979) and Novak (1981).

8. *Doctors and Lawyers are Worse*

Compared to doctors and lawyers associations, unions are pikers. It is *explicitly* illegal, not merely implicitly so through labor legislation, for a person who is not licensed by them to do so to compete with a member of either profession. That is, practicing medicine without a license is a criminal offense, and a non-lawyer who attempted to defend a client would be thrown summarily out of court. Powers of this sort are simply unavailable to union organizations. Why criticize the latter, then, and not the former?

The defense against this criticism is straightforward. Attorneys and physicians associations are but examples of *unions*.[23] In criticizing the latter we are also doing so for the former. Remember, we are defining unions as organizations that use coercive force, whether of the blue- or white-collar variety it matters not, to mulct higher than market wages[24] out of long-suffering employers, and, indirectly, consumers. If this does not aptly describe these two professions, then nothing does.

9. *A Miscellany of Questions*

Is there any libertarian distinction worth making between now illegal "closed shops" (in which membership can be compelled) and "agency shops" (in which workers cannot be compelled to join the union)?

The key point for the libertarian is, does an institution necessarily initiate violence against innocent non-aggressors? If so, it is illicit; if not, then licit. There is nothing in principle improper about a closed-shop union that does not rely on labor legislation nor beat up scabs, provided that the employer wishes to hire people only on the precondition that they join. Similarly, on these assumptions, an agency shop is incompatible with libertarianism, if the owner wants to hire only union members.[25] The word "compelled" in the above statement is problematic. Am I "compelled" to pay for the candy bar I have just taken? Well, yes, in the sense that if I do not I am a thief. But, no, in that this purchase of mine is strictly a voluntary one.

Should unions be considered illegal even if, they do not benefit from labor legislation of the Wagner Act variety and any use of violence is forbidden and the law is enforced? Yes. But then I would characterize such organizations not as "unions" but as workers associations, or some such, to make the distinction between these theoretical entities (which do embody rights of free association) and actual labor organizations.

Do unions violate rights even if all members support the union? Yes, they do violate the rights of scabs.

---

[23]Friedman (1962, Chapter 9) went so far as to characterize the American Medical Association as the nation's most powerful union. No truer words have ever been printed.

[24]For the case of doctors, see Hamowy (1984).

[25]It is all a matter of contract. For a defense of the Yellow Dog contract, see Chapter 8 in this volume.

Can unions enhance rights when they exist inside a political system that exceeds its proper boundaries (e.g., one with a powerful state). After all, they check state power. My answer (Block, 2005) to this is a qualified "Yes." It then becomes a strategic or prudential matter, not one of principle: The libertarian would support whichever is the weaker party. As that is usually the union, not the government, this would be a case for support of even coercive labor unions.

10. *Potential*

Every group, and, indeed, every individual, has the *potential* to commit violence. Why pick on unions alone?

It is because *only* unions, and *no* other groups apart from government itself, have the *legal right* to initiate violence. Yes, *every* person or group of persons can potentially commit violence. But only unions have been granted by government permission to share in its rights of force majeure.[26]

Let me end with a joke, a very pertinent one. A husband and wife are out on a rowboat, fishing, or at least the husband is. The wife abhors fishing, and is reading a book on the boat, merely keeping him company. He swims away, leaving his fishing rods in the boat. The boat drifts to a part of the lake where fishing is illegal. A sheriff then comes upon the wife blissfully reading her book, and announces he is going to arrest her for fishing in restricted waters. She objects that she was only reading, not fishing. The officer of the law replies, "But you have all the equipment necessary to fish." The wife replies, "If you arrest me for fishing when I was only reading a book, I shall accuse you of rape." "Rape?" replies the sheriff. "Why, I made no threats, and took no action that could be construed in any such manner." Her retort: "But you have all the necessary equipment."

We *all* have the "equipment" with which to initiate violence. But only unions, apart from government, have the legal *right* to do so. This right is demonstrated on the blue-collar front by the purposeful failure of the police to stop strike violence of the type that, had it occurred in any other context[27] would have called for swift and immediate police counter-action. It occurs in the white-collar context on the basis of laws such as the Wagner Act.

---

[26] http://en.wikipedia.org/wiki/Force_majeure

[27] Sometimes the police refuse to intervene when minority group members are creating mayhem, out of fear that the violence will only escalate if they do. To the extent this is true, we have to modify our claim that only unions have a legal right to initiate violence against non-aggressors.

References

"Anarchy, State and Rent Control." *The New Republic* Dec. 22 1986 pages 20–21.
http://www.daft.com/~rab/liberty/Miscellaneous/Nozick-article

Baird, C (1980). *Rent Control: The Perennial Folly.* Washington DC: The Cato Institute.

Block, W (1991/1976). *Defending the Undefendable.* New York: Fox and Wilkes.

Block, W (1996). Road Socialism. *International Journal of Value-Based Management,* 9, 195–207.

Block, W (14 February 2000). *Watch Your Language.*
http://www.mises.org/fullarticle.asp?control=385&month=17&title=Watch+Your+Language&id=19
[22 August 2007].

Block, W (1 April 2000). *Taking Back the Language.*
http://www.mises.org/fullarticle.asp?control=406&month=19&title=Taking+Back+the+Language&id=19
[22 August 2007].

Block, W (20 April 2000). *Word Watch.*
http://www.mises.org/fullstory.asp?control=414&FS=Word+Watch [22 August 2007].

Block, W (2002). A critique of the legal and philosophical case for rent control. *Journal of Business Ethics,*
40, 75–90. http://www.mises.org/etexts/rentcontrol.pdf [22 August 2007].

Block, W (2004). Libertarianism, positive obligations and property abandonment: Children's rights.
International Journal of Social Economics, 31(3), 275–286.

Block, W (14 September 2005). *The Evil of Unions: In the public as well as the private sector.*
http://www.lewrockwell.com/block/block54.html [22 August 2007].

Block, W and R Whitehead (2005). Compromising the uncompromisable: A private property rights approach
to resolving the abortion controversy. *Appalachian Law Review,* 4(2), 1–45.

Bohm-Bawerk, E (1959/1884). *Capital and Interest* (GD Hunke and HF Sennholz, Trans.). South Holland,
IL: Libertarian Press. [Original work published 1884].

Friedman, M (1962). *Capitalism and Freedom.* Chicago: University of Chicago Press.

Gordon, D (1994). Justice and redistributive taxation: James Buchanan vs. Ludwig von Mises. *The Review of
Austrian Economics,* 8(1), 117–131.

Grampp, WS (April 1950). Some effects of rent control. *Southern Economic Journal.* 425–426.

Hamowy, R (1984). *Canadian Medicine: A Study in Restricted Entry.* Vancouver: The Fraser Institute.

Hayek, FA (1959). Unions, inflations, and profits. In *The Public Stake in Union Power,* PD Bradley (ed.), pp.
46–62. Charlottesville, VA: University of Virginia Press.

Hayek, FA (1981). The repercussions of rent restrictions. In *Rent Control: Myths and realities,* W Block
(ed.), Vancouver: The Fraser Institute.

Hessen, R (1979). *In Defense of the Corporation.* Stanford, CA: Hoover Institution Press.

Hoppe, H-H (2001). *Democracy — The God That Failed: The economics and politics of monarchy,
democracy, and natural order.* New Jersey: Transaction Publishers.

Hutt, WH (1973). *The Strike Threat System: The economic consequences of collective bargaining.* New
Rochelle, NY: Arlington House.

Hutt, WH (1989). Trade unions: The private use of coercive power. *The Review of Austrian Economics,* III,
109–120, http://www.mises.org/journals/rae/pdf/rae3_1_7.pdf

Johnson, MB (ed.) (1982). *Resolving the Housing Crisis: Government policy, decontrol, and the public
interest.* San Francisco: The Pacific Institute.

Novak, M (ed.) (1981). *The Corporation: A theological inquiry.* Washington DC: American Enterprise
Institute for Public Policy Research.

Nozick, R (1974). *Anarchy, State and Utopia.* New York: Basic Books.

Petro, S (1957). *The Labor Policy of the Free Society.* New York: Ronald Press.

Raimondo, J (2000). *An Enemy of the State: The life of Murray N. Rothbard*. Amherst, NY: Prometheus Books.

Rothbard, MN (1998/1982). *The Ethics of Liberty*. Atlantic Highlands, NJ: Humanities Press.

Sabrin, M (1995). *Tax Free 2000: The Rebirth of American Liberty*. Lafayette, LA: Prescott Press.

Salins, PD (1980). *The Ecology of Housing Destruction: Economic effects of public intervention in the housing market*. New York: New York University Press.

Sechrest, LJ (1999). Rand, anarchy, and taxes. *The Journal of Ayn Rand Studies*, 1(1), 87–105.

Tucker, W (1990). *The Excluded Americans: Homelessness and housing policies*, Washington DC: Regnery Gateway.

LABOR RELATIONS, UNIONS AND COLLECTIVE BARGAINING: A POLITICAL ECONOMIC ANALYSIS

Walter Block
College of the Holy Cross, Amherst

It is not difficult to document the fact that many segments of our society extol the virtues of unionism, as commonly practiced. Some people defend unions as a means of promoting employment. Others feel that "social justice" is a sufficient warrant for this curious institution. So deeply embedded in our folkways is the concept that unions are legitimate institutions that many mainline religious organizations have even gone so far as to invite them to organize their own Church employees – on what they see as moral grounds.

The simple fact is that in the minds of most pundits, unions have a legitimate role to play in our society. How else can we account for the fact that gangs of organized laborers who have engaged in violent strikes not only still remain at large, but are widely applauded for their courage and convictions? Were any other group of people to have interfered with the lives and property of others in a similar manner, they would have been summarily clapped into jail, and been considered proper objects of fear, loathing, ridicule and pity, the reaction elicited in most people by activities criminal.

Complexity

Contrary to the popular notion, however, unionism is a complex phenomenon, which admits of a voluntary and a coercive aspect. The philosophy of free enterprise is fully consistent with voluntary unionism, but is diametrically opposed to coercive unionism. What do all varieties of unionism, both coercive and voluntary, have in common? Unions are associations of employees, organized with the purpose of bargaining with their employer in order to increase their wages.[1]

---

[1] Since money wages are funds which the employees take home, and working

What, then, is the distinction between invasive and non-invasive unions? The latter obey the libertarian axiom of non-aggression against non-aggressors; the former do not. Legitimate unions, in other words, limit themselves to means of raising wages which do not violate the rights of others; illegitimate unions do not so inhibit themselves.

Some pundits have declared their "full support for the principle of free and voluntary association in labor unions." If this constitutes moral approval of voluntary unions, and condemnation of the coercive type, well and good. But if it is intended to apply to extant labor organizations, this statement is disingenuous. It is not even a rough approximation of how organized labor has operated – and still continues to operate – in the modern world.

Coercion

Let us be absolutely clear on this distinction, for it is at the root of any accurate assessment of unionism. There are those labor organizations which do all they can to raise their members' wages and working conditions – except violate the (negative) rights of other people by initiating violence against them. These can be properly called "voluntary unions". But then there are those which do all they can to promote their members' welfare both by legitimate non-rights-violative behavior as well as by the use of physical brutality aimed at non-aggressing individuals.

With regard to the activity of "coercive unions" defined in this manner, Ludwig von Mises has stated:

---

conditions embody funds which are spent, at least in part, in behalf of the employees while on the job, there are really two desiderata here. One, the total of money wages and working conditions, and two, the allocation between them. On the free market, the employer has a great incentive to allocate these two sorts of wage expenditures in accordance with the desires of his employees. If, for example, the workers in his plant prefer most of their wages in the form of take-home-pay, and very little in the form of expenditure for amenities on the job site, the employer who ignores this desire (or, equivalently, fails to ferret out this information) will suffer higher quit rates – or else he will have to increase his total wage package, in order to compete with other employers who are better able to discern employee tastes in this matter.

LABOR RELATIONS, UNIONS AND COLLECTIVE BARGAINING 479

... labor unions have actually acquired the privilege of violent action. The governments have abandoned in their favor the essential attribute of government, the exclusive power and right to resort to violent coercion and compulsion. Of course, the laws which make it a criminal offense for any citizen to resort – except in case of self-defense – to violent action have not been formally repealed or amended. However, actual labor union violence is tolerated within broad limits. The labor unions are practically free to prevent by force anybody from defying their orders concerning wage rates and other labor conditions. They are free to inflict with impunity bodily evils upon strikebreakers and upon entrepreneurs who employ strikebreakers. They are free to destroy property of such employers and even to injure customers patronizing their shops. The authorities, with the approval of public opinion, condone such acts ... In excessive cases, if the deeds of violence go too far, some lame and timid attempts at repression and prevention are ventured. But as a rule they fail ... What is euphemistically called collective bargaining by union leaders and 'pro-labor' legislation is of a quite different character. It is bargaining at the point of a gun. It is bargaining between an armed party, ready to use its weapons, and an unarmed party under duress. It is not a market transaction. It is a dictate forced upon the employer ... It produces institutional unemployment.

The treatment of the problems involved by public opinion and the vast number of pseudo-economic writings is utterly misleading. The issue is not the right to form associations. It is whether or not any association of private citizens should be granted the privilege of resorting with impunity to violent action.

Neither is it correct to look upon the matter from the point of view of a 'right to strike.' The problem is not the right to strike, but the right – by intimidation or violence – to force other people to strike, and the further right to prevent anybody from working in a shop in which a union has called a strike. (Ludwig von Mises, 1966, pp. 777-79).

And in the view of Friedrich Hayek:

> It cannot be stressed enough that the coercion which unions have been permitted to exercise contrary to all principles of freedom under the law is primarily the coercion of fellow workers. Whatever true coercive power unions may be able to wield over employers is a consequence of this primary power of coercing other workers; the coercion of employers would lose most of its objectionable character if unions were deprived of this power to exact unwilling support. Neither the right of voluntary agreement between workers nor even their right to withhold their services in concert is in question. (F.A. von Hayek, 1960, p. 269.)[2]

Given that there are legitimate and illegitimate forms of labor organization, it follows that sound public policy consists of defending the former and eliminating the latter. In legal terminology, this reduces to a call for the repeal of legislation that promotes invasive action, and for an expansion of the legal protections for non-invasive ones. In the just society, a union may do anything that individual citizens have a right to do, and must refrain from all activities prohibited to other citizens. The labor code, in other words, ought be nothing more than the ordinary rule of law (Hayek, 1973; Leoni, 1961, pp. 59-76), applied to management-labor relations.

This leads us to the $64,000 question. Which arrows in the quiver of organized labor are invasive, and which are not? Let us start off by mentioning several legitimate techniques utilized by organized labor, and then look at the panoply of illegitimate actions engaged in by unions.

Legitimate Unionism
Mass Walkout
First is the mass walkout: threatening, or organizing, a mass

---

[2]Says Morgan O. Reynolds, 1984, p. 50: "Hitting a person over the head with a baseball bat is much less likely to be treated as criminal if the person wielding the bat is an organized (i.e. unionized) worker in a labor dispute." See also Hutt, 1989.

LABOR RELATIONS, UNIONS AND COLLECTIVE BARGAINING 481

walk out, unless wage demands are met.[3] This is not an infringement of anyone's rights, since the employer, in the absence of a contract, cannot compel people to work for him at wages they deem too low. Nor is it any valid objection to this procedure that the workers are acting in concert, or in unison, or in collusion, or in "conspiracy." Of course they are. But if it is proper for one worker to quit his job, then all workers, together[4], have every right to do so, en masse.[5] All conspiracy laws ought to be repealed, provided only that the agreement is to do something that would be legal when undertaken by a single individual.

There are numerous conservatives, as opposed to libertarians, who take the view that anti-trust and anti-combines law ought to be applied to unions.[6] Thus, even what we have been describing as voluntary unions would be for them illegitimate, because they claim that "collusive actions" on the part of unions "'exploit' the community as a whole,"[7] in their violation of consumers' sovereignty.[8] But this only shows that there is all the world of difference between economists who support the system of laissez-faire capitalism, on the one hand, and those who favor a system of national or state capitalism on the other.

---

[3]This is on the assumption that there is no valid employment contract in effect at this time which prohibits such an act.

[4]This follows directly from a defense of voluntary socialism, vis a vis coercive Voluntary unionism is merely one facet of the former. For an elaboration of this point, see Walter Block, 1990.

[5]This is not a violation of the law of composition, or an instance of this fallacy. The only serious challenge to the textual statement is the case where harms can be additive. For example, the scenario where if one person touches another, slightly, it is not a rights violation, because no harm is done, whereas if a million persons do so, the victim can indeed be harmed, and thus there is a rights violation. The difficulty with this line of argument, though, is that even the first slight touching, done by only one person, is an illicit act, even though the harm is slight, or even non existent, provided only that the victims person has been interfered with. See (to be supplied).

[6]In contrast, libertarians take the view that anti-trust and anti-combines legislation ought not be applied to anyone, neither unions nor business firms. See Armentano (1972, 1982).

[7]W. H. Hutt (1973, p.3, 1989); Schmidt (1973); Simons (1948). In sharp distinction, for a libertarian analysis which *defends* the right of organized labor to threaten or to quit in unison, see Petro (1957); Reynolds (1984).

[8]For a critique of Hutt, see Murray N. Rothbard (1970, pp. 561-566).

Back to Work Legislation

Again, libertarians would disagree with many "right-wing" conservatives on the question as to whether it is improper for governments to enact legislation forcing unions back to work where a union strike threatens to disrupt broad segments of the economy and to harm innocent parties not involved in the dispute. The libertarian viewpoint holds that the government does not have such a right, and that this follows from the basic libertarian premise of self-ownership. In the words of Murray Rothbard (1978, pp. 83, 84):

> On October 4, 1971, President Nixon invoked the Taft-Hartley Act to obtain a court injunction forcing the suspension of a dock strike for eighty days; ... It is no doubt convenient for a long suffering public to be spared the disruptions of a strike. Yet the 'solution' imposed was forced labor, pure and simple; the workers were coerced, against their will, into going back to work. There is no moral excuse, in a society claiming to be opposed to slavery and in a country which has outlawed involuntary servitude, for any legal or judicial action prohibiting strikes – or jailing union leaders who fail to comply.

Conventional conservatives tend to place the national good above the good of individuals, so there is a basic disagreement between right-wing conservatives and libertarians on this issue.

Boycott

Another activity held to be legitimate by libertarians is the boycott, whether primary or secondary. A boycott is simply the refusal of one person to deal with another.[9] All interaction in

---

[9]Thus, all anti-discriminatory laws are incompatible with the libertarian legal code. For an analysis which shows that such legislation is itself a rights violation, and that the free marketplace is the best protector of liberties, see Friedman (1985), Sowell (1983), Williams (1982).

It is logically inconsistent to maintain that people do not have the right to discriminate against one another, and that they do have the right to boycott, since the boycott is merely an orchestrated discrimination against certain individuals or groups

LABOR RELATIONS, UNIONS AND COLLECTIVE BARGAINING 483

a free society must be on a mutual basis, but there is no presumption that any particular interaction must take place. It is part and parcel of the law of free association that any one person may refuse to associate with another for any reason that seems sufficient to him. Since a boycott is merely an organized refusal to deal with another, and each person has a right to so act, then people may act in this way in concert. A "hot edict," whereby a union declares the handling of certain products to be prohibited by organized labor, is a special case of the boycott. Provided that there is no contract in force which is incompatible with such a declaration, it, too, is an entirely legitimate activity. Says Rothbard (1983, p.131) in this regard:

> A boycott is an attempt to persuade other people to have nothing to do with some particular person or firm – either socially or in agreeing not to purchase the firm's product. Morally, a boycott may be used for absurd, reprehensible, laudatory or neutral goals. It may be used, for example, to attempt to persuade people not to buy non-union grapes *or* not to buy union grapes. From our point of view, the important thing about the boycott is that it is purely voluntary, an act of attempted persuasion, and therefore that it is a perfectly legal and licit instrument of action ... a boycott may well diminish a firm's customers and therefore cut into its property values; but such an act is still a perfectly legitimate exercise of free speech and property rights. Whether we wish any particular boycott well or ill depends on our moral values and on our attitudes toward the concrete goal or activity. But a boycott is legitimate *per se*. If we feel a given boycott to be morally reprehensible, then it is within the rights of those who feel this way to organize a counter boycott to persuade the consumers otherwise, or to boycott the boycotters. All this is part of the process of dissemination of information and opinion within the framework of the rights of private property.

---

of people.

Furthermore, 'secondary' boycotts are also legitimate, despite their outlawry under our current labor laws. In a secondary boycott, labor unions try to persuade consumers not to buy from firms who deal with non-union (primary boycotted) firms. Again, in a free society, it should be their right to try such persuasion, just as it is the right of their opponents to counter with an opposing boycott.

Sorenson

An illustration of this principle took place in Canada. Alderman Bill Sorenson of North Vancouver City had voted to contract out the municipal garbage collection services to private enterprise. And, to add insult to injury – at least in the eyes of Local 389 of the Canadian Union of Public Employees (CUPE) – he also voted for a wage freeze covering all city employees.

The union didn't take long to strike back.

As it happens, Sorenson was the operations manager for the North Shore Community Credit Union, a local banking facility. As it also happens, Local 389 of CUPE holds deposits with this credit union. In response to Alderman Sorenson's votes on city council, the Union withdrew $25,000 of its funds from the bank which employed Sorenson.

Now this decision to withdraw funds was no mere coincidence. It was motivated by spite – an attempt to get back at a part-time politician by attacking him in his capacity as a private citizen.

As a result of this act, Mr. Sorenson resigned his seat on the city council – it isn't clear whether he was forced to do this to keep his job.

According to pundits, this sorry spectacle was a threat to democracy. Said one editorialist, "It was a mean, cheap tactic on the part of a trade union, and no credit to the labour movement as a whole."

Mean? Yes. Cheap? Yes. Petty? Again, yes. But let's put things into perspective. The union, and all other depositors for that matter, have every right in the world to withdraw funds at any time they wish, for whatever reason seems sufficient to them. That, after all, is the meaning of a demand deposit. Such

LABOR RELATIONS, UNIONS AND COLLECTIVE BARGAINING 485

an arrangement is the embodiment of a contract between two mutually consenting parties, the depositor and the lending institution. In choosing to withdraw $25,000, even for this spiteful reason, CUPE Local 389 was thus completely within its moral and legal rights. The $25,000 is owned by the union. It and it alone has the sole right to determine its place of investment. Neither Mr. Sorenson nor the credit union for which he works has any right to determine where, how or whether this money shall be invested. Certainly their rights have not been abridged by the decision of the proper owners[10] to withdraw the money from the care of the bank.

Not only has the union every right to withdraw its funds for this reason, but other groups in society act in the same way – without the wailing and gnashing of teeth visited upon CUPE.

Does anyone really doubt that corporations deposit and withdraw their funds in accordance with what they perceive as their own best interests? Certainly, church groups and others have publicly withdrawn holdings from banks which have invested in South Africa, or which support firms which are not "ecologically sound." And do not consumers continually pick and choose amongst the stores they will patronize, partially on the basis of boycotting merchants who displease them, sometimes on the most subjective of grounds? Why should unions be singled out for opprobrium for stewardship of their own money?

Then there is the difficulty of legally prohibiting such behavior. How could government stop this practice without dictating how to spend and invest private property? Any attempt to stop such practices would surely involve us in the scenario warned against so eloquently by George Orwell, in his book *Nineteen Eighty-Four*.

Contrary to the political commentators, this act of boycott was a moderate response by the union, certainly when com-

---

[10]We are assuming for the moment, in effect, that CUPE is a legitimate or non coercive union organization. Unfortunately, this is not at all the case. Their illegitimacy stems, however, not from their decision to boycott Sorenson's bank; it is a result of their failure to renounce initiatory violence as a means of conducting business.

pared to the acts which are customary to organized labour. Canadian unions, and also those in the U.S., as a matter of institutional arrangement are commonly allowed to invoke the coercive power of government in order to pursue their own commercial goals. This is their defining characteristic, for organized labour is one of the few institutions in our society permitted to use the threat of fines and/or jail sentences to prohibit competition.

This is a reference, of course, to the manner in which non-union workers who compete with unions for jobs are treated. They are branded as scabs and pariahs. Canadian law forces the employer to "bargain fairly" with the union, and thus prohibits him from dealing with those who would compete for the jobs of organized workers. Even though mutually agreeable contracts could be made between employers and "scabs" for the jobs and pay-scales rejected by striking workers, labour legislation forbids such an occurrence.

So there we have it. On the one hand, a union boycott which violates no rights, but which is roundly condemned by commentators. On the other hand, the union practice of restricting entry to employment which is a patent violation of the rights of every non-union would-be competitor for these jobs. And yet this immoral practice is condemned by practically no one, and even enjoys the prestige and protection of modern law.

A greater travesty of justice can scarcely be imagined.

Illegitimate Unionism
Picketing

Now let us consider several illegitimate union activities. These are acts which coercive unions engage in, but which non-coercive unions totally eschew. Picketing, for example, is morally illicit, and therefore should be outlawed, because it is equivalent to a threat or an initiation of physical force.[11] This

---

[11]In the typical legal analysis of this subject, only secondary picketing (which is not directly aimed at the employer, but rather at third parties, in order to in this way more effectively impact the employer) is even discussed. Implicit in this analysis is the understanding that primary picketing is a legitimate activity. See for example Gall

LABOR RELATIONS, UNIONS AND COLLECTIVE BARGAINING 487

activity must be clearly distinguished from a boycott. In picketing, the object is to coerce and often physically prevent people who would like to deal with the struck employer (suppliers, customers, competing laborers – "scabs," or strikebreakers) from so doing. In a boycott, in contrast, the aim is to mobilize those who already agree with the strike to refrain from making the relevant purchases. True, one may try to convince neutral parties, but in a boycott the means of doing so are strictly limited to non-invasive techniques. Once physical encroachments are resorted to, a boycott becomes converted into picketing.

There are those who characterize picketing as merely "informational." In order to see the problematic nature of such a claim, try to imagine what our response would be were McDonald's to send its agents, hundreds of them, carrying big sticks with signs attached to them (picket signs), to surround the premises of Burger King, or Wendy's, in order to give "information" to their customers or suppliers. In like manner, we do not allow Hertz to picket Avis, or General Motors to picket Ford. There is absolutely no doubt that such activities would be interpreted, and properly so, as an attempt to intimidate. If these firms wish to convey information, they have other avenues open to them: advertising, direct mail, contests, give-aways, bargains, etc. And the same applies to a union. If it wishes to communicate, it must restrict itself to these activities.

Nevertheless, it is continually asserted that the pickets are only at a job site in order to impart the information that a strike is in progress; however, it is "conceded" that the picketers become enraged if they see anyone engaging in commercial endeavors with the struck employer. The attempt, here, is to claim that these "interferences" (people going about their ordinary business, attempting to ignore the strike) are responsible for the violence which is endemic on a picket line. But one cannot have it both ways. Either there is only knowledge being given out, or there is not. If there is, then how do we account

---

(1984).

for the typicality with which violence arises on the picket line? Are its members particularly "sensitive?"

But this is all beside the point. Even if violence was never associated with picket lines, this would only prove they were so successful in their intimidation that none was necessary. The libertarian non-aggression axiom precludes both the actual initiation of violence as well as the threat thereof; thus, even picketing which is (so far) non-violent is a threat to all would-be crossers of the picket line.

A more accurate interpretation of picketing (whether primary or secondary) is as a nuisance or harassment. This is precisely how it would be regarded were it to take place in any other commercial or personal arena.

Suppose, that is, that a person vacates the premises of landlord A, and patronizes landlord B instead. Surely the courts would cast a baleful eye on A, if he, together with his family, cronies, and business associates, began to picket the tenant for being "unfair." Or take another case. Suppose that a man divorces his spouse, and then along with all his friends "pickets" the home of his ex-wife, warning off possible suitors. Would this be considered an informational exercise in free-speech rights? Hardly. On the contrary, it would be clearly seen for the harassment it is, and be summarily prohibited by any court in the land.

Can we afford any less rigorous a definition of justice in labour-management relations?

There is one complication, however. It concerns the legal status of the area on which the picketing occurs. If the picket line operates on private property, the analysis from the libertarian perspective is clear and straightforward: this activity may properly occur only with the permission of the owner. Otherwise, as we have seen, it must be interpreted as oppressive. Unfortunately, in a series of cases concerning the right of picketing and leafletting, the courts have undermined the private property status of streets and thoroughfares in shopping malls, by a finding that these areas are "public places." But they were privately built, are privately operated and maintained, and therefore ought to be considered as part of the private sector –

LABOR RELATIONS, UNIONS AND COLLECTIVE BARGAINING 489

their use to be determined by their owners.

It is far more difficult to determine the proper use of public streets and sidewalks. For the libertarian theorist, these areas are a conundrum. Given that it is morally improper and economically inefficient for the government to have nationalized them in the first place (Block, 1979), it is difficult to determine whether or not picketing should be allowed on the public sidewalk, for example, right in front of the employer's premises. The determination of whether to allow any public assembly (e.g., a parade) to disrupt the normal traffic patterns on government streets is essentially an arbitrary one. It depends upon public pull, not on philosophically determined rights.

Perhaps the best course of action in this moral vacuum is to treat the picketers as if they were merely offering information, as they so vociferously claim. In this case, the best analogy is the man who walks up and down the street with sandwich board placards advertising for a local merchant. Would the court allow one or even two such moving billboards? Certainly, provided that they kept some distance between themselves, and did not interfere with passersby. Would the court allow dozens of tightly packed sandwich board carriers who impeded the normal traffic flow? Certainly not. We conclude from these considerations that striking unionists who use "public property" should be treated exactly like any other group of people attempting to advertise information. If the courts would allow one or two sandwich boarders the use of the public sidewalk, they should extend the same right to informational union picketers. And where they would deny this right to dozens or hundreds of sandwich boarders, they must act in the same way with regard to organized labor.

Scabs

Who are the innocent persons against whom coercive union violence is commonly directed? These are the people at the bottom of the employment ladder, the least, last, and lost of us, the individuals after whose welfare we should take particular concern if we have any regard for the poor. They are, in a word, "Scabs."

490   **JOURNAL OF SOCIAL, POLITICAL & ECONOMIC STUDIES**

Now scabs have had a very bad press. Even the appellation ascribed to them is one of derogation. But when all the loose and inaccurate verbiage is stripped away, the scab is no more than a poor person, oft-times unskilled, uneducated, under- or unemployed, perhaps a member of a minority group, who seeks nothing more than to compete in the labor market, and there to offer his services to the highest bidder.

In fact, it is no exaggeration to consider the scab the economic equivalent of the leper. And we all know the treatment with regard to lepers urged upon us by moral and ecclesiastical authorities.

In their pro (coercive) union stance, defenders of organized labor expose themselves as untrue to the morally axiomatic principle of the preferential option of the poor, which was adumbrated by both the U.S. and Canadian Conferences of Catholic Bishops.[12] The "poor," in this case, are not the princes of labor, organized into gigantic, powerful and coercive unions. Rather, they are the despised, downtrodden and denigrated scabs.[13]

Violence

A strange adventure recently befell Patrick McDermott, the 27 year old son of Canadian Labor Congress president Dennis McDermott. Young Patrick was innocently riding a bus in suburban North York, in Ontario, when he witnessed a beating in the street. Dianne McIntyre, aged 42, was being assaulted by a man – whereupon our hero jumped off the bus, came to the rescue of the damsel in distress, and for his pains was wrestled to the ground by four other men, colleagues of the hoodlum battering Mrs. McIntyre, and was kicked and punched while he was down.

"No big deal" you say? "Happens every day?" Well, yes,

---

[12]For a critique of these documents, see Block (1986, 1983).

[13]Nothing said here mitigates against the legitimacy of voluntary unions, those which restrict themselves to mass walkouts and other non-invasive activity. The only difficulty is that at present, such entities are non existent, at least in North America, to the best of the present author's knowledge.

LABOR RELATIONS, UNIONS AND COLLECTIVE BARGAINING 491

unfortunately; street violence seems to be part and parcel of modern day life, not only in the U.S., but increasingly in Canada as well.

But this case was exceptional. The victimized woman was crossing a picket line at the main Visa credit card center for the Imperial Bank of Commerce, and the five bully boys were bank workers, engaged in a labor strike against this financial institution. What a position to be in for Patrick McDermott, a staunch union supporter in his own right, and son of the outgoing president of the C.L.C.!

Mr. McDermott the younger tried to remain loyal to his principles. That is, to both of them: chivalry and defense of innocent persons against assault and battery on the one hand, and unionism on the other. Although suffering from an arm injury, bruised ribs and a split lip in his confrontation with the minions of organized labour, he stated that he still believes "in the strike and the cause, but when it comes to goons hitting defenseless women, it's got to stop. That guy should be thrown out of the union."

This, however, is too facile, by half. Unionism as practiced in the Western democracies is intrinsically a violent, confrontational and physically aggressive institution. Young Mr. McDermott cannot have it both ways. He must either renounce the "cause," or give up on his principle that goons should not be able to beat up innocent persons.

Why is this? How can it be that a widely respected institution, organized labour, *necessarily* initiates violence against non-aggressing people?

The reason is straightforward. Actual union practice, and the labour codes of the land which underlay it, are predicated on the assumption that competition, no matter how well it works elsewhere in the economy, is simply inappropriate for the labour market. And not only inappropriate, but deserving of legal penalties as well. Labour enactments commonly mandate that the employer "bargain fairly" with a union, when what he may want to do most of all is ignore his striking employees entirely, and hire competing workers (i.e., "scabs") in their place. Some Canadian provinces (e.g., Quebec) prevent management from hiring temporary replacements for the duration of the labour dispute; others allow this, but insist that the firm not deal more favorably with these laborers than with its unionized work force. If the employer declines to be bound by these restrictions, he is liable to fines or even jail sentences – which is certainly equivalent to visiting violence against a person, the employer, for doing no more than encouraging competition in the labour market.

It is perhaps for this reason that the police and courts turn a blind eye – or even a sympathetic one – to situations where union violence is directed against the employer, or, in the case of Mrs. McIntyre, against those who support scab workers by crossing picket lines. "If the government will physically prohibit labour market competition anyway, why penalize organized labour for doing the same thing?" seems to be the prevailing opinion.

A moment's reflection will convince us that this practice – union violence *or* government violence practiced against employers and/or scabs – is completely unjustified. The non-employed competing workers (scabs) have every bit as much right as the striking unionists to compete for jobs offered by the employer. Any other conclusion would set up two classes of people – unionists and scabs – with different types of rights. But all people have the *same* human rights to compete for employment, without being victimized by physical violence, whether from unionists *or* policemen.

As for the assault and battery perpetrated on Patrick McDermott and Dianne McIntyre, a union spokesman termed the incident "minor," and said there were no plans for disciplinary action against the pickets who injured them! And of course the police did nothing to quell this violence in our streets, even though they and all citizens would have been outraged had this situation occurred in any context other than that of a labour strike.

Breakdown of Law and Order

One way to understand this phenomenon of the widespread acceptability of union violence is to focus on the role of the

LABOR RELATIONS, UNIONS AND COLLECTIVE BARGAINING 493

police. They are, after all, supposedly society's fail-safe mechanism against violence. The problem, however, is that this institution, too, has been beset by the virus of accepting unions as legitimate.[14]

According to Mr. Bob Stewart, Chief of Police of Vancouver, one of Canada's largest cities, the use of violence by his constables is inappropriate in a labour dispute. Happily, this man is not a complete pacifist; this view only applies, it would appear, with regard to union unrest. Addressing a meeting of the Atlantic Police Chiefs, he stated "the role of the police officer is to maintain peace and order and not be seen as partisan."[15] The reason for this low profile, it was contended, is that a labour dispute is really a contract dispute between two parties, and not a dispute with police.

It is easy to understand the motivation behind this stance. Canada sees itself as a very stable, polite and civil society, and union-management confrontations are potential tinder boxes. The last thing desired is to fan the flames of violence that have so unfortunately erupted in other corners of the globe.

Nevertheless, there are grave flaws in such a view. Were it to have come from someone else, who did not occupy such an exalted place in the country's law enforcement hierarchy, it could be easily dismissed. But when it is stated by a high ranking police official, it has great capacity to do harm.

First of all, there is the danger that strikes will become more violent, not less. If the police announce beforehand that they will not energetically quell labor violence, this may encourage hotheads to give vent to their more base instincts.

Secondly, it is the very rare case indeed when a person picks a fight directly with a policeman (except, perhaps, when the officer of the law is disguised.) Typically, the services of the police officer are called upon when there is a dispute between two parties, neither of whom is engaged in a direct altercation with the police. But when two men are fighting in a public

---

[14]For a moral and religious defense of unionism, see Novak (1984), U.S. Bishops (1984); for a critique, see Block (1986).

[15]*Vancouver Sun*, July 8, 1987.

street, or when one is assaulting and battering another, we expect the policeman to intervene, with force if necessary, even if the dispute does not directly concern him. After all, we the citizens supposedly pay taxes for police protection, and we expect these services when *we* are attacked, not only when *they* are.

Thirdly, this philosophical position is woefully ignorant of what actually takes place during a strike. Superficially, it is a confrontation between employer and union, who are, or in some cases once were, parties to a labour contract. But it is only in the rare instance that the unionized workers attack their employer's plant, or their employer; after all, they work there, and when the dispute is solved, they typically prefer to have a plant in which to return back to work.

On the contrary, a strike is almost always a dispute between parties who are unrelated by contract. That is, between organized labour and replacement workers, or strike breakers. The union brands these individuals as "scabs," and then initiates violence against people who are guilty only of daring to bid for the jobs currently claimed by the unionists.

Further, it is not really important whether or not the two disputants are contractually linked. Even if they are, it is still the sworn duty of the police to stop – by force if necessary – either side from initiating violence against the other.

That a Canadian Police Chief purposefully wishes to take a "low profile" under such circumstances only indicates he does not really understand the purpose or significance of his job.

Job Ownership

Another defense of picketing and attendant violence concedes that this is a physically aggressive activity, but asserts that it is not an initiation of coercion, but rather a defense of private property rights, namely the jobs of the striking unionists. There is a certain superficial plausibility in this rejoinder. However, the "scab" is not stealing the job of the striking coercive unionist. A job, by its very nature, cannot be owned by any one person. Rather, it is the embodiment of an agreement between *two* consenting parties. In the case of the strike, organized labor is

LABOR RELATIONS, UNIONS AND COLLECTIVE BARGAINING 495

unsatisfied with the offer of the employer. It is publicly re-
nouncing this offer. It therefore cannot be said that these
workers still "have" these jobs.[16] Under laissez faire, *all* people
are allowed to compete for jobs in a free labor market. It is a
vestige of the guild system to think that there are two groups of
people with regard to employment at any given plant: the
coercive unionists, who own the jobs, or have a right to them,
and all other people, who must refrain from bidding for them.

To some extent we are fooled by the very language we use
in order to describe this situation. We speak of "my" job, or
"your" job, or "his" job, or "her" job; this use of the possessive
pronoun does seem to indicate real possession, or ownership.
We also speak of "my" tailor, or "my" employee, or "my"
customer, and yet it would be nothing short of grotesque to
assign ownership rights to any of these relationships. All of
them are based on mutuality, not ownership on the part of
either person. This use of the term "my" does not imply
ownership. If it did I could forbid "my" employee from quitting
his job. If it were "my" customer, I could prevent him from
taking his business elsewhere, to a competitor. And if it were
"my" tailor, it would be a violation of my rights if he moved to
another city, retired, or entered a new occupation.

A job is an embodiment of an agreement between two
consenting parties – employee and employer. It cannot be the
possession of only one of them. A worker no more owns "his"
job than does a husband own "his" wife. A striking union which
forcibly prevents the employer from hiring a replacement is like
a husband who divorces his wife – and then threatens to beat
her up, and any prospective new suitor as well – if she tries to
remarry. Just as one spouse may now divorce the other for any
reason or for none at all, the employer should be able to fire an
employee without being compelled to show "cause." Our laws do

---

[16]We must assume that there is no longer a valid employment contract in force
between the employer and employees. If there is, then the workers do indeed "own"
these jobs, but only because of the contract (assuming that it was initially agreed upon
without duress), not because of any superior status they may claim as members of the
union caste.

not force the worker to justify a decision to quit his job, and the employee-employer relationship should be an entirely symmetric one.

Sweat Shops

What of the claim that without picketing, coercive unions would be rendered virtually powerless, and in the absence of strong coercive labor organizations, the working people would be forced back into the "sweat shops." First of all, even if this claim were true, picketing would still be unjustified, and a violation of the basic libertarian premise against the initiation of violence. Secondly, even if coercive unions were all that stood between the sweat shop and present living conditions for their members, it still does not follow that the lot of *working people* would be improved by picketing. For this activity is aimed not so much at the employer as at the competing worker, the strike breaker. The major aim of the picket line, as we have seen, is to prevent alternative workers from attaining access to the job site. Indeed, the very terminology employed by coercive unionists to describe him, "scab," is indicative of the extreme denigration in which he is held. But these people are *working people* too. Further, as we have noted, they are almost always poorer[17] than the striking coercive unionists. This is seen by the fact that the "scabs" are usually more than happy to take the offer spurned by the strikers. So if there is anyone who needs to be protected from the specter of the "sweat shop," it is not the coercive unionist, but the scab.

Thirdly, it is profoundly mistaken to believe that the modern level of wages depends upon coercive union activity. As any introductory economic textbook makes clear[18], wages depend,

---

[17]The Canadian and U.S. bishops are on record as supporting the "preferential option for the poor." Yet, inconsistently, they support coercive unionism as against the "scabs," who are their major victims. However, the scab may be considered as the economic equivalent of the leper. But the ecclesiastical and biblical authorities urge upon us the kindest of treatment with regard to lepers. Therefore, their own analysis of the scab is illogical.

[18]Even those written by authors who are far from sympathetic to the free enterprise system. See for example Samuelson (1970, chapter 29.)

LABOR RELATIONS, UNIONS AND COLLECTIVE BARGAINING 497

to the contrary, on the productivity of labor. If wages are bid above productivity levels, bankruptcy and consequent unemployment will tend to result.[19] If wages somehow find themselves below the rate of marginal revenue productivity, other employers can earn profits from bidding these workers away from their present employers – by continually improving the job offer until wages and productivity levels come to be equated.

There is abundant evidence to support the view that coercive unionism cannot be credited with the explosion of wages and living standards. For one thing, the modern coercive labor movement has only been with us in this century, and only gained much of its power (in the U.S.) with the advent of special legislation in the 1930s, when its share of the labor force rose from 5% to 20% (Rothbard, 1978, p. 84). And yet wages, welfare and standards of living have been on the increase for hundreds of years before that. For another, the economies of countries of southeast Asia such as South Korea, Taiwan, Hong Kong, Singapore, have been burgeoning in the last several decades, in the virtual absence of unionism, coercive or voluntary (Novak, 1987). As well, there have been sharp wage increases in industries – within countries with a strong labor movement – which are completely unorganized. Examples include banking, computers, even house cleaners.

The comparison between the U.S. and Canada is also instructive. In 1960, the (coercively) unionized sector in both countries was about 30%; by 1988, labor organizations represented over 40% of the Canadian work force, but less than 15% in the U.S. If the union-as-the-source-of-all-prosperity hypothesis were correct, we would have noted a slippage toward sweat shop labor conditions in the U.S., and an era of extreme affluence in Canada. Needless to say, that has not at all been the

---

[19]This was the fate of West Virginia, which fell victim to the activities of John L. Lewis, and organized labor in the coal fields.

case.[20]

Homework

A man's home may be his castle, but not as far as working there is concerned – at least according to legislation which restricts commercial activity in one's own domicile.

Originally, such laws were placed on the books in order to support legislation concerning child labor and compulsory minimum wages. As well, the unions protested vociferously that homeworkers would be very difficult to organize, and the result would be a return to sweatshop conditions.

In the modern era, however, the people who wish to work at home are more likely to be reasonably well off women who wish to earn a bit of extra pin money. For example, there was a "kerfluffle" over several hundred women in the New England states who were knitting snow mittens and ski caps, and who justified this practice on the grounds of "freedom of enterprise." And, as if to show that politics does make strange bedfellows, they also defended themselves on the basis of womens' liberation. Being able to work at home was the only way that many of them could work at all – while continuing to watch over their children.

The debate over home knitters is really only a tempest in a teapot. At most, it involves several thousand seamstresses in an industry that has been on the verge of being supplanted by technology for many years now. Of far greater statistical significance will be the likely move of clerical workers from office to home. This is now just beginning to be made possible

---

[20]Grubel and Bonnici (1986, pp. 40-43). As well as the differing unionization rates, the two countries also experienced widely divergent unemployment insurance policies. In 1970, the U.S. and Canada both spent about 0.9% of their G.N.P. on unemployment insurance benefits; by 1983, the U.S. had maintained its previous level of 0.9%, but Canada's had risen to 3.4%, an increase of 277%! (pp. 44-47). These two events had a profound effect upon the unemployment rates of the two North American neighbors. Traditionally, U.S. and Canadian unemployment rates have moved together within a narrow range. In 1963 for example, they were both slightly less than 6%. But as the disparate unionization and unemployment policies began to take effect, the Canadian rate began to exceed that for the U.S. In the early 1980s a gap of some 4% opened up (p. 2).

LABOR RELATIONS, UNIONS AND COLLECTIVE BARGAINING 499

by technological breakthroughs in computers and word processors, and has thus so far amounted to only a trickle. If present trends continue, however, it is possible that this small stream could turn into a tidal wave.

If this occurs, the union argument that cottage industry is synonymous with sweat shop conditions will be given even wider publicity. It is incorrect, and public policy based on its truth will, as a result, be counterproductive. We can no longer countenance the idea that unionization is all that stands between the laborer and the sweat shop. Thus, there is simply no case for interfering with the institution of home work, no matter how big it becomes.

And there is every moral reason for allowing this new form of industrial organization. People have a natural right to do whatever they please, provided only that their actions do not infringe on the rights of others to do exactly the same. Those who favor both unionism and women's liberation will have to make a choice: one or the other. As this example shows, they cannot have it both ways.

Unequal Bargaining Power

A major reason given by some commentators for their unseemly support of unionism is that employers frequently possess greater bargaining power than do employees in the negotiation of wage agreements. Such unequal power may press workers into a choice between an inadequate wage and no wage at all, it is alleged.

But this rather seriously misconstrues the process of wage determination. In a free labor market, wages are basically set by the marginal revenue productivity of the employee – not on the basis of bargaining power, scale of enterprises, or size of labor units. Were the bargaining power explanation for wage rates correct, remuneration would be negatively correlated with the concentration ratio; that is, industries with fewer employers would pay lower wages than ones with many – and pay would be unrelated to measures of productivity such as educational attainment. Needless to say, no evidence for this contention exists.

The typical reason for supposing that there is unequal bargaining power[21] is that there are more employees than employers.[22] If so, this is hardly sufficient to establish the case. Let us assume that bargaining power is defined in such a way that when there is a difference of opinion over wages, or a dispute about them, the person with the greater bargaining power is more likely to attain his goal than is the person with the lesser bargaining power[23]. But in actual point of fact, the likelihood of attaining one's goal in a bargaining situation depends almost entirely on whether the wage is above, below, or equal to equilibrium, e.g., productivity levels (Hutt 1973, ch. 5). In the first case, the employer will have more "bargaining power," as wages will tend to fall in any case; in the second case, the employee will have more "bargaining power," as the market will dictate an increase in wages. One may say, if one wishes, that in the third case "bargaining power" is equal, since wages will tend not to change. But on the basis of Ocham's Razor it would be more scientific to dispense with the concept of bargaining power[24] entirely, and confine our purview to basic

---

[21]For a particularly unsophisticated version of this view, see Weiler (1980, p. 26), who states: "...workers realized that they had no real leverage in dealing with their employer on an individual basis. True, any one employee might threaten to quit if her pay was not raised. But any sizable employer, let alone a national bank, could always get along without that single employee, who ability and contribution is fungible and who is easily replaced if and when she makes her exit. By contrast the employee will find that she cannot make do without her employer, since she needs a job to earn a living, and jobs may not be too plentiful."

[22]Other attempted justifications of this thesis are that employers are typically more wealthy than employees, and that it is easier for the former to replace the latter than the inverse.

[23]To define bargaining power in the opposite manner (so that the person or group with greater bargaining power would tend to lose disputes over wages) would be to render the argument ludicrous.

[24]There are more customers than merchants (and more whites than blacks, more right handed persons than southpaws, more brunettes than blondes). Does this mean that the latter have more "bargaining power" than the former whenever the two are embroiled in competition, or in a dispute over the terms of trade? Not a bit of it. Customers have more "bargaining power" than merchants when prices are presently above equilibrium, that is, when goods are in surplus, because prices tend to be fall in such cases. Likewise, merchants have more "bargaining power" than consumers when prices are below equilibrium, i.e., when there is a shortage of the good in question,

LABOR RELATIONS, UNIONS AND COLLECTIVE BARGAINING 501

supply and demand analysis of the labor market.

The bargaining power notion is also erroneous in that it disregards the basic economic tenet that in a free market wages tend to be equated with productivity levels. If wages are higher than worker productivity, the enterprise tends to become bankrupt; if lower, the firm suffers a high quit rate, as employees are enticed away by other employers. It is only when wages and productivity are equal that there is no automatic market impetus for change (Hazlitt, 1979).

Weiler sneeringly rejects this as "sophisticated economic analysis," and thus a "somewhat romantic notion," that is somehow out of step with what "has always seemed intuitively clear to workers – and to their employers." Continues Weiler (1980, pp. 26,7):

> In real life, labour markets are notoriously imperfect. There is no central clearinghouse to set an auction price for labour. Workers are poorly informed about alternative jobs and comparative compensation. Once the average employee has invested a significant part of his working career in a single job, he faces tangible and psychological barriers to moving on. Thus employers have the effective ability to quote the price they will pay for labour and to make that price stick.

On closer inspection this latter statement sounds more like the ravings of a Marxist than the sober commentary of a scholar of labor markets. Weiler goes on to assert that "typically, employers (do not) set those wage rates at exploitative levels," but this only compounds the fallacy. Why, if they have the power to do so, do supposedly profit-maximizing firms refrain from "exploiting" labor?

Nor do the other parts of this analysis withstand scrutiny. No central clearinghouse for labor is necessary for the smooth functioning of labor markets, nor is worker information

---

because prices tend to rise in such cases.

required. As long as there is competition between employers, and knowledge of wages on at least one side of the market – for example that of the employer – the market operates inexorably to equate compensation and productivity levels. And further, to the extent that long-term employees face psychological costs in job switching, they are earning psychic profits by remaining with their present employer. If they are reluctant to leave, this is because they are earning non-monetary income over and above their actual salaries by remaining precisely where they are.[25]

Labor Legislation

It follows from our analysis of coercive unionism that much of our present labor legislation is mischievous and misguided. If voluntary association and mutual consent are the only legitimate foundations of employment; if it should be strictly forbidden for one group of workers to forcibly prevent another ("scabs") from competing for jobs; then it follows that government-made laws which are inconsistent with these principles are incompatible as well with the libertarian legal code. For example, there should be no laws which compel the employer to "bargain in good faith" with any one set of employees; he should be allowed to deal with anyone he wishes. Further, all legislation prohibiting an employer from firing striking workers, and hiring replacements on a permanent basis, should be repealed. Says Rothbard (1973, pp. 84-85):

> It is true that the strike is a peculiar form of work stoppage. The strikers do not merely quit their jobs; they also

---

[25]Weiler (1980, p. 27) maintains, without benefit of citation, that "empirical investigation of labour markets in the absence of collective bargaining discloses a remarkable dispersion of wage rates paid to workers with comparable skills in comparable jobs and in comparable industries and regions – all contrary to the hypothesis of competitive markets, which are supposedly marred by trade unionism." But how large is "remarkable"? Who is to determine that the skills, industries and regions are truly comparable? Ivory tower researchers? Nor can these unnamed studies take into account non monetary psychic on-the-job earnings, attained, as Weiler himself postulates, by long tenure on the job.

LABOR RELATIONS, UNIONS AND COLLECTIVE BARGAINING 503

assert that somehow, in some metaphysical sense, they still 'own' their jobs and are entitled to them, and intend to return to them when the issues are resolved. But the remedy for this self-contradictory policy, as well as for the disruptive power of labor unions, is not to pass laws outlawing strikes; the remedy is to remove the substantial body of law, federal, state, and local, that confers special governmental privileges on labor unions. All that is needed, both for libertarian principle and for a healthy economy, is to remove and abolish these special privileges.

These privileges have been enshrined in federal law – especially in the Wagner-Taft-Hartley Act, passed originally in 1935, and the Norris-LaGuardia Act of 1931. The latter prohibits the courts from issuing injunctions in cases of imminent union violence; the former compels employers to bargain 'in good faith' with any union that wins the votes of the majority of a work unit arbitrarily defined by the federal government – and also prohibits employers from discriminating against union organizers .... Furthermore, local and state laws often protect unions from being sued, and they place restrictions on the employers' hiring of strikebreaking labor; and police are often instructed not to interfere in the use of violence against strikebreakers by union pickets. Take away these special privileges ...

It is characteristic of our statist trend that, when general indignation against unions led to the Taft-Hartley Act of 1947, the government did not repeal any of these special privileges. Instead, it added special restrictions upon unions to limit the power which the government itself had created .... The government's seemingly contradictory policy on unions serves, first, to aggrandize the power of government over labor relations, and second, to foster a suitably integrated and Establishment-minded unionism as junior partner in government's role over the economy.

Conclusion

It is an important aspect of public policy-making to examine extant labour codes with a view to revising them. In the past, such attempts have been superficial; they have placed bubble gum, band aid, and scotch tape solutions on a corpus in need of major surgery. Our legislative representatives must go to the heart of the matter this time out, for economic justice, the rule of law and the health of the economy depend upon it.

In the field of labour relations, the most important issue is the strike. Actually, this is misnomer, as it refers not to one act, but to two. A strike is, first, a withdrawal of labour in unison from an employer, on the part of the relevant organized employees. To this, there can be no objection. If a single individual has a right to withdraw labour services, or to quit a job, he does not lose it merely because others choose to exercise their rights, simultaneously.

There is a second aspect of the strike, however. This element is pernicious, insidious and entirely improper: the union practice of making it impossible for the struck employer to deal with alternative sources of labour, who are anxious to compete for the jobs the strikers have just vacated.

A properly revised labour code, then, would allow strikes in the sense of mass refusals to work, or quits in unison. It would entrench this behavior, as a basic element of the rights of free men. But is would *limit* union activity to this one option. It would thus prohibit, to the full extent of the law, any and all interferences with the rights of alternative employees ("scabs") to compete for jobs held by union members. It would end, forevermore, all picketing, and other such forms of threatened or actual violence.

Although many people think that pickets are aimed at the struck employer, they are actually an attack on competing workers ("scabs"). And just as our laws should not allow business firms to picket the premises of suppliers, competitors or customers, no group of workers should be able, by picketing, to forcibly prohibit another group of workers – almost always poorer – from bidding for jobs. A proper labour code would thus define a "legitimate union" as one which strictly limited its actions to organizing mass resignations. A "legitimate union" would eschew picketing, violence, and all other special advantages – legislative or otherwise – vis-a-vis its non-unionized

LABOR RELATIONS, UNIONS AND COLLECTIVE BARGAINING 505

competitors. This would end, once and for all, the legal fiction that workers who have left their job can yet retain any right to employment status in those positions.

We must conclude that the key distinction in any analysis of unions is between those that engage in coercion – whether directly or through the intermediation of unjust laws. And that sound public policy, in the first best sense, consists not of allowing illegitimate union activities, coupled with restricting them by the imposition of secret ballots, etc., but rather of stripping them of all coercive powers. The only just unions are those which limit their activities to boycotts, mass walkouts and other such activities that any one person has a right to engage in. When labor organizations transcend these limitations, they must be reined in, if economic justice is to prevail.

REFERENCES

Armentano, Dominick T.,
   1972   The Myths of Antitrust, New Rochelle, N.Y.: Arlington House. Armentano, Dominick T., *Antitrust and Monopoly: Anatomy of a Policy Failure*, New York, Wiley, 1982.

Block, Walter
   1979   "Free Market Transportation: Denationalizing the Roads," *Journal of Libertarian Studies*, Vol. III, No. 2, Summer 1979, pp. 209-238.
   1976   *Defending the Undefendable*, New York: Fleet Press.
   1983   *On Economics and the Canadian Bishops*, Vancouver, the Fraser Institute.
   1986   *The U.S. Bishops and Their Critics: An Economic and Ethical Perspective*, Vancouver, the Fraser Institute, 1986.
   1990   "Private Property, Ethics and Wealth Creation," *Toward an Ethic of Wealth Creation*, Peter L. Berger, ed., San Fransisco: Institute for Contemporary Studies, 1990.

Friedman, Milton
   1977   *On Galbraith and on Curing the British Disease*, Vancouver: The Fraser Institute.
   1985   "Capitalism and the Jews," *The Morality of the Market: Religious and Economic Perspectives*, Walter Block, ed., Vancouver: The Fraser Institute.

Gall, Peter A.
   1984   "Regulation of Picketing under the B.C. Labour Code: Some Cracks in the Institutional Foundation," *The Labour Code of British Columbia in the 1980s*, Joseph M. Weiler and Peter A. Gall, eds.,

Vancouver: Carswell Legal Publications.
Grubel, Herbert G. and Josef Bonnici
  1986   *Why is Canada's Unemployment Rate So High?*, Vancouver, the Fraser Institute.
Hayek, Friedrich A.,
  1960   *The Constitution of Liberty*, Henry Regnery Company, Chicago.
  1973   *Law, Legislation and Liberty*, The University of Chicago Press, Chicago.
  1989   *The Fatal Conceit : The Errors of Socialism*, Chicago, The University of Chicago Press.
Hutt, W. H.
  1973   *The Strike Threat System: The Economic Consequences of Collective Bargaining*, New Rochelle, N.Y., Arlington House.
  1989   "Trade Unions: The Private Use of Coercive Power," *The Review of Austrian Economics*, Vol. III, 1989, pp. 109-120.
Leoni, Bruno
  1961   *Freedom and the Law*, New York: Van Nostrand.
Mises, Ludwig von,
  1966   *Human action*, Chicago: Regnery.
Novak, Michael
  1987   *Will it Liberate? Questions about Liberation Theology*, New York, Paulist Press, 1987.
Novak, Michael et. al.,
  1984   Toward the Future: Catholic Social Thought and the U.S. Economy - A Lay Letter, New York: Lay Commission on Catholic Social Teaching and the U.S. Economy, 1984.
Ohashi, T.M. and T.P. Roth, eds.,
  1980   *Privatization: Theory and Practice*, Vancouver: The Fraser Institute.
Petro, Sylvester
  1957   *The Labor Policy of the Free Society*, New York, Ronald Press.
Reynolds, Morgan O.
  1984   *Power and Privilege: Labor Unions in America*, New York: Manhattan Institute for Policy Research.
Rothbard, Murray N.
  1970   *Man, Economy and State*, Los Angeles, Nash.
  1973   *For New Liberty*, New York, Collier.
  1983   *The Ethics of Liberty*, Atlantic Highlands, N.J. Humanities Press.
Samuelson, Paul
  1970   *Economics*, New York, McGraw-Hill, 8th edition, chapter 29.
Schmidt, Emerson P.
  1973   *Union Power and the Public Interest*, Los Angeles, Nash.
Simons, Henry C.
  1948   *Economic Policy for a Free Society*, Chicago, University of Chicago Press.

LABOR RELATIONS, UNIONS AND COLLECTIVE BARGAINING 507

Sowell, Thomas
1983   *The Economics and Politics of Race*, New York, Morrow.
U.S. Bishops' Pastoral Letter on Catholic Social Teaching and the U.S.
Economy, Origins: NC Documentary Service, November 15, 1984,
Vol. 14, No. 22/23.
Vonnegut, Kurt
1982   "Harrison Bergeron," *Discrimination, Affirmative Action, and Equal Opportunity*, Walter Block and Michael Walker, eds., Vancouver: The Fraser Institute.
Walker, Michael ed.,
1976   *The Illusion of Wage and Price Control*, Vancouver: The Fraser Institute.
1988   *Privatization: Tactics and Techniques*, Vancouver: The Fraser Institute.
Weiler, Paul
1980   *Reconcilable Differences*, Toronto: Carswell.
Williams, Walter E.
1982   *The State Against Blacks*, New York: McGraw-Hill.

Labour Union Policies: Gains or Pains?

Jason Evans and *Walter Block*

The Authors

Jason Evans, Economics Department, University of Central Arkansas, Conway, AR 72035, USA.

Walter Block, Harold E. Wirth Eminent Scholar Endowed Chair in Economics, 6363 St. Charles Avenue, Box 15, Miller 321, College of Business Administration, Loyola University New Orleans, New Orleans, LA 70118, USA.

The popular saying "you cannot see the forest for the trees" is often used to refer to a situation in which the main theme is ignored or misunderstood due to excessive focus on its components. Supporters of labour unions and their policies could be said to suffer from this distorted point of view. Union leaders have preached the benefits of their organisations to society for so long that many people take them at their word. They claim to help the underdog worker get his fair share of the economic pie with better wages, hours, and other conditions (Reynolds, 1987, p.11). Organised labour also claims to make gains for its members such as setting standards to increase workers' levels of skill and competence, shortening the work week from seventy to forty hours, and increasing their level of health and welfare.

It is important, however, to look not only at the effects of labour union policies in the short-run (trees), but in the long-run as well (forest). When this approach is taken, the real effects of these organisations becomes clear. Their policies distort wages and prices, curb production, decrease nonmonetary rewards, hinder labour/management relations, and have an overall detrimental effect on the economy. So who benefits from this labour agenda? The union leaders, who collect billions of dollars in dues every year from the "underdog workers" they represent.

There is a very popular myth in the United States that labour unions are beneficial to society because of the wage "gains" received by their members. But these wage "gains" are nothing more than wage rates that are pushed above the level that a competitive free market would have brought through supply and demand. This is oftentimes achieved through a strike, or even the threat of a strike (Hazlitt, 1973, p.2). The strikers would like to focus attention on the plight of the so-called underpaid workers and keep the public from asking one important question - who will pay for these higher wages? The union answer is typically that the excess profits of the company involved will cover the increase. That is, the union is playing the role of a contemporary Robin Hood, taking from the rich and giving to the poor.

Volume 9 Number 1 2002   73

But the real answer is that virtually everyone pays for these higher wages in one way or another. These short-run "gains" quickly turn into long run "pains" for the entire society. By striking, the union is basically announcing that its workers are more important than everyone else and no one else can do their jobs.[1] By refusing to work and preventing others from taking the vacated jobs, the union is obviously excluding non-union workers. The latter pay in the form of unemployment. Other employees suffer because the "gains" of the union workers cause a decrease in their real wages. This is true because of the effects that higher wages have on the producers of the goods and services the population must buy. Higher production costs force producers to increase prices in order to make a profit. Basically, while union public relations would have everyone believe that they are hurting only the producers by digging into their profits, they are actually damaging consumers in the form of higher prices of goods and services. These "gains" have actually been found to hurt some union members themselves because the higher wage rates in a particular industry will generally lead to less employment there (Hazlitt, 1973, p.3).

No, the only way to increase real wages is to increase productivity. With greater output per man hour, unit costs of goods are lower, thus increasing the amount of product that can be bought with the same amount of wages. Labour unions work assiduously against increasing wealth in this way. One example is their opposition to capital investment and modernisation. Since both have a positive effect on productivity and efficiency, the effect is to retard the lowering of production unit costs. Unions are also notorious for undermining management's ability to do its job, which is to increase efficiency. Perhaps the most well known example of this is the union practice of featherbedding, or make-work projects, which provides their rank and file with unnecessary jobs that are of no benefit to the employer or to the final consumer. So, despite union claims to the contrary, it is evident that unions cannot raise real wages, only distort them. The only way to increase real wages is to increase productivity (Hazlitt, 1973, p.1).

Unions claim to increase productivity by reducing turnover. More specifically, they claim to nearly wipe out the problem of voluntary quits (Rees, 1989, p.131).[2] According to union leaders, lower turnover results in a more experienced workforce with reduced training costs, thus increasing productivity. Unions can also, at times, call for their workers to increase individual productivity if there is a competitive threat in a particular industry which leads the union leaders to believe that jobs could be lost due to the closing of a plant. On the surface, it seems as though they have the best interests of their own workers in particular and of society in general in mind. One hint that this is an erroneous claim, is that the motivation behind these types of actions can best be summed up by a simple equation: "Lost jobs equal lost dues." These increases in production are yet more "trees" used in labour union propaganda to cover up the "forest" of the more harmful effects of unions on productivity.

For if organised labour can pump up productivity on a temporary basis in order to respond to a perceived threat to its hegemony, this is proof positive that in the more typical absence of such specific outside competition they are *reducing* product per man hour. It is as if a slacker were to become unusually busy when and only when his boss is looking at him with a jaundiced eye. This is surely evidence of less than a full effort when the manager is focusing elsewhere.

On the other hand, in industries not earmarked by restrictions on entry (Hamowy, 1984), there is *always* outside competition, breathing down the neck of any specific firm. That the union increases productivity on a temporary basis to ward off a given threat indicates that at other times, its reduction of productivity puts the business concern at undue risk.

Perhaps the most widely known way that labour unions curb productivity is through the use of subdivisions of labour and make-work rules. Subdivision of labour is basically a narrowing of the tasks of each individual worker to the point where he does only one job, or one part of it when the whole process could easily be done by him in its entirety. This results in delays and jurisdictional disputes over who does which tasks.[3] Take for example, a mason, who under the rules of subdivision of labour must build a wall and do nothing else. If the workers who carry the bricks and mix the mortar fall behind and need help to catch up, the mason will not be able to help them even though he is physically fit and able to do so. Closely related to subdivision of labour is the use of aforementioned make-work rules in which workers are employed to do unnecessary jobs. Commonly referred to as "featherbedding" or "overstaffing," these rules are designed for the sole purpose of having more union members employed. Once again, more union jobs means more dues in the pockets of the union leaders.

Another infamous practice of organised labour is opposition to technological advances and machinery that could replace labourers. For example, unions have gone as far as to require an extra man on trains in the early twentieth century to shovel coal into the engine, called a fireman (it is very important to note that this took place after the invention of the modern engine when coal-powered engines became obsolete and no longer in use). While these technological advances would cause some personnel to lose their jobs, they would greatly increase production as well as product quality levels. Innovation would also free workers to participate in other industries which would benefit from the increased production and lower costs from the unionised industry.

In the view of Eleanor Roosevelt, the Hilary Clinton of her day, in a syndicated newspaper column in 1945, "We have reached a point today where labour saving devices are good only when they do not throw the worker out of his job" (cited in Hazlitt, 1979, p.54). But the whole *point* of labour saving technology is to "unemploy" workers in jobs that no longer need doing, so that they can be freed up to undertake tasks impossible to accomplish under the old technology.

At one time in our history, all elevators were manually operated. Automatic mechanisms have long since replaced these archaic devices. One way to interpret this is as unemployment perpetrated by vicious capitalists. But this Marxian notion cannot come to grips with the fact that had elevators continued to be manually operated, this would have limited the size of buildings to say nothing of depriving us all of the goods and services that can now be provided by all the people no longer needed to operate these conveyances.

Labour unions also oppose payment on the basis of output or efficiency, preferring that all workers in a specific field be paid a standard wage. This artifically reduces workers' incentives to put forth their best effort. Promotion by seniority fosters the same lackadaisical effort because no matter how hard a worker tries he will not be promoted before his time. This greatly limits the American workforce by keeping talented, hard workers from moving up to positions of power. An outright deliberate curbing of productivity is union initiated "slowdowns" in which the labour barons advise the rank and file to cut back production and oppose their fellows who produce more than they (Hazlitt, 1979, p.150)[4]

To further demonstrate the relationship between productivity and wages it is necessary to focus on the effects of the former on the latter. Because of the raise in production costs brought about by the higher wages of union workers, cutbacks in production often occur. Companies are in business to make money so it makes perfect sense that paying artificially high wages will have negative repercussions. Since many union members are protected from being let go from their jobs, the only option is to reduce the hours used to produce goods or services. For marginal producers, who are just barely able to stay in business, these union "gains" can force them to stop production altogether. This causes goods or services that would have been produced to not be brought into existence (Branden, 1963, p.2).

In addition to monetary rewards, such as wages, pensions, and bonuses, employers pay workers with fringe benefits such as insurance, locational convenience, free parking, air conditioning, music, low noise levels, and on-the-job training, just to name a few (Reynolds, 1987, p.67). Under capitalism, market forces will determine the level at which these benefits are allocated to the workers, and employers will adjust the amounts of these fringe benefits to match productivity. That is, for example, if the employees would prefer to be able to listen to music while on the job and the benefits to the employer outweigh the costs, the employer would most likely allow the workers to have a radio. If not, he will face a higher than optimal quit rate, as workers migrate to more musical pastures. The employer, of course, wants to minimise labour costs while maximising productivity. In most cases these fringe benefits will increase product per man hour because the employees are happier and willing to work harder. However, if the non monetary payments hinder production, for example, if the workers would like a million dollar surround sound stereo system installed so they can dance all day instead of work, the employer will not likely go along with the idea. Although that is a rather extreme example, it demonstrates what labour unions want for their workers. Leaders not only want higher wages for their members, they also want to squeeze every resource of the employer in order to make their union more attractive to prospective dues payers. It is clearly a case of labour bureaucrats trying to do everything they can in order to sign up more members.[5]

In any study of the negative effects of unions, it is necessary to include the negative implications for labour/management relations. The relationship between management and unions is strained at best because of their positions at opposite ends of the labour spectrum. Management on the one hand wants to maximise productivity, efficiency, and profits. The labour union at the other end, wants to maximise the number of workers, wages, and fringe benefits. It is a relationship where there must be a balance in order to communicate effectively. When unions force wages up, management must respond by cutting back on production in order to minimise costs (Reynolds, 1987, p.66). This can be accomplished through layoffs and normal turnover, or through a substitution of labour-saving machinery in addition to closer supervision of workers.

All of these devices, however, are in conflict with the basic goals of the unions, exacerbating relations between the two parties. Management expects more output from the highly paid union workers and would prefer to increase monitoring, impose more strict quality standards, and entertain a general expectation of increased productivity to meet this end (Reynolds, 1987, p.66). The union, of course, fights management for control of working conditions. The result is often a communication breakdown in which there is no teamwork between the management team and the union leaders and workers. According to P.J. Sloane, this conflict is maximised when both sides are nearly equal in terms of bargaining power (Sloane, 1978, p.31). This is true because without the power to easily defeat the other side, both must work hard to try to win the conflict. This is a very inefficient and costly way to settle problems which should never exist in the first place. If left to market forces, wages and worker efficiency would balance each other (see Block, 1991, 1996a), Kauffman (1992), Petro (1957), Poulson (1982), Reynolds (1982, 1984, 1987)). Surveys of on-the-job satisfaction report that union workers have a lower level of job satisfaction than non-union workers[6] (Reynolds, 1987, p.66). The struggle for power between organised labour and management is a major factor contributing to worker dissatisfaction.[7] Union workers also often find it more difficult to communicate with management on a personal or team level, which also stems from poisoned relationships. Government laws mandating that management must work with the union tie the hands of business and slap it in the face, to add insult to injury. The "invisible hand" of Adam Smith cannot do its job if it is tied behind one's back (Smith, 1776).

The overall effects of labour union policies on the economy are highly detrimental. The unnatural and forced raising of wages causes a misallocation of resources and an inefficient workforce (Reynolds, 1987, p.80). This brings about a lower gross domestic product because of the inability to utilise all of the assets available to the economy. Organised labour is also responsible for higher levels of unemployment than would normally exist in a free-market situation, causing yet more losses to the economy in the form of lost production. The opposition to modernisation also slows down economic progress by hindering production and efficiency. With technological advances come more jobs doing other productive work that can benefit the whole country. As stated by Henry Hazlitt, "Work creates work, and there is no limit to the amount of work to be done" (Hazlitt, 1979, p.150). Opposition to technology also reduces capital investment, further placing production in jeopardy. Economic growth is reduced by socialistic interferences with labour markets. Perhaps the most wasteful effect of union policies is the direct costs of strikes, strike threats, labour consultants, negotiating costs, grievance costs, bureaucratic costs, and government spending which come straight out of the pockets of every worker (Reynolds, 1987, p.82). These and all the other negative effects of labour union policies make them, in Henry Hazlitt's words "The chief antilabour force" (Hazlitt, 1973, p.1). In order to reach the maximum level of production and income, the market must be left alone to allow the natural market forces to prevail.

Despite the foregoing analysis of the negative aspects of unions, many people are likely to conclude that these costs must be borne somehow because of the importance of collective bargaining. Their idea is that without this guarantee, wages would be determined solely by management, and would thus shrink them to that level pertaining in sweat shops, or even perish the thought, to medieval or third world levels.

Nothing could be further from the truth. Wages are determined by productivity, not at the discretion of an employer. Any firm paying a worker worth $20 per hour only $5 per hour would earn a pure profit of $15 during this time period. This would attract hordes of competitors willing to surpass such a low wage,[8] so as to be able to take advantage of this labour "exploitation." But this process ensures that wages will be bid up to productivity levels, or, rather, not deviate from them in the first place.[9]

Evidence for this contention abounds. Wages are high in the non unionised fields of banking, insurance and computers; and in virtually non unionised countries such as Japan, South Korea, Hong Kong and Singapore. Real wages were rising for dozens of decades before the advent of unions in the early 20th century. As for the market process of raising wages to productivity levels, eloquent testimony to the importance of this phenomenon is given by the California and Oregon growers who travel hundreds of miles to attract Mexican workers being paid far less in that country than their productivity levels in the US.

No. Individual "bargaining" will do quite well for raising wages to productivity levels. Workers need collective bargaining like a fish needs a bicycle.

Endnotes

1. This draws our attention to the union claim of job ownership. They attempt to protect jobs in much the same way that one would protect private property. According to Walter Block, "A job, by its very nature, cannot be owned by any one person. Rather, it is the embodiment of an agreement between *two* consenting parties" (Block, 1991, p.494). How is it, then, that these workers could have any more right to a certain job than any other competitor seeking that particular employment?

2. For a critique of the labour economics of Rees, see Block (1996b).

3. These jurisdictional disputes are often bitter and quite costly, with a complex system set up to handle arbitration (Rees, 1989, p.133).

4. These workers are treated rather cruelly, being denounced, asked to quit, or sometimes even beaten (Hazlitt, 1979, p.150).

5. Once again, it is necessary to mention the love the union leaders have for membership dues.

6. This is true in spite of their higher wages and all other benefits of union membership.

7. These policies include make-work rules which increase monotony (Reynolds, 1987, p.67).

8. We much incorporate the costs of finding such workers into the analysis, or implicitly assume there are so many such "underpaid" employees that unearthing them would repay search costs.

9. In real world markets, of course, there is only a tendency toward equilibrium, not the certainty that we will be at this point at any given time.

References

Block, Walter (Winter 1991) "Labor Relations, Unions and Collective Bargaining: A Political Economic Analysis," *Journal of Social Political and Economic Studies*, Vol. 16, No. 4, pp.477-507.

Block, Walter (1996a) "Comment on Richard B. Freeman's 'Labor markets and institutions in economic development," *International Journal of Social Economics*, Vol. 23, No. 1, pp.6-16.

Block, Walter (1996b) "Labor Market Disputes: A Comment on Albert Rees' 'Fairness in Wage Distribution," *Journal of Interdisciplinary Economics*, Vol. 7, No. 3, pp.217-330.

Branden, Nathaniel (1963) "Labor Unions" *http://www.nathanielbranden.net/que/que31.shtml*

Hamowy, Ronald (1984) *Canadian Medicine: A Study in Restricted Entry*. Vancouver: The Fraser Institute.

Hazlitt, Henry (1973) "Conquest of Poverty," Chapter 13. http://www.hazlitt.org/e-texts/poverty/ch13.html

Hazlitt, Henry (1979) *Economics in One Lesson*. Arlington House Publishers, New York.

Kauffman, Bill (1992) "The Child Labor Amendment Debate of the 1920s; or, Catholics and Mugwumps and Farmers," *The Journal of Libertarian Studies*, Vol. 10, No. 2, Fall, pp.139-170.

Petro, Sylvester (1957) *The Labor Policy of the Free Society*. New York, Ronald Press.

Poulson, Barry W. (1982) "Substantive Due Process and Labor Law," *The Journal of Libertarian Studies*, Vol. VI, No. 3-4, Summer/Fall, pp.267-276.

Rees, Albert (1989) *The Econmics of Trade Unions*. The University of Chicago Press, Chicago.

Reynolds, Morgan O. (1995)*Economics of Labor*. South-Western College Publishing.

Reynolds, Morgan O. (1987) *Making America Poorer: The Cost of Labor Law*. Washington DC: Cato Institute.

Reynolds, Morgan O. (1984)*Power and Privilege: Labor Unions in America*. New York: Manhattan Institute for Policy Research.

Reynolds, Morgan, O. (1982) "An Economic Analysis of the Norris-LaGuardia Act, the Wagner Act and the Labor Representation Industry," *The Journal of Libertarian Studies*, Vol. VI, No. 3-4, Summer/Fall, pp.227-266.

Sloane, P.J. (1978) *Trade Unions*. The Institute of Economic Affairs. Lancing, England.

Smith, Adam (1776/1779) *An Inquiry into the Nature and Causes of the Wealth of Nations*. Indianapolis, IN: Liberty Fund.

The Yellow Dog Contract: Bring It Back!

Walter Block

September 5, 2005

The Yellow Dog Contract is an honorable contract. It states that one of the conditions of employment is that the worker agrees not to join a union. It is no different, in principle, from the requirement that if you come visit someone in his house at his invitation, you must wear a funny hat, or agree not to associate with anyone he specifies, say, an enemy of his. In each case, that of the firm, and the private home, the one making this "demand" is exercising his rights of free association. He is saying, in effect, "if you want to associate with me, you must do thus and such." You are perfectly free to refuse to do so, but then, he will not associate with you. He will not allow you into his home, nor agree to employ you.

But suppose you need a job? Is it not "unfair" that the employer will not hire you unless you give up, in advance, your rights to organize with others for better wages and working conditions? No; this situation is no more illicit than if you need to go to a party, and hate to wear a funny hat but the host insists upon it. Or demands, as the price of entry, that you not consort with his enemies. After all, it is *his* house.

We all assert our rights of free association through implicit "threats" of this sort. The wife says to the husband, "If you gamble away our money, I'll leave you." The husband may say to the wife, "If you run around with other men, I'll divorce you." The customer says to the shop-keeper, "If you give me a bad product, you'll never see me in here again." The restaurant owner says to his diners, "If you can't behave yourself, I'll have to ask you to leave."

Free association is a crucially important element of liberty. Without the right to associate with those we (mutually) choose, we are in effect and to that extent, slaves. The *only* thing wrong with slavery was that the slave could not quit. He was forced to "associate" with the slave master against his will.

But the principle is the same with the Yellow Dog Contract. When it is banned, the employer is compelled to associate with a potential union member. To force him to do so is to coerce an innocent man. It is to violate his right to freely associate with others on a voluntary basis.

The argument for doing so is that without unions, wages would plummet to whatever low levels the "generosity" of employers would yield. But nothing could be further from the truth. Wages are, rather, determined on the basis of (marginal revenue) productivity, which in turn stems from how hard and smart we work, and with the cooperation of how much and of what quality capital equipment. Wages were rising long before the advent of unions in the early part of the 20th century, in industries (computers, banking, insurance, baby sitting, lawn mowing) and countries (Hong Kong, Singapore, Taiwan) without any such labor organizations. Labor unions reached their apex decades ago, and now account for *single digits* in the private sector; during this sharp decrease in the proportion of workers "protected" by organized labor, wages catapulted.

No, labor organizations are instead an economic tape-worm, infesting firms and eating away their substance. It is no accident that what is now the "rust belt" in the northeastern quadrant of the country is the most heavily unionized sector of the nation, and suffered the greatest degree of plant closings when they were trying to suck the blood of entrepreneurs. Nor is it any puzzle that the south, the least unionized part of the country, became one of the fastest growing areas. Parasitical labor organizations in the coal-fields were also responsible for the economic devastation of West Virginia.

Are unions per se illegitimate? No. If all they do is threaten mass quits unless their demands are met, they should not be banned by law. But as a matter of fact, not a one of them limits itself in this manner. Instead, in addition, they threaten the person and property not only of the owner, but also of any workers (they call them "scabs") who attempt to take up the wages and working conditions spurned by the union. They also favor labor legislation that compels the owner to deal with the union, when he wishes to ignore these workers and hire the "scabs" instead.

The Yellow Dog Contract, in addition to safeguarding employer and employee rights of free association, also serves as a remedy against union inflicted economic disarray and violence against innocent people and their property. Long live the Yellow Dog Contract. Bring it back. Now.

An Economic and Ethical Analysis of Unions

Walter Block

Editor's Note: This debate over unions and wages took place in the pages of The Maroon, *the student newspaper of Loyola University New Orleans between Walter Block of the economics department and Boyd Blundell, of the religion department. This debate is in five parts. The first four appeared in* Maroon *as follows: Block on 9/8/06, Blundell on 9/15/06, Block on 9/22/06 and Blundell on 10/6/06. The* Maroon *declined to publish Block's follow up reply, but we include this as the fifth entry in the debate, which appears below. The sixth part of the series is a report on a debate that took place between Walter Block and Fr. David Boileau, professor of philosophy at Loyola University, on this very topic.*

Part I

Unions don't guarantee fair employee wages
Walter Block
9/8/06

According to some, the reason we need unions is because without them employers would grind employees into the ground. Were organized labor to disappear, wages would plummet; workers would have to work on Sundays, tip their hat to their bosses and suffer all sorts of other indignities — including losing virtually all improvements in working conditions made over the last century.

This is all wrong. Wages and working conditions aren't set by firms. Rather, they depend upon the productivity of labor. This can be defined as the extra amount of revenue brought in by adding one more person to the payroll. For example, if there were 1,000 workers creating an item that sold for X dollars and then the 1,001st employee came on board and the firm's sales rose to $X + $7, then the marginal revenue productivity of the last person hired would be $7 per hour.

111

Wages can't (long be) higher than this amount, or the company will lose money on every worker it hires. For example, if compensation is $10, and revenue taken in due to the efforts of the worker is $7, then the firm loses $3 every hour the man is on the shop floor.

On the other hand, a situation can't long endure where wages are lower than this amount. For example, suppose pay was $2 per hour, while productivity remained at the $7 level we are considering. Then, the employer would earn a pure profit of $5 every hour. This can't last. Other companies would have incentive to hire such a worker away from his employer. Assuming that the productivity of the latter would be the same $7 everywhere, a competitor could offer, say, $2.25. This would be a substantial increase over and above the present salary of $2 and yet would allow the newcomer to earn a profit of $7 – $2.25 = $4.75. But if this would work, so would a bid of $2.50, $2.75, $3.00, etc. Where would this process end? As near to $7 as allowed by the costs of finding such "underpaid" workers and convincing them to switch jobs for higher pay. This doesn't mean that under free enterprise there will be no deviations from this amount. But there is an inexorable tendency for wages to continually move in the direction of this equilibration.

If wages were really set by employers, why is it that employees such as Shaquille O'Neal, Brad Pitt and Brittney Spears all earn mega bucks? Generosity? No, the reason they do is because their productivity (ability to fill seats in sports arenas, movie theaters, concerts) is very high. If their present employers did not pay them in accordance with productivity others would gladly jump in and do so.

When unions artificially boost wages above this stipulated $7 productivity, they look good in the short run. But in the long run they create business failures and rust belts. It's impossible for any substantial length of time to maintain wages above productivity levels.

What determines the level of productivity and hence the wages? This is based on how hard and how intelligently people work and the amount and sophistication of the tools and capital equipment they are given by their employer to work with. This, in turn, depends upon how much saving occurred in the previous periods and even before that, how economically free and law abiding is the populace. The more reliance on private property rights and free enterprise, other things equal, the better in this regard.

If organized labor is really the only institution that stands between the workingman and abject poverty, how is it that real wages have been increasing while the rate of unionization has been declining over the last half century? Why is it that some industries that have never come within a million miles of unions (computers, banking, accounting) pay very high wages, often in excess of that earned by the rank and file? Given that they are at the mercy of the capitalist pigs, should they not have been ground into the dust? How can it be that the south, which is the least unionized part of the country, is the fastest growing? What accounts for the fact that countries where Western-style unionism is all but unknown (Hong Kong, Singapore, Japan) there are economic powerhouses, with standards of living envied in many places on the globe?

Walter Block is a professor of business administration and Wirth chairman of economics.

Part II

Block's stance on unions not even wrong
Boyd Blundell
9/15/06

In physics, when a theory fails to be verified in experiments or is found to have errors, it is considered wrong. But if the theory is so incoherent, so disconnected from reality that no sense can be made of it, physicists say that it is not even wrong. Such a theory doesn't deserve to be judged alongside real theories.

Walter Block's column last week on unions and wages is not even wrong.

The essay has more fundamental errors than paragraphs, so only the most offensive can be addressed. The first paragraph sets up a strawman, the mysterious "some" who think that were unions to disappear, working conditions would revert to the horrors of the 19th century industrial factories.

Let's be clear: nobody says this. It's a figment of Block's imagination. The reason nobody says this is that the gains in working conditions and wages that were made by unions have now been codified in law. It is the laws that guarantee basic workplace safety, minimum wages and so on. Where these laws vary, so do working conditions.

The one thing Block's "some" do in fact say is that wages would drop if unions disappeared. He thinks this is all wrong, that there is an "inexorable tendency" toward an equilibrium between wages and productivity. He argues that a situation in which a worker is paid $5 an hour less than his productivity "can't last," because such a productive worker could be hired away.

Now it pains me to have to instruct the eminent Dr. Block on such a basic fact of economics, but wages are driven primarily by bargaining power as dictated by supply and demand; productivity is virtually irrelevant (except as a soft ceiling). If an owner needs five workers but there are five hundred workers who need work, the owner will offer the lowest possible wages needed to attract five suitable workers in order to maximize profits. The more workers available, the lower that wage will be. If there are more workers than work, wages tend to go down; more work than workers, wages tend to go up.

What unions provide in all this is collective bargaining. Workers, especially lower skilled workers, generally need their jobs more than the company needs them, so in negotiating wages and benefits, the company has more leverage. Companies can, and do, punish individuals who ask for wage and workplace improvements, which keeps wages down. But if the workers bargain collectively, they gain leverage, because the company can ill afford to lose all of them at once. The more leverage the workers gain, the better they can do in the free negotiation of contracts. It's called capitalism.

Block's citation of stars like Shaq and Brad Pitt, who are in no way representative, only goes to prove this. Shaq and Pitt have uncommon leverage (their skills are in great demand), and yet both are still members of powerful unions (NBAPA and SAG). Why? Because they are aware that their equally productive forbears were not nearly as well compensated for or secure in their work because they could not bargain collectively.

Then things get truly bizarre. Block asks about the decline of union power and the rise of real wages over the past fifty years, ignoring both the strengthening of wage laws and the disconnect between the median real wage and corporate profitability. He then asks (somewhat incoherently) about how some non-unionized industries nonetheless pay very well. Answer: the workers' skills are in demand, so they have more leverage.

He then asks why the South, "the least unionized part of the country, is the fastest growing?" This is just confusing. The topic under discussion is wages, which means Block is claiming the South is the fastest growing in wages. That's simply false. There is actually a decline in median real wages in the South over the last six years, and the South is doing worse than any other region. The citing of Singapore is both false (they do have unions) and misleading (only poor countries envy it).

Boyd Blundell is an assistant professor in the religious studies department.

Part III

Blundell's argument misses point
9/22/06

Dear Editor:

I am sorry to learn that my column "Unions Don't Guarantee Fair Employee Wages," in *The Maroon's* Sept. 8 issue, was "so incoherent" that it didn't even rise to the level of being wrong, in the view of Professor Boyd Blundell who responded in the Sept. 15 issues of *The Maroon*.

This professor of religious studies charges that I am guilty of a "straw man" argument; but there are indeed "some" who think that without unions, and the labor legislation they have engendered, we would now revert back to the horrors of 19th century industrial factory wages and working conditions. But many people believe this fallacy. Blundell himself is a case in point.

He opines that our relatively high wage levels (including working conditions, safety, etc.), are guaranteed by being codified in law. If so, why do not the governments of economically backward countries such as North Korea, Cuba, Chad, etc., immediately place laws on their books ensuring the same kind of benefits? Is it really so "incoherent" to claim they cannot do so since real wage levels stem not from government edict but rather from how hard and smart employees work, and with the cooperation of how much and of what quality of capital goods, and that this in turn depends upon the level of economic freedom prevailing in a country?

Prof. Blundell maintains that worker "productivity is virtually irrelevant" to the setting of labor's compensation. Rather, it is driven by "bargaining power." But the latter depends almost entirely on the former. He thinks it is possible that there can be "more workers than work." But if so there would be either be no scarcity (and hence no need for any labor in the first place), or, due to government setting artificially high wages, thus creating this very unemployment. Blundell is evidently thinking of the Great Depression; but this was caused by government meddling, not capitalism.

My faculty colleague misunderstands the institutions of capitalism. Yes, they are compatible with unionized orchestrated mass quits, or threats thereof, but certainly not with threatening picketers, beating up scabs or modern labor legislation.

The stratospheric salaries paid to Shaq and Brad Pitt are not at all a result of their union membership. Other equally unionized actors and athletes earn a fraction of their pay. Presumably, it is "a flat earth hypothesis" on my part that these stars are very productive and thus can attract masses of fans, while their lesser-known and lesser-able colleagues are unable to do so. Computer nerds, bankers, lawyers, stockbrokers and financial advisors earn high salaries without benefit of unions. Yes, they have "bargaining power," but this is due to their productivity.

Professor Blundell concedes that unions are not necessary since numerous industries without them "pay very well." But this was pretty much my entire point in the article he so derisively dismisses. He misreads me as saying that the South is the fastest growing "in wages." I said, "fastest growing," period. This is because for the past few decades industries have been running away from the heavily unionized rust belt Northeast, and bringing their jobs to the South.

Sincerely,
Walter Block, Ph.D.

Part IV

Block's argument still off point
10/6/06

Dear Editor:

I will try to conserve my energy by concentrating on the original issue. I am continuing primarily because I think that this university's commitment to critical thinking is important, and that it is good for the students to see this in action.

Professor Block claims that "the stratospheric salaries paid to Shaq and Brad Pitt are not at all a result of their union membership," but rather their productivity. It is worth noting first that the example is not an honest one. Block was originally making a point about the productivity and wages of the average "worker" on the "shop floor" and then responded to an imaginary challenge by appealing to the very top performers in labor markets that have nothing to do with a shop floor.

So let's talk about professional athletes. Professional athletes generally have a unique set of skills that are very valuable to their employers but are not transferable to other jobs. A quality NFL player now makes millions of dollars a year. The questions is, what would that player do if he did not play NFL football? His unique skills do not make him productive at an elite level in any other job.

This is the dilemma that faced each player individually before unionization. The owners employing them knew those players had no options and thus offered them compensation that was only a small fraction of their productivity. If a player refused, the owner faced only a small decline in quality, as there were other players almost as good with similarly limited options. (This is what is meant by having "more workers than work.") The player, on the other hand, faced a massive plunge because he had no other marketable skills. All the bargaining power was with the owner.

Note that, unlike Block, I am not simply proposing theories — this is what actually happened. NFL players were, for example, forbidden to have agents, required to pay for their own equipment, and not paid when injured. Nor was this peculiar to football. In the National Hockey League, injured players were even forced to work the concession stands and parking lots during games, and the most productive player in NHL history, Gordie Howe, retired with a pension of $800 a month. Only when the players in the major professional sports organized to bargain collectively did any of these conditions begin to change. Productivity did not change; bargaining power did.

Remarkably similar stories can be told regarding the relationship between movie studios and actors, which led to the formation of the Screen Actors Guild. So Professor Block's central claim that compensation "depends almost entirely" on productivity is demonstrably false, even using the unreasonable examples he chose.

If we actually look at the average worker on the "shop floor," the point only becomes more painfully obvious. Professor Block wishes a formal debate with me on this issue, I am, in the spirit of modeling critical thinking, at his disposal.

Sincerely,
Boyd Blundell, Ph.D.
Assistant Professor
Religious Studies

Part V

Block, Walter. "Blundellian Economics"

It is a pleasure to once again dialogue with Professor Boyd Blundell on the economics of wages, unions, productivity, government intervention and free enterprise.

The issue that divides us is how to explain and account for the existing pattern of wage payments. According to introductory economic textbooks, this depends, basically, on marginal revenue productivity (MRP): on how much the additional worker contributes to total productivity. This sets an upper bound. You can't get blood out of a stone. If the workers, together, are paid more than what they all produce, the employer loses money and must eventually be forced into bankruptcy.

This is also the lower bound, but with three exceptions. Why will a situation not long endure where a worker who produces, say, $15 per hour is nevertheless paid, only, $5, for example? This is because a pure profit of $10 would be earned by the firm on his labor, and this profit opportunity would be like a magnet, attracting other companies to get in on this good thing. How could they do so? Someone else will offer $5.01, and "exploit" this employee to the tune of $9.99. But then another will offer $5.02, and profit from his labor by $9.98. Where will this process end? Ultimately, in equilibrium, there are zero profits and the wage will thus rise to $15.

But this assumes, first exception mentioned above, that there are no transactions costs involved; that workers paid less than their MRP can be found easily, will change jobs to earn even one cent per hour more, there are no personal losses from job switching to either side, the wage contracts can be written costless, etc. In the real world, when the wage in our example gets very close to $15 the bidding will tail off. For instance, it may not be worthwhile for a company to search for those with this skill set earning $14.90, unless, perhaps, there are a lot of them out there. Plus, productivity levels are continually changing, pay scales are based on estimates, so there is little reason to expect exactitude. However, if wages and productivity levels diverge, the forces mentioned above will be brought to bear to reduce incipient gaps between them, in either direction.

The second exception is that the MRP of the worker holds true in at least one other, and presumably many other venues. We implicitly assume that a carpenter, for example, who can produce $15 worth of services for Firm A, can also do so for B, C, D.... Some introductory microeconomic and labor economics texts, but all intermediate ones, will mention that if this is not the case the rule of wage equals MRP will have to be modified. Now, the economic axiom will be that wages will equal productivity in the second best alternative, or some such. Why? Well, suppose that the employee can produce $15 for Firm A, but, due to some sort of heterogeneity, can account for only $10 worth of goods at Firms B, C, D.... We can readily see why wages in such a case will be bid up from an initial $5, should they start there, to $10. But no one, apart from A, would pay him any more, and thus A would have no incentive to go any higher, to match these other non-forthcoming offers. In such a case, economic theory maintains that the wage will be indeterminate; it will lie somewhere between $10 and $15. A MRP of $10 will bring wages up to that level in equilibrium; the rest will be determined by Prof. Blundell's favorite explanation: bargaining power. But don't think even in such a case that the wage will not rise above $10. Remember, Firm A benefits by $15 by hiring such a person. If they pay him $12 for example, they still earn a pure profit of $3. If they do not hire him at all, they lose this opportunity.

But this example is extremely rare in the real world. It certainly does not apply to Shaq and Brad Pitt. There are numerous firms for which these two can place rear ends in seats at about the same rate. Shaq's productivity would be roughly the same for the Knicks, the Lakers (his previous team), the Hornets or his present team, the Heat. Apart from that he could play in a European, Asian or South American league, with only a slight reduction in (second best) MRP, and hence wages. If unlikely in the extreme, all of these firms tried to pay Shaq and his fellow athletes far less than that, they could always borrow a leaf from the old American Basketball Association, and set up their own new league in competition with the NBA and all these others.

Let us get back down out of the stratosphere and into the realm of the more ordinary worker, the carpenter. Can the union help him raise his wage above his assumed MRP of $15, and still remain within the bounds of morality? No. If they were but to limit themselves to a mass quit, and/or the threat thereof, then the answer would be yes. For there are serious transactions costs in replacing, say, 500 workers, all at one fell swoop. There is thus a bit of bargaining power, between, perhaps, the $14.90 the market might otherwise settle at, and the $15. (This gap would be due to the fact that the process necessary to raise the wage to the full $15 do not come for free.)

But the union does more, far more, than restrict itself to mass quit threats. In addition it attempts to preclude competitors from bidding for these jobs. It restricts entry by beating up scabs (a blue collar technique) and getting the government to do their dirty work for them via labor legislation (white collar criminality). Organized labor cannot justify these barbaric practices by claiming ownership over these jobs. Rather, employment is the embodiment of an agreement between employer and employee. It can be owned by neither.

What of the argument that productivity plays no role in wage determination because the process of bidding up wages to productivity levels ($15 in our case) will be short-circuited by unemployment? This is predicated on the "lump of labor" fallacy. Here, there is only so much work to be done in an economy, and there are too many willing workers to do it all. No. Employment opportunities stem from the primordial fact of scarcity: we always want more than we have. If I invent a cure for all tooth problems and thereby unemploy the nation's dentists I am a hero, not a villain. I free up thousands of highly skilled people to undertake other tasks that before my "invention" simply could not be done.

From whence then springs the unemployment we see in the world, then? It stems from subsidizing unemployment (the unemployment insurance system) and from artificially boosting wages above MRP levels: harmful minimum wage legislation, which attacks low-productivity workers such as teenage black males, and, you guessed it, unions which also compel higher than productivity wages levels, and create havoc in the form of rust belts.

The third exception stems from the fact that some employees work in a sector of the economy producing goods which will not reach the final consumer for many years (basic or heavy industry), and others create items which will reach him relatively quickly (for example, retail markets). If they are to receive the same wages, the marginal productivity must be higher in the former cases than in the latter, due to discounting. Thus, the real theory is based not on marginal revenue productivity (MRP) but rather on *discounted* marginal revenue productivity (DMRP). For more on this rather complex issue, which elementary treatments often, with good reason, ignore, see Chapter 4.

Prof. Blundell has expressed interest in a formal debate on these matters. I am happy to comply. The economics club (motto: free pizza and laissez faire capitalism) is certainly interested in arranging a meeting for this purpose.

Part VI

Economics, ethics debated
Alethia Picciola
10/6/06

Loyola professors the Rev. David Boileau, S.J., and Walter Block on Tuesday afternoon brought to life the university's ideals of thinking critically while debating and acting justly in terms of economic and political policies.

At the economics club meeting both Block and Boileau agreed that everyone is generous at heart and that no one really wants people to starve. They had very different ideas, however, about employee and employer relationships, as well as how to solve the problem of poverty.

Block, an economics professor, argued that "unions are just a criminal gang," when on-strike workers interfere with the rights of potential workers either by physical violence or by getting laws passed that force their employers to retain existing workers.

"The only way we should deal with each other is on a voluntary basis," said Block. He compared involuntary interaction at its worst with slavery, referring to the role of a boss who is forced to continually associate with his existing workers who cannot quit their jobs.

Boileau, who teaches philosophy, focused more on the ethical side than on economics. He said that Block's point of view describes a man who is outside of his environment.

The positive outcome of unions is that they provide equity and safety to its employees, traits that don't come cheap, Boileau said.

He also spoke at length about the concept of corporate responsibility. He said that while employees have a responsibility to their employer to produce goods as quickly and efficiently as possible, the employer has a responsibility to his workers to provide them with living wages.

Loyola wasn't off limits as Boileau questioned the ethics of the university.

"We're supposed to be a social justice institution. We had buses out here (earlier this year) taking kids to Wal-Mart. Those sandals you got on, buddy, were built by a Chinese girl for $1 an hour," he said to a student attending the meeting.

Not having enough money to pay living wages was a problem for the unions to solve, said Boileau. "It's an economics problem, it's a political problem, it's no longer a union problem."

The long-debated question of who product capital belongs to came up several times during the debate, as well.

Boileau argued that the capital belongs not only to the employers, but also to the employees. If the employee is making the product, the employer has a moral responsibility to give that employee a living wage, Boileau said.

Block disagreed, saying that "any worker can get up on his hind legs and declare himself a boss," by saving money and hiring an employee. The employers own the capital because they are the ones making investments and taking risks on products, Block said.

Employer responsibility was also addressed at the debate in addition to owner responsibilities. "The living wage comes from productivity, not from the generosity of the employer," Block said.

Wages cannot go up, though, unless productivity goes up, Block said. One way to raise productivity is to have high-quality capital equipment, which, according to Block, is achieved through the freedom of companies to obtain capital from whomever they can.

November 10, 2006

International
Journal of Social
Economics
23,1

6

Comment on Richard B. Freeman's "Labor markets and institutions in economic development"

Walter Block

College of the Holy Cross, Worcester, MA, USA

Freeman (1993) has a two-part thesis. The first, borrowed faithfully from Coase (1992), is that economic analysis cannot exist in a vacuum; rather, it must be embedded in, and appreciative of, the prevailing institutional arrangements. The second part, with which Coase may be expected to have somewhat less sympathy, is that when our analysis incorporates successfully the actual institutional setting in which economic activity takes place, a whole host and variety of conservative shibboleths, or pro free enterprise findings, can no longer be maintained. Among these are five fairly typical presumptions within the profession, he tells us. They are:

(1) Bias towards urban élites in underdeveloped countries retards economic development (Freeman (1993) calls these nations "Developia", but the description used in the text describes more accurately what is actually going on in these countries, as opposed to what men of good will might wish were occurring there).

(2) Pro union and minimum wage legislation are economically counter-productive.

(3) Markets outperform central planning, or industrial policy, and that "clear property rules and privatization are necessary for rational economic behavior" (Freeman, 1993, p. 403).

(4) Countries which feature military dictatorships and suppress labour organizations increase income inequality.

(5) People are usually satisfied with economic progress.

Before considering each of these claims in turn we must begin by noting that Freeman adopts the usual debater's ploy of stating his opponent's views in a highly exaggerated manner. This "straw man" technique is seen, for example, in the dare "… if you think that military dictatorships that suppress labor *necessarily* produce high income inequality, the income distributions of Korea and Taiwan should give you pause" (emphasis added, Freeman, 1993, p. 403). The point is, no reputable economist, or anyone else for that matter, has ever asserted that there is any such necessary connection; the contention, on the contrary, is only that this seems to be a statistical or empirical reality. By

International Journal of Social
Economics, Vol. 23 No. 1, 1996,
pp. 6-16. © MCB University Press,
0306-8293

putting excessive claims in the mouths of his intellectual opponents, Freeman does indeed enhance his ability to "prove them wrong". But this is merely a Pyrrhic victory, as he only succeeds in undermining views they do not actually hold. Let us now turn to a consideration of each of his allegations in detail.

Rural-urban differentials

Freeman states: "A principal tenet of the orthodox view of labor in development is that urban wages are excessively high relative to rural wages" (p. 403). Although Freeman does indeed cite one set of authors (Harris and Todaro, 1970) in support of this declaration, he realizes that this is only a proxy for an underlying causal explanation: the aggrandizement of the public sector at the expense of the private. And it is this, not some anti urban bias, that is the explanation of economic stagnation in the third world offered by a whole host of commentators (Ayittey, 1991, 1992; Bauer, 1981, 1984; Bauer and Yamey, 1957; Louw and Kendall, 1986; Williams, 1989).

Freeman points to a number of studies which show that urban/rural relative wages have been falling; but this is only an approximation of public-private compensation. Only two (Colclough, 1991; Lindauer *et al.*, 1988) address themselves directly to this issue, but leave open the question as to whether economic growth took place as a result. A reasonable expectation, *ceteris paribus* conditions holding, is that if public sector wages fell relative to those in the private, this would tend to boost the economy, given that – apart from a contestable[1] small but necessary "nightwatchman" function for the state – the main role of government bureaucrats is not to promote wealth creation, but rather to batten off it.

Of course, this change affects only relative wages. In absolute terms, even if real governmental wages declined, they could still be far in excess of those available in the rural sectors, and in excess of those needed for the growth enhancing minimal state.

Labour rigidities

Freeman's treatment of this subject is marred with a serious equivocation. In his first paragraph he starts off by characterizing the traditional view as one which sees "institutional interventions in the labor market (such as) minimum wages, unions", etc. as "major impediments to growth and employment" (Freeman, 1993, p. 404).

One would think that after such an introduction he would proceed to show that these interferences have not slowed down the rate of growth or job creation. Instead, and here is the equivocation, he tries to show that in the event, these policies have not actively been pursued by governments in the Third World. Here, he refers to the ILO study indicating that the "real value of minimum wages plummeted", or "were enforced weakly" (p. 404). Now surely this is a very different issue. It is one thing to assert that minimum wages, unions, etc., if enacted, retard the economy. It is an entirely different matter to show that they have not been pursued; that they have been, in effect, dead letter laws. The one

International
Journal of Social
Economics
23,1

8

has no implication for the other. Surely, a strictly enforced law setting one cent per hour as the minimum wage, or even one for $50 per hour with absolutely no penalty for violation, cannot be expected to have much of a negative effect on economic development[2].

Freeman (1993) avers that "the optimal policy may be to enforce minima in large (multinational) firms with low elasticity of demand for labor and where the supply of foreign capital is inelastic" (p. 404). But politicians who follow this advice risk discouraging the golden (multinational) goose from locating in their countries. Nor is it easy to rescind wage minima once they have been inaugurated. Has the recent French experience with minimum wages taught us nothing[3]? Then there is his claim that:

> Taiwan enacted a fairly high minimum in 1988 and increased it substantially in 1992. One interpretation of this pattern is that the minimum is an endogenous "luxury policy", imposed when it helps workers at little economic cost but dropped when it risks serious economic harm (Freeman, 1993, p. 404).

It would be hard to pack more economic fallacies into a shorter statement. One problem is the implication that the Taiwanese growth rate persisted during these years. If so, it accomplished this goal despite these enactments, not because of them. Freeman, who seems long on historical examples but short on economic theory, has no way to distinguish between these two very different states of affairs. Another problem concerns the notion of "luxury policy". Why is it that one and the same law – establishing wage minima – can help workers at times, and at other times "risks serious harm"? The straightforward answer is that this enactment can help workers only in the very short run, before the firm has had a chance to substitute other factors of production for the one which has suddenly increased in price[4]. But just as soon as this can be humanly accomplished, the employer moves along his isoquant to a new optimal point. And with this new allocation of resources, the unskilled labourer will find himself priced out of the market. The difficulty with this author is that he offers no alternative explanation; nor does he appear to accept this one, otherwise he would not speak of the "help" to be afforded by this law.

To conclude this section, Freeman's is indeed a strange defence of minimum wage laws: they are not so bad when not enforced. With friends like him, this law hardly needs enemies.

A similar analysis pertains to unions. Where once we might have expected a ringing endorsement of this curious institution (and probably would have received it, had this piece been written 30, 20, or even ten years ago) from this quarter, we are told instead that the defanged version is really not too harmful. It cannot be denied that non-aggressive unions, such as the Christian Labour Association of Canada, which specifically abjure violence, are vastly to be preferred to the ordinary garden variety labour organizations which have, as their main stock in trade, the legal ability to preclude entry of competing workers through threats of violence. But even this moderate version of the union case is not, strictly speaking, correct. Just as a dead letter minimum wage law will still have some pernicious effects, even though far less than the virulent strain, so are moderate, "well behaved" unions still problematic, economically speaking. This is because unions are *per se* illegitimate organizations (Block, 1991a). Either they themselves threaten ("picket line") violence, or they struggle to enact "labour" legislation which mobilizes the state to act on their behalf in this regard. For example, laws which compel the employer to "bargain in good faith" with unionized employees, when he would just as soon fire and replace with "scabs", are an infringement on his rights of free association.

Needless to say, these laws have strong and negative impacts on economic development, *ceteris paribus*. Freeman (1993) holds, to the contrary, that "unions ... are associated with greater training, increased fringe benefits, and higher productivity" (p. 405). And in this, there is no gainsaying that he is correct. But he has hit on the truth accidentally, and for reasons he will not appreciate. First, he is correct in using the phrase "associated with", and not "caused by"; second, in maintaining that organized workers are among the cream of the crop. Labour organizations comprise the most highly skilled, but they do not create any of these benefits. Rather, the direction of causation is the other way around. These benefits enable them to be better organized than the "scabs" with lesser skills who are precluded from bidding for work slots with lower wage offers. If anything, however, the effect of unionism – apart from entry restrictions, which do indeed raise their wages – is to lower productivity; this is dissipated in work to rule, grievance procedures, jurisdictional disputes with other unions, featherbedding, etc.

Whether or not it is "justifiable ... to dismantle such institutions" (Freeman, 1993, p. 405) is of course a normative issue. It is justifiable if one believes in stamping out the initiation of violence; it is not if not. From a positive point of view this issue raises the question of whether unions enable poor countries to improve their lot. And from this perspective the answer is clear: both economic theory and history suggest the negative.

In the absence of unionism, rising profits typically mean greater investment, still more earnings and jobs and, hence, additional economic development. In their presence, the employer is diverted from this entrepreneurial task. His energy must be spent in protecting his profits from avid and unscrupulous unionists. Even when he succeeds in these defensive measures, he must pay the opportunity costs of these energies in terms of additional business, profits, etc. And when he fails, it is even worse: it is at the cost of potential bankruptcy.

Freeman (1993) concludes this section on a straw man note: "don't look to government or union labour-market interventions as prime causes of trouble" (p. 405). One need not look to minimum wages or unionism as the prime culprit explaining underdevelopment without seeing these phenomena as part of the problem, not part of the solution. "Government", of course, is entirely a different matter, one not addressed by Freeman in this section, and perhaps tacked on to

International
Journal of Social
Economics
23,1

10

this sentence to make it less of a straw man argument than otherwise it would appear.

Ownership and government

Exhibit "A" in Freeman's case against the market is China. Here is a country, as communist as can be, and yet, in the teeth of massive governmental ownership and control it is one of the fastest growing in the entire world. "Take that!", advocates of capitalism, Freeman seems to be saying. Actually, he does say this, only in more scholarly language:

> The message to the transition economies of Eastern Europe has been that the road to successful economic performance requires privatization of enterprises and a clear residual claimant. The failure of reform communism in Hungary, Poland, and other Eastern European countries seemed proof that only private ownership could create the incentives for an effective market economy (Freeman, 1993, p. 405).

And then along comes China, an utter refutation of all this pro market bias, in his view:

> Did China privatize its state firms or clearly define property rights for its various enterprises as part of its economics reforms? No. Chinese ownership rights (i.e., the particular residual income claimant) are complex at best (Freeman, 1993, p. 405).

There are several difficulties here, however. First of all, there is nothing in the Adam Smithian paradigm which requires a simple and clear residual claimant. A complex one, as occurs in China according to our author will do just fine. Second, it is not true that this country now functions completely in the absence of a marketplace. There are, after all, numerous privatizations and strings of free enterprise zones spread out all across this nation. After the passing of Mao, deregulation has hit the country like a storm[5]. In addition, China functions in a world economy, and as such enjoys the advantages of market prices generated elsewhere[6]. Third, there is the entrepreneurial spirit of the Chinese people (Sowell, 1983, 1994). Everywhere they go, wherever unfettered by Communism, they have made a vast economic success of their lives. Whether it is in the USA, or Malaysia, or Australia, people of Chinese extraction work hard, become highly educated, and wealthier on average than their "host" groups, even though often discriminated against by them[7]. It was only in China – under Communism – that the Chinese failed in their economic pursuits. Is it any wonder then, that when this country relaxed some of its economic regulations and restrictions – of which Freeman can find nothing negative to say – and set up privatization and free enterprise zones to a modest extent, that its economy would start to flower like parched grass too long kept away from water?

But is all this because of or in spite of the move towards economic freedom? Here, Freeman is at his absolute weakest. Unaccountably, he shows no evidence of being aware of any theory which attempts to account for economic success. In his intellectual arsenal is only the ability to look around the world, cite some countries that are developing and some that are not, and then argue that anything (seemingly anything) that characterizes the former can account for their success. In fact, he goes so far as to consider the possibility that socialism, communism, central planning, government industrial policy, can actually lead to economic prosperity. This is neither the time nor the place to rehearse the reasons why this is not so (we can acknowledge, however, the raw coverage it takes to say these things in 1993, four years after the crumbling of the Marxist system), why this cannot be so. Suffice it to say that without markets, and market prices, economic calculation is impossible (Hoppe, 1989; Mises, 1966, 1981; Salerno, 1990, 1993); that be the economic planners ever so smart, and armed with all the modern computers beyond the dreams of avarice, the economy does not benefit from the automatic weeding out process[8] provided for by the profit and loss system (Hazlitt, 1979); that the market serves as a vast communication and informational providing network, without which central planning is like a chicken running around with its head cut off (Hayek, 1948).

This is why the presumption is always in favour of markets, not bureaucratic control, notwithstanding Freeman's contention that he has "shifted the burden of proof from those who see a positive role for a development-oriented government to those who argue that interventions are harmful" (Freeman, 1993, p. 406).

If China was Freeman's exhibit "A" in his search for a successful centrally planned economy, Singapore, clearly, is exhibit "B". Now this is more than passing curious, in view of the fact that numerous studies have shown the economy of this city state to be among the most free in the world (Block, 1991b; Easton and Walker, 1992; Walker, 1988). Talk about biting off more than you can chew. It is undoubtedly true, as Freeman points out, that Singapore deviates from the pure vision of the classical liberal nightwatchman state of *laissez-faire* capitalism (e.g. Nozick, 1974) with government owned enterprises, payroll taxes, state investment, wage dictations, etc. But by world standards, this does not at all deserve to be characterized as "massive state intervention". One must bring a sense of perspective to matters of this sort.

And it is the same with exhibits "C" and "D", Taiwan and Korea. Freeman (1993) reports gleefully that the central governments of these countries have "initiated new sectors, directed private investments into some areas, and regulated foreign trade in ways that are inconceivable in the United States or United Kingdom" (p. 406). Again, true enough. But Taiwan and Korea are still relatively free, from an economic perspective, at least when compared to most nations of the world.

By the way, Freeman is labouring under a geographical delusion: there is no such country on the entire planet named "Korea" as he seems to think. As it happens, there are only North Korea and South Korea. And this opens up an interesting comparison, for these two serve almost as a laboratory experiment. They have in them the same people, with the same genetic pool, the same ethnic stock, the same culture, the same just about anything anyone would ever care to mention – with the exception of one small detail. The North

Richard B.
Freeman's
"Labor markets"

11

International
Journal of Social
Economics
23,1

12

functions under a system of which one would expect Freeman to give his enthusiastic support, while the South labours under a cloud of relative economic freedom[9]. Needless to say, we are not offered this as a case in point by our author.

Freeman (1993) concludes this section by tasking Hong Kong with "institutional intervention ... because the government owns the land, and the business élite may operate as a small directing cartel" (p. 406). Again, this is true enough, but there is a certain perspective missing. Hong Kong is usually rated the freest country on the globe (Block, 1991b; Easton and Walker, 1992; Walker, 1988), despite its admitted shortcoming on the land ownership front. One would think that Freeman would ignore this case, since it is a prime example of great economic growth for many decades, despite a continual influx of very poor people fleeing the People's Republic of China. As for allowing cartels, this is an example of economic freedom, not its absence (Armentano, 1972, 1982; Armstrong, 1982; Rothbard, 1962). It is compatible with the free enterprise system that co-operation between businessmen be legalized. Rather than a slur on the market credentials of Hong Kong, it is a diminution of that of those countries in the west with antitrust laws on their books.

Income inequality

The Freeman (1993) thesis in this section is that "it is widely believed that" (p. 406) under dictatorial regimes, particularly those headed by army officers, the poor suffer, both absolutely and relatively, whether or not growth occurs. The military dictatorships in South America particularly in Brazil and Colombia, seem to bear this out. However, and here comes his "refutation" of orthodox economics, this does not apply to Taiwan and Korea [*sic*], both of whose army high commands have suppressed independent unions and yet have "attained low levels of income inequality" (p. 406).

Again, an interesting hypothesis, but one which does not escape unscathed. First of all, it is pejorative to refer to "attaining" low levels of income inequality. There is nothing in all of value free economics which can serve as the basis for the claim that greater income inequality is "worse" than less[10]. One can make such normative proclamations if one wishes, but unless they are backed up by some sort of economic or philosophical justification, they cannot be allowed to stand uncontested[11].

Second, there is no consensus among economists (Block and Walker, 1988; Frey *et al.*, 1984) that military regimes necessarily reduce income equality. Certainly, there is no economic theory which holds this to be true. That it would describe practice in South America, but not Asia, should not be the occasion, then, of any intellectual embarrassment.

With regard to Freeman's contention to the effect that unions are instrumental in lowering income disparities, however, here there is a presumption on the part of economics – one which is at some variance with his views. Consider what we know of this relationship. The main business of unions is to restrict entry; their *raison d'être* is to reduce the competition for jobs they want. But who are the people so restricted? They are typically lower skilled workers, labelled "scabs", who are almost always poorer than their better organized union counterparts. But if the main effect of unions is to raise their wages at the expense of those lower down in the income distribution, their effect on equality is precisely the opposite of that contemplated by Freeman. In order to see this, one need only consider the economic effect of unions on the lowest income decile, or quintile, where the "scabs" are likely to be over represented.

Richard B.
Freeman's
"Labor markets"

13

Progress and satisfaction

This section is also problematic. Here, Freeman (1993) seemingly turns on its head the usual expectation that, *ceteris paribus*, economic progress would lead to human satisfaction. He gives as an example economically expanding Korea[12], "where surveys of attitudes and revealed behavior portray an utterly disgruntled labor force … with high levels of dissatisfaction". Naturally, this is a not so thinly disguised call for a greater role for unionism: "Growth without decent treatment of workers of the sort that free unions normally guarantee is evidently not the best path to a satisfied population" (p. 407).

It would be interesting to contemplate a study comparing the satisfaction of unemployed and underemployed "scab" labour in countries with "free unions" of the sort advocated by Freeman, with fully employed unskilled workers in those without this institution. The clear expectation is that those at the bottom of the labour skills pyramid would be far happier in the absence of unions than in their presence.

Even apart from Freeman's union advocacy, attitude surveys are notorious for their inaccuracy as a benchmark of satisfaction. Further, there is the well known "revolution of rising expectations". This refers to people who have been exploited for many years and now benefit from a slight improvement in living standards. Instead of being happy with their new lot in life they become more enraged; now, for the first time, they appreciate just how bad their condition was, how unjust, and have the ability to change things, and thus are determined to do something about their plight. We must take with a grain of salt, then, the surveys that Freeman cites. They need not at all indicate deterioration in condition; an improvement is a realistic alternative interpretation.

Conclusion

We have traced the Freeman exegesis through five separate but not unrelated topics. He sees his analysis as embodying Coase's call for sensitivity to institutional arrangements. A competing perspective views it instead as a cry for more socialism, more interventionism, more coercive unionism – not a very pretty picture, particularly as half the world is now involved in the difficult process of extricating itself from these very abuses.

International
Journal of Social
Economics
23,1

14

Notes

1. For the limited government side of this debate see: Friedman, M. (1962, 1975, 1987); Friedman, M. and Friedman, R. (1980, 1983); Nozick (1974); Smith (1776/1965); for the alternative see: Friedman, D. (1979, 1989); Hoppe (1989, 1993); Rothbard (1970, 1973, 1982); Woolridge (1970).

2. They will have some effect, however, and this will be negative. Postulate two communities, otherwise equal in every way, except that one has enacted a "dead letter" minimum wage law, and the other has not. Is there any doubt that new investment will tend to concentrate in the latter? Nor is there any doubt about the reason: it is easier to raise the level of wages, or enforcement, of a law already on the books, than to attain this *de novo*.

3. There an attempt was made to introduce a lower teenage minimum wage, not even to discontinue the policy entirely. Even this moderate attempt at reform had to be withdrawn due to riots in the streets. It is therefore perhaps irresponsible to urge politicians to flirt with such dangerous legislation.

4. We should not lose sight of the fact that the worker is only helped in the short run at the expense of the firm which employs him. Apart from being akin to "killing the goose that lays the golden eggs", this state of affairs is problematic from the point of view that there is thus nothing in all of welfare economics which can support the view that this "help" is of benefit to society as a whole. All we know is that the worker is "helped", the employer hurt. Unless we are willing to utilize illicit interpersonal comparisons of utility, no such conclusion may be reached.

5. Horner (1994, p. 35) speaks of "decisions by China's government to allow capitalism an unprecedented sway throughout the country". He maintains that "there is little Maoism in today's China. Indeed, almost immediately upon Mao's death in 1976, his successors began to undo his work".

6. Why is it that the economy of the Soviet Union was able to function – after a fashion – for 70 years? One of the reasons was "The Sears Roebuck Catalogue", e.g. the fact that the USSR could avail itself of prices generated in true markets in the west.

7. In this they resemble the Jewish people, according to Sowell; they, too, do better than average, economically speaking, in every country in which they have settled – with the exception of Socialist Israel – even though they, too, have borne the brunt of discrimination, and worse.

8. It is undoubtedly true that central planners are far more sophisticated than businessmen, and have far better credentials. Neither Ray Kroc, Sam Walton, nor Bill Gates has earned a PhD. All were school dropouts. Nevertheless, the market singled out these three men, and hundreds more like them, and did not so distinguish their public sector counterparts.

9. A similar "laboratory experiment" was provided by East and West Germany, and is by Taiwan and China, neither considered – not too surprisingly – by Freeman in his world wide search for examples comparing markets and central planning.

10. For a refutation of the philosophical claim that this is true by Rawls (1971), see Nozick (1974).

11. A similar criticism applies to the oft heard phrase: "perverse income transfers", which denotes taking money from the poor and giving it to the rich. This, too, is unwarranted, because unless there is some proof of, or evidence on behalf of the notion that there is too much income inequality, such transfers need not be considered "perverse".

12. He presumably refers to South Korea, not the economic basket case to its north where, presumably, people are more happy with their lot in life, since they are immeasurably poorer.

References

Armentano, D.T. (1972), *The Myths of Antitrust*, Arlington House, New Rochelle, NY.

Armentano, D.T. (1982), *Antitrust and Monopoly: Anatomy of a Policy Failure*, Wiley, New York, NY.

Armstrong, D. (1982), *Competition vs. Monopoly*, The Fraser Institute, Vancouver.

Ayittey, G.B.N. (1991), *Indigenous African Institutions*, Transnational Publishers , New York, NY.

Ayittey, G.B.N. (1992), *Africa Betrayed*, St Martin's Press, New York, NY.

Bauer, P.T. (1981), *Equality, the Third World and Economic Delusion*, Weidenfeld & Nicolson, London.

Bauer, P.T. (1984), *Reality and Rhetoric: Studies in the Economics of Development*, Harvard University Press, Cambridge, MA.

Bauer, P.T. and Yamey, B.S. (1957), *The Economics of Under-developed Countries*, The University of Chicago Press, Chicago, IL.

Block, W. (1991a), "Labor relations, unions and collective bargaining: a political economic analysis", *Journal of Social, Political and Economic Studies,* Vol. 16 No. 4, Winter, pp. 477-507.

Block, W. (Ed.) (1991b), *Economic Freedom: Toward a Theory of Measurement*, The Fraser Institute, Vancouver.

Block, W. and Walker, M. (1988), "Entropy in the Canadian economics profession: sampling consensus on the major issues", *Canadian Public Policy*, Vol. XIV No. 2, June, pp. 137-50.

Coase, R.H. (1992), "The institutional structure of production", *American Economic Review,* Vol. 82 No. 4, September, pp. 713-9.

Colclough, C. (1991), "Wage flexibility in sub-Saharan Africa", in Standing, G. and Tokman, V. (Eds), *Towards Social Adjustment*, International Labor Organization, Geneva, 1991, pp. 211-34.

Easton, S.T. and Walker, M.A. (1992) (Eds), *Rating Global Economic Freedom*, The Fraser Institute, Vancouver.

Freeman, R.B. (1993), "Labor markets and institutions in economic development", *American Economic Review*, Vol. 83 No. 2, May, pp. 403-8.

Frey, B.S., Pommerehne, W.W., Schneider, F. and Gilbert, G. (1984), "Consensus and dissension among economists: an empirical inquiry", *American Economic Review*, December, Vol. 74 No. 5, pp. 986-94.

Friedman, D. (1979), "Private creation and enforcement of law: a historical case", *Journal of Legal Studies*, Vol. 8, pp. 399-415.

Friedman, D. (1989), *The Machinery of Freedom: Guide to a Radical Capitalism,* 2nd ed., Open Court, La Salle, IL.

Friedman, M. (1962), *Capitalism and Freedom*, University of Chicago Press, Chicago, IL.

Friedman, M. (1975), *There's No Such Thing as a Free Lunch*, Open Court Publishing Co., La Salle, IL.

Friedman, M. (1987), *The Essence of Friedman*, Hoover Institution Press, Stanford, CA.

Friedman, M. and Friedman, R. (1980), *Free to Choose*, Harcourt Brace Jovanovich, San Diego, CA, New York, NY and London.

Friedman, M. and Friedman, R. (1983), *Tyranny of the Status Quo,* Harcourt Brace Jovanovich, San Diego, CA, New York, NY and London.

Harris, J.R. and Todaro, M.P. (1970), "Migration, unemployment and development: a two sector analysis", *American Economic Review,* Vol. 60, March, pp. 126-41.

Hayek, F.A. (1948), "Socialist calculation I, II, and III", *Individualism and Economic Order,* University of Chicago Press, Chicago, IL.

Hazlitt, H. (1979), *Economics in One Lesson,* Arlington House Publishers, New York, NY.

International Journal of Social Economics 23,1

16

Hoppe, H.-H. (1989), *A Theory of Socialism and Capitalism: Economics, Politics and Ethics,* Dordrecht, Boston, MA.

Hoppe, H.-H. (1993), *The Economics and Ethics of Private Property: Studies in Political Economy and Philosophy,* Kluwer, Boston, MA.

Horner, C. (1994), "Losing China again", *Commentary,* Vol. 97 No. 4.

Lindauer, D.L., Meesook, O.A. and Suebsaeng, P. (1988), "Government wage policy in Africa: some findings and policy issues", *World Bank Research Observer,* Vol. 3, January, pp. 1-25.

Louw, L. and Kendall, F. (1986), *South Africa: The Solution,* Amagi Publications, Bisho, Ciskei.

Mises, L. von (1966), *Human Action,* Regnery, Chicago, IL.

Mises, L. von (1981), *Socialism,* Liberty Fund, Indianapolis.

Nozick, R. (1974), *Anarchy, State and Utopia,* Basic Books Inc., New York, NY.

Rawls, J. (1971), *A Theory of Justice,* Harvard University Press, Cambridge, MA.

Rothbard, M.N. (1962), *Man, Economy and State,* Nash, Los Angeles, CA.

Rothbard, M.N. (1970), *Power and Market: Government and the Economy,* Institute for Humane Studies, Menlo Park, CA.

Rothbard, M.N. (1973), *For a New Liberty,* Macmillan, New York, NY.

Rothbard, M.N. (1982), *The Ethics of Liberty,* Humanities Press, Atlantic Highlands, NJ.

Salerno, J.T. (1990), "Ludwig von Mises as social rationalist", *Review of Austrian Economics,* Vol. 4, pp. 26-54.

Salerno, J.T. (1993), "Mises and Hayek dehomogenized", *Review of Austrian Economics,* Vol. 6 No. 2, pp. 113-46.

Smith, A. (1776/1965), *An Inquiry into the Nature and Causes of the Wealth of Nations,* Modern Library, New York, NY.

Sowell, T. (1983), *The Economics and Politics of Race: An International Perspective,* Morrow, New York, NY.

Sowell, T. (1994), *Race and Culture: A World View,* Basic Books, New York, NY.

Walker, M.A. (1988) (Ed.), *Freedom Democracy and Economic Welfare,* The Fraser Institute, Vancouver, BC.

Williams, W.E. (1989), *South Africa's War against Capitalism,* Praeger, New York, NY.

Woolridge, W.C. (1970), *Uncle Sam the Monopoly Man,* Arlington House, New Rochelle, NY.

III. The Minimum Wage

A Primer on Jobs and the Jobless

Walter Block

3/9/2004

With the economics of employment and unemployment constantly discussed on the business pages and political campaigns, let us turn our attention toward fundamentals and root out some fallacies.

If the media tell us that "the opening of XYZ mill has created 1,000 new jobs," we give a cheer. When the ABC company closes and 500 jobs are lost, we're sad. The politician who can provide a subsidy to save ABC is almost assured of widespread public support for his work in preserving jobs.

But jobs in and of themselves do not guarantee well-being. Suppose that the employment is to dig huge holes and fill them up again? What if the workers manufacture goods and services that no one wants to purchase? In the Soviet Union, which boasted of giving every worker a job, many jobs were just this unproductive. Production is everything, and jobs are nothing but a means toward that end.

Imagine the Swiss Family Robinson marooned on a deserted South Sea island. Do they need jobs? No, they need food, clothing, shelter, and protection from wild animals. Every job created is a deduction from the limited, precious labor available. Work must be rationed, not created, so that the market can create the most product possible out of the limited supply of labor, capital goods, and natural resources.

The same is true for our society. The supply of labor is limited. We must not allow government to create jobs or we lose the goods and services which otherwise would have come into being. We must reserve precious labor for the important tasks still left undone.

Alternatively, imagine a world where radios, pizzas, jogging shoes, and everything else we might want continuously rained down like manna from heaven. Would we want jobs in such a Utopia? No, we could devote ourselves to other tasks — studying, basking in the sun, etc. — that we would undertake for their intrinsic pleasure.

Instead of praising jobs for their own sake, we should ask why employment is so important. The answer is, because we exist amidst economic scarcity and must work to live and prosper. That's why we should be of good cheer *only* when we learn that this employment will produce things people actually value, i.e., are willing to buy with their own hard-earned money. And this is something that can only be done in the free market, not by bureaucrats and politicians.

But what about unemployment? What if people want to work, but can't get a job? In almost every case, government programs are the cause of joblessness.

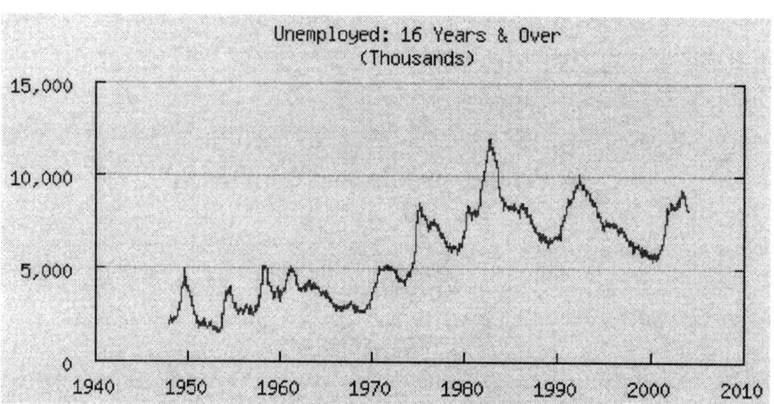

Minimum Wage. The minimum wage mandates that wages be set at a government-determined level. To explain why this is harmful, we can use an analogy from biology: there are certain animals that are weak compared to others. For example, the porcupine is defenseless except for its quills, the deer vulnerable except for its speed.

In economics there are also people who are relatively weak. The disabled, the young, minorities, the untrained — all are weak economic actors. But like the weak animals in biology, they have a compensating advantage: the ability to work for lower wages. When the government takes this ability away from them by forcing up pay scales, it is as if the porcupine were shorn of its quills. The result is unemployment, which creates desperate loneliness, isolation, and dependency.

Consider a young, uneducated, unskilled person, whose productivity is $2.50 an hour in the marketplace. What if the legislature passes a law requiring that he be paid $5 per hour? The employer hiring him would lose $2.50 an hour.

Consider a man and a woman each with a productivity of $10 per hour, and suppose, because of discrimination or whatever, that the man is paid $10 per hour and the woman is paid $8 per hour. It is as if the woman had a little sign on her forehead saying, "Hire me and earn an extra $2 an hour."

This makes her a desirable employee even for a sexist boss. But when an equal-pay law stipulates that she must be paid the same as the man, the employer can indulge his discriminatory tendencies and not hire her at all, at no cost to himself.

Comparable Worth. What if government gets the bright idea that nurses and truck drivers ought to be paid the same wage because their occupations are of "intrinsically" equal value? It orders that nurses' wages be raised to the same level, which creates unemployment for women.

Working Conditions. Laws which force employers to provide certain types of working conditions also create unemployment. For example, migrant fruit and vegetables pickers must have hot and cold running water and modern toilets in the temporary cabins provided for them. This is economically equivalent to wage laws because, from the point of view of the employer, working conditions are almost indistinguishable from money wages. And if the government forces him to pay more, he will have to hire fewer people.

Unions. When the government forces businesses to hire only union workers, it discriminates against non-union workers, causing them to be at a severe disadvantage or permanently unemployed. Unions exist primarily to keep out competition. They are a state-protected cartel like any other.

Employment Protection. Employment protection laws, which mandate that no one can be fired without due process, are supposed to protect employees. However, if the government tells the employer that he must keep the employee no matter what, he will tend not to hire him in the first place. This law, which appears to help workers, instead keeps them from employment. And so do employment taxes and payroll taxes, which increase costs to businesses and discourage them from hiring more workers.

Payroll Taxes. Payroll taxes like Social Security impose heavy monetary and administrative costs on businesses, drastically increasing the marginal cost of hiring new employees.

Unemployment Insurance. Government unemployment insurance and welfare cause unemployment by subsidizing idleness. When a certain behavior is subsidized — in this case not working — we get more of it.

Licensing. Regulations and licensing also cause unemployment. Most people know that doctors and lawyers must have licenses. But few know that ferret breeders, falconers, and strawberry growers must also have them. In fact, government regulates over 1,000 occupations in all 50 states. A woman in Florida who ran a soup kitchen for the poor out of her home was recently shut down as an unlicensed restaurant, and many poor people now go hungry as a result.

When the government passes a law saying certain jobs cannot be undertaken without a license, it erects a legal barrier to entry. Why should it be illegal for anyone to try their hand at haircutting? The market will supply all the information consumers need.

When the government bestows legal status on a profession and passes a law against competitors, it creates unemployment. For example, who lobbies for the laws which prevent just anyone from giving a haircut? The haircutting industry — not to protect the consumer from bad haircuts, but to protect themselves against competition.

Peddling. Laws against street peddlers prevent people from selling food and products to people who want them. In cities like New York and Washington, D.C., the most vociferous supporters of anti-peddling laws are established restaurants and department stores.

Child Labor. There are many jobs that require little training — such as mowing lawns — which are perfect for young people who want to earn some money. In addition to the earnings, working also teaches young people what a job is, how to handle money, and how to save and maybe even invest. But in most places, the government discriminates against teenagers and prevents them from participating in the free enterprise system. Kids can't even have a street-corner lemonade stand.

The Federal Reserve. By bringing about the business cycle, Federal Reserve money creation causes unemployment. Inflation not only raises prices, it also misallocates labor. During the boom phase of the trade cycle, businesses hire new workers, many of whom are pulled from other lines of work by the higher wages. The Fed subsidy to these capital industries lasts only until the bust. Workers are then laid off and displaced.

The Free Market. The free market, of course, does not mean Utopia. We live in a world of differing intelligence and skills, of changing market preferences, and of imperfect information, which can lead to temporary, market-generated unemployment, which Mises called "catallactic." And some people choose unemployment by holding out for a higher paying job.

But as a society, we can insure that everyone who wants to work has a chance to do so by repealing minimum wage laws, comparable worth rules, working condition laws, compulsory union membership, employment protection, employment taxes, payroll taxes, government unemployment insurance, welfare, regulations, licensing, anti-peddling laws, child-labor laws, and government money creation. The path to jobs that matter is the free market.

THE MINIMUM WAGE: DOES IT
REALLY HELP WORKERS?

Paul McCormick
HOLY CROSS COLLEGE
WORCESTER, MA

and

Walter Block
UNIVERSITY OF CENTRAL ARKANSAS
CONWAY, AR 72035

ABSTRACT

There is perhaps no greater cognitive dissonance than that which exists between the view that economists and non economists have about the minimum wage law. According to the latter, most recently articulated by President Clinton in an attempt to raise the level of coverage from $5.15 to $6.15, this law is all that stands betwen poorer working Americans and a continued loss in the purchasing power of their salaries. However, as the present paper points out, this is a snare and a delusion. The minimum wage actually creates unemployment for the unskilled. Far from benefitting them, it is a positive harm to those at the bottom of the labor market.

"More and more Americans are working hard without a raise. Congress sets the minimum wage. Within a year the minimum wage will fall to a forty-year low in purchasing power. But millions of Americans and their children are trying to live on it. I challenge you to raise their minimum wage"[1], President Clinton said to Congress in his State of the Union address on January 23, 1996. Yet, does the minimum wage really raise the wages of workers working below the minimum wage? Can wages be legislated? Are most minimum wage workers really struggling to live on the minimum wage? Both economic theory and empirical evidence suggest that the minimum wage does nothing to raise wages and actually increases unemployment, especially among teenagers and minorities.

One of the most widely perpetrated myths is that a wage is different from a price. In fact, no difference exists between wages and prices. Quite simply, a wage is the price of labor that the employees agrees to work for and the employer agrees to pay.[2] Therefore, the laws of supply and demand govern wages, and a minimum wage is nothing more than a price floor which inevitably causes a surplus of labor. At a minimum wage of $4.25 an hour, workers whose productivity is worth $1.00 an hour will most likely be unemployed. Employers who would have hired workers worth $1.00 an hour to do various jobs will now be forced to hire workers at $4.25 an hour for the same jobs, or else the jobs will be eliminated. As Henry Hazlitt states, "You cannot make a man worth a given amount by making it illegal to offer him anything less. You merely deprive him of the right to earn the amount that his abilities and situation would permit him to earn, while you deprive the community even of the moderate services he is capable of rendering. In brief, you substitute a low wage for unemployment."[3]

To exaggerate, consider the effects of a minimum wage of $100,000 a year on the unemployment rate of high school and undergraduate college students. Who would hire many of these workers at that rate of pay? These students do not possess the skills to perform these jobs worth $100,000 a year. Therefore, they will likely all be unemployed.

It is possible that the wages of many workers whose productivity is close to a newly-imposed minimum wage but below that minimum wage will receive a raise in the short run, simply because it will be too costly to fire workers immediately without hiring replacements. However, over time the employees whose productivity was close to the minimum wage will be fired and replaced by either more productive labor or more productive machinery. The unemployment caused by the minimum wage can be disguised, but it cannot be avoided.

Proponents of the minimum wage might say that the substitution of more productive machinery for less productive labor is beneficial, reasoning that technological improvements are always good. But if the technology was really so beneficial, why would it take a new minimum wage law to force the employers to implement the technology? If the employer could profit regardless of a new minimum wage law from the substitution of technology for labor, why not substitute the technology in the first place?[4]

Does the minimum wage always cause unemployment? No. One exception occurs when an employee is earning less than he is worth. However, these situations are quite rare and are better dealt with through the market.[5] For the usual practice is for employees being paid less than their productivity levels would warrant is to seek other jobs at higher remuneration. Alternatively, such workers are likely to become targets for "raiding" competitive employers. If there are such underpaid workers, as there are, say, in Mexico, firms can be counted upon to send trucks and buses there, to scoop up these people, take them to the farms of California and Texas, at higher wages than available at home.

The employer may also compensate for the enactment of minimum wage legislation by taking fringe benefits or other non-wage benefits from the employee by paying the cost of the fringe benefit to the employee in the form of wages.[6] While the employee will not benefit, he may stand to lose much if he values the fringe benefit more than its cost given to him in the form of wages. The resulting surplus of labor from the imposition of the minimum wage will also decrease the incentive of employers to offer fringe benefits and other non-wage benefits. Now the employer has more worker to choose from and thus has the luxury of selectivity.[7]

In any of the above cases, the minimum wage does not in the long run increase the compensation the worker receives for his work. In fact, the minimum wage often harms the workers even if they do manage to keep their jobs. Nevertheless, it is important to note that the workers in the above cases will at first appear to have benefited by an increase in the minimum wage. They will still be employed, and many will assume that their wages or total compensation package will increase in the long run through an increase in the minimum wage. Unfortunately, this is not the case. In short, Congress cannot give people more by legislating a minimum wage. "You cannot make a man worth a given amount by making it illegal for anyone to offer him anything less.[8]

Many advocates of a higher minimum wage argue that the present minimum wage is insufficient to support a family. Besides the fact that the minimum wage law hurts those same families, empirical evidence suggests that the typical minimum wage worker is not supporting a family. "In 1992 less than 10% of the workers earning the minimum wage were heads of families below the poverty level ... Most (62 percent) of the minimum wage workers were employed only part-time. More than one-third were teenagers. The typical minimum wage worker is a spouse or teenage member of a household with an income well above the poverty level. Therefore, even if the adverse impact of a higher minimum wage on both employment and non-wage forms of compensation is ignored, a higher minimum wage will exert little impact on the income of the poor,"[9] says James D. Gwartney and Richard L. Stroup, economists from Florida State University and Montana State University, respectively.

If one wishes to examine empirical evidence of the effects of the minimum wage on employment, teenagers are an important group to study. Besides the fact that teenagers make up a large portion of the workers either having or desiring minimum wage jobs, teenagers also tend to be some of the least skilled and inexperienced workers in the labor force. Minority teenagers are also an important group to study. Latin Americans, have less skill, experience, and education than white teenagers, and consequenty receive less in wages than a white teenager without the minimum wage law. Therefore, the minimum wage will tend to drive blacks and Latin Americans into the ranks of the unemployed to a greater extent than white teenagers.[10]

When considering various statistical evidence on the effects of the minimum wage on the rate of unemployment, it is important to make some key observations. For one, employers are able to compensate for some of the unemployment effects of the minimum wage as seen in some of the cases outlined above. However, the minimum wage exerts other harmful effects besides unemployment when employers are forced to compensate in other ways for the higher price of labor caused by a new minimum wage law. An adverse effect of the minimum wage which is especially harmful to teenagers is the fact that the minimum wage eliminates the chance of teen-agers taking certain low-wage jobs, gaining experience, and eventually obtaining better careers in the future. The minimum wage makes it difficult for employers to afford to give teenagers jobs with training.[11]

Secondly, many teenagers upon having difficulty finding a job will become discouraged and leave the labor force entirely. These workers will not be counted in the ranks of the unemployed, although many would take a job if they were offered one. Lastly, it is important to realize that the minimum wage in the United States does not apply to all jobs.[12] As a result, some low-skill workers will still be able to get jobs in the sector not covered by the minimum wage.

In graphs issued by the Bureau of Labor Statistics, the effects of the minimum wage on various groups are shown. The first graph illustrates the effects of the minimum wage on teenage and non-teenage (adult) males. The first year on the graph is 1948, which is three years after the minimum wage was first raised in 1945 to $.40 an hour. The last year on the graph is 1964. The unemployment rate in 1948 for teenage males was 8%. By 1964, with a minimum wage of $1.25 an hour, the unemployment rate for teenage males was about 13%. Conversely, the unemployment rate fluctuated from about 2% in the 1952-1954 period to a high of 5% in 1958 for adult males. By 1964, the unemployment rate for adult males was below 4%.[13]

On another graph, the unemployment rate for white male teenagers and nonwhite male teenagers is compared to increases in the minimum wage. In 1948, the unemployment rate for both groups was around 8%. By 1964, when the minimum wage was $1.25 an hour, white male teenage unemployment was at 12%, while nonwhite male teenage unemployment was at a whopping 22%-23%.[14]

Perhaps the most interesting minimum wage increase to study is the 1956 increase from $.75 to $1.00 an hour, a 33% increase. Within a few years, the unemployment rate rose for all groups. From 1956-1958, the unemployment rate for adult males rose from less than 4% in 1956 to 5% in 1958, while the aggregate teenage male unemployment rate rose from about 10% to almost 16% in the same period. From 1956-1958, the white male teenage unemployment rate rose from about 10% to around 14%, while nonwhite male teenage unemployment rose from about 13% to a whopping 24% in that

period.[15] The majority of the nonwhite teenagers are black. Both Yale Brozen and Milton Friedman deny that discrimination or the migration of blacks from the South to the cities of the North caused a wide gap to emerge between the unemployment rate of white male teenagers and nonwhite male teenagers after the 1956 minimum wage increase. "If (the wage gap) had been caused by increasing prejudice or increasing migration, you would expect a gradual widening of the spread. But it wasn't a gradual widening. It occurred quite suddenly,"[16] said Brozen.

Most studies done on the effects of the minimum wage on unemployment have shown that "a 10 percent increase in the minimum wage reduces teenage unemployment by 1 to 3 percent."[17] The effect of the minimum wage of teenage unemployment is relatively low largely because of its effect of eliminating non-wage compensation such as job training programs.[18] However, most of these studies do not incorporate the period of the 1980s, a unique period in which the minimum wage was only raised once to $3.35 an hour in 1981. Due to inflation, the real value of the minimum wage decreased approximately 20% from 1981-1989. Therefore, the rate of teenage unemployment should have decreased because of the substantial devaluation of the minimum wage. A study by Alison J. Wellington found that if the period of the 1980's was incorporated, a 10% increase in the minimum wage leads to a less than 1% decrease in teenage unemployment.[19]

In short, President Clinton's analysis of the minimum wage is wrong. The minimum wage does not positively affect the long-run earnings of Americans. Neither economic theory nor carefully-scrutinized empirical evidence indicate that this is the case. And, contrary to what the President implies, most minimum wage earners are not supporting families. In fact, the minimum wage for the most part does not help those it intends to help. In reality, the minimum wage causes unemployment and other problems.

144

ENDNOTES

[1] "Prepared Text for the President's State of the Union Message." *New York Times* 24 Jan. 1996 p. A13.

[2] Henry Hazlitt, *Economics in One Lesson* (New York: Crown Publishers, 1979) p. 134.

[3] Hazlitt p. 135.

[4] Hazlitt p. 29.

[5] Hazlitt p. 135.

[6] Alison J. Wellington, "Effects of the Minimum Wage Law on the Employment Status of Youths: An Update," *The Journal of Human Resources* vol. 26 (Winter 1991): p. 33.

[7] James D. Gwartney and Richard L. Stroup, *Economics: Private and Public Choice* (New York: Hardcourt Brace and Company, 1995) p. 697.

[8] Hazlitt p. 135.

[9] Gwartney and Stroup p. 697.

[10] Gwartney and Stroup p. 697.

[11] Gwartney and Stroup p. 696.

[12] Wellington p. 30.

[13] Yale Brozen and Milton Friedman, *The Minimum Wage: Who Really Pays?* (Washington, D.C.: The Free Society Association, 1966). p. 15.

[14] Brozen and Friedman p. 12.

[15] Brozen and Friedman p. 12, 15.

[16] Brozen and Friedman p. 13.

[17] Gwartney and Stroup p. 696.

[18] Gwartney and Stroup p. 696.

[19] Wellington p. 27.

BIBLIOGRAPHY

1. Brozen, Yale, and Milton Friedman. *The Minimum Wage Rate: Who Really Pays?* Washington, D.C.: The Free Society Association, Inc., 1966.

2. Gwartney, James D. and Stroup, Richard L. *Economics: Private and Public Choice.* New York: Hardcourt Brace and Company, 1995.

3. Hazlitt, Henry. *Economics in One Lesson.* New York: Crown Publishers, 1979.

4. "Prepared Text for the President's State of the Union Message." *New York Times* 24 Jan. 1996, p. A13.

5. Wellington, Alison J. "Effects of the Minimum Wage Law on the Employment Status of Youths: An Update." *The Journal of Human Resources* vol. 26 (Winter 1991): pp. 27-46.

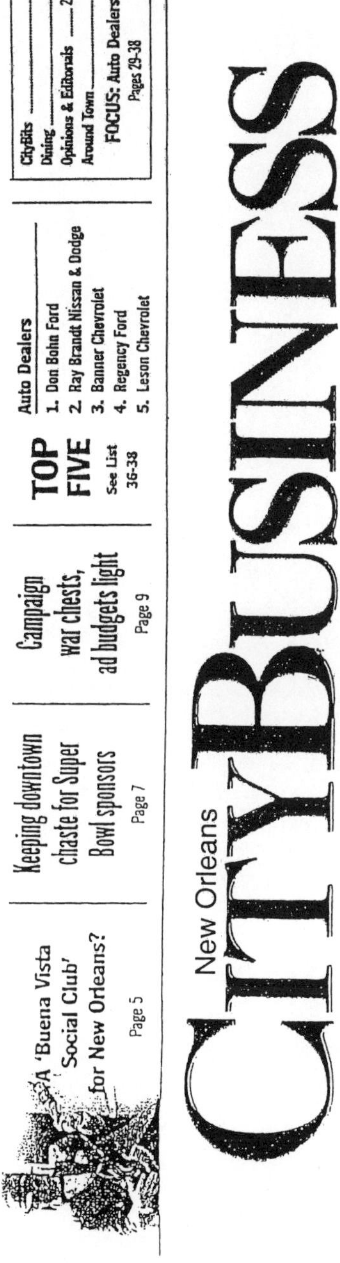

Guest Perspective Walter Block

Delusions of rising wages

NEW ORLEANIANS are scheduled to vote on whether to peg the city's minimum wage $1 per hour higher than whatever level is called for in federal law. Right now, this is $5.15 per hour, so if this initiative passes Feb. 2, the local minimum wage would be $6.15.

One argument being bruited about in favor of this initiative is that it is hard to make ends meet on less than $11,000 per year for a 40 hour a week job, the amount presently stipulated by legislation. Another is that it is simply unfair to allow those at the bottom of the economic pyramid to suffer with so little, while others have so much more.

The opposition, so far, is all but limited to the New Orleans business community. It complains that the wage increase would drive some firms into bankruptcy and others into neighboring parishes; all, particularly in the hospitality industry, would have to cut back on their number of employees, fringe benefits and on-the-job training.

However far apart they are on some issues, both sides agree to one thing: A rise in the minimum wage level called for by law will actually increase wages. There was even a headline in a local newspaper expressing this view: "Election set on raising wages in N.O."

This is economic illiteracy.

Contrary to popular opinion, there is no necessity for a rise in the mandated minimum wage to increase pay. Strictly speaking, this piece of legislation does not at all stipulate that the wages of even one person be raised. Rather, it states only that it shall be illegal to pay anyone less than a certain amount. These are two very different things.

> **Contrary to popular opinion, there is no necessity for a rise in the mandated minimum wage to increase pay.**

Think of the high-jump bar in track and field. If it is raised, say, from 60 inches to six feet, this does not mean that any athlete will be able to rise above the new height. Very much to the contrary, some competitors who were capable of attaining the previous goal will fall short in the face of the new one. A high jumper whose personal best was 65 inches could sail over the previous level with room to spare, but his rear end would knock down the bar at the new height.

Similarly, a worker with marginal revenue productivity (the amount an employee can add to his employer's revenues) of $5.50 would be fully employable at the old minimum wage level. His boss could even earn a profit off of him. If the employee brings in $5.50 per hour, then, minus $5.15, the employer would net 35 cents per hour. But at the new pay of $6.15, the firm would lose 65 cents per hour, hardly a paying proposition.

The minimum wage is not an employ-ment law; it is an unemployment law. It states that anyone with productivity of less than the amount predetermined by the politicians cannot get any job at all from a profit-earning enterprise.

If we could really legislate ourselves out of poverty — the implicit premise of both sides of this "debate" — then why should we stop at $5.15 or even $6.15? Why not fix wage minimum at $10,000 per hour? The reason is, it would be clear as crystal that virtually everyone — with the possible exception of Madonna and Mike Tyson — would be unemployed at that point. But the same principle is at work in the present case.

This law will greatly worsen the economic condition of local unskilled workers. It will do so in two ways, the first, directly, the second, indirectly. First, the new higher minimum wage will raise the bar over which such potential employees will have to jump in order to get a job. Second, this law will harm the New Orleans business community.

Crescent City firms will be put at a competitive disadvantage vis-a-vis competitors in neighboring parishes who can pay less for workers. Youngsters and other unskilled workers rendered all but unemployable will have to travel farther to find jobs. Many will not be able to do so, and will instead enter the soul-destroying welfare rolls.

Local firms will be hurt, but will be able to substitute away from the labor factor of production newly priced out of the market (the unskilled) either by automating, or through customer efforts. This is why we now have lots of self-serves, both in filling stations and restaurants; before the advent of this pernicious legislation, these facilities were manned with entry-level workers.

If the minimum wage law is such an utter disaster for those at the bottom of the income scale, what is the real impetus behind it?

Organized labor is always and ever in competition with inexperienced youth and others who typically hold low-paying jobs. What better way of eliminating the competition than by pricing them out of the market? Unions can shed crocodile tears about the plight of the unskilled while fixing thing so that the latter are forever unemployed. Their dupes among the clergy and the do-gooding community, totally ignorant of even the rudiments of economics, stand on the sidelines and cheer. This ploy ("unskilled wages rose, so should our own") also serves as leverage for union wage increases; but, as organized labor is not itself a licit institution, this can hardly be counted as a positive.*

Dr. Walter Block holds the Harold E. Wirth Eminent Scholar endowed chair in economics at the Loyola University College of Business Administration. He can be reached by phone at 504-864-7934 or by e-mail at wblock@loyola.edu

The Minimum Wage Once Again

by Walter Block*

There is a great cognitive dissonance between what is commonly taught in economics and what passes for sound public policy analysis within the beltway of Washington D.C. Nowhere is a more stark example of this to be found than in the case of the minimum wage law. Many students know that a minimum wage law exacerbates unemployment, particularly for low-skilled, youthful, and minority males. Card and Krueger (1994, p. 792) even characterize this claim as "the central prediction of the textbook model of the minimum wage," but attempt to undermine it their article.

In what might be called "the political economy" of minimum wage legislation, we must take note of the fact that instead of following the received wisdom of the economics profession in this case1, the Clinton Administration's recent legislative proposals have seen fit to follow the lead of Card and Krueger (CK). Why? Although this can only be speculative, one reason for President Clinton's support of the CK study may be that it is consonant with the position of organized labor. Unions favor minimum wages since they are always in competition with low wage labor; when unskilled labor costs are legislatively increased, this factor of production becomes less competitive with union's own supply of labor to the market. That is, suppose a union were to demand an increase in their wage from $17 to $20 per hour. The natural reaction of the employer would be to decrease its demand for this now more expensive factor of production, skilled union labor, and to replace it with a substitute factor, unskilled labor.[2] And, without a minimum wage law, this would always be a potent threat, limiting the demands for wage increases on the part of unions. However, with a hefty minimum wage level, particularly an increasing one, this potential danger to organized labor is eliminated. For example, a firm mat easily substitute away from skilled labor if it can hire unskilled workers at $4 per hour. At $6 per hour, the competitive advantage of resorting to the lesser skilled sector of the labor market may be entirely dissipated.[3]

---

* Walter Block, Harold E. Wirth Eminent Scholar Chair in Economics, College of Business Administration, Loyola University New Orleans, New Orleans, LA 70118. http://www.cba.loyno.edu

[1]See Frey, et. al. (1984) and Block and Walker (1988) in this regard.

[2]This substitution, of course, could not take place on anything like a one for one basis. By definition, no one unskilled laborer can do the job of a skilled one. But two or three or four unskilled workers, could, at the margin, provide enough labor power so that total production need not be compromised.

[3]No one, not even the most rabid advocates of minimum wages, urges that their level be raised without limit. For example, were the law to make it illegal to pay less than $1000.00 per hour, there is no economist who would deny that this would wildly exacerbate unemployment, as there are very few market participants with a productivity at or above this level. State Card and Krueger (1999, p. 2): "Our analysis of this new policy intervention is consistent with the conclusion that modest changes in the minimum wage have little systematic effect on unemployment (emphasis added by present author)." But if a minimum wage law mandating $1000.00 per hour would devastate unemployment prospects since few people can produce at this level, it would appear to follow logically that a more modest minimum wage law of, say, $5.00 per hour would ravage the employment prospects for people with productivity in or especially below this range - precisely the contention of most economists. See also Card and Krueger (1995). The one possible exception to this general rule might be Galbraith (2000), who characterizes as fallacious five contentions he attributes to mainstream economics, including the claim that "Rising minimum wages cause unemployment." Stated in so stark a manner, without the CK qualifier as to "modest changes," Galbraith might well be interpreted as claiming that any rise in the minimum wage is economically harmless. He continues: "A furious fight on this issue ensued as recently as 1995 when two distinguished researchers, Alan Krueger of Princeton and David Card of the University of California, Berkeley, broke ranks to declare that the evidence contradicted this thesis. Since then, the minimum wage has gone up twice, and unemployment has continued to decline. Card and Krueger were right--and so was their fundamental criticism of basic labor market theory." The point is, Card and Krueger cannot reasonably be interpreted as making a fundamental criticism of basic labor market theory. For them to do so, they would have to claim that any increase in the minimum wage, no matter how large, would not have deleterious effects on employment. Instead, as I read CK, they are making the far more modest and thus less radical or fundamental claim that this is true only for modest increases.

What is the case in behalf of minimum wages offered by CK? New Jersey raised its minimum wage floor from \$4.25 to \$5.05 per hour on 1 April 1992. Card and Krueger attempt to assess the unemployment effects of this change as of 5 November - 31 December 1992, roughly seven and a half months later, and were unable to find any.[4] So, they conclude that there were none. In response, I discuss several errors of omission and commission, on theoretical and empirical grounds; I maintain, as a result, that their dismissal of conventional economic theory was premature.

CK's sample is the employment and wage experience of the employees of 399 fast-food restaurants in New Jersey and Pennsylvania -- 321 in the former state that were subject to the new wage minimum and 78 in the latter state that were not affected by the increase. They make several comparisons.

- One is between the pay and unemployment patterns in their New Jersey sample versus their Pennsylvania control group.

- Another contrasts the state of affairs prevailing in New Jersey before and after the imposition of the new minimum wage level (\$5.05) on 1 April 1992. Their first wave of interviews took place during the period 15 February and 4 March 1992. Their second set occurred between November 5 to December 31, 1992.

- A third comparison was between low- and high-wage employers in New Jersey, the latter presumed not to be affected by the increase in the mandated wage level.

What did CK find? No matter which of the three ways the experiment was conducted, there was no reduction in employment in the affected restaurants in New Jersey. On the contrary, there was a small but statistically significant increase in jobs in these cases. CK attempt to eliminate other alternative explanations for this phenomenon. These deserve scrutiny.

Reliability. First, and most obvious, is data reliability. CK (1994, p. 778) base their case for reliability on the ground that there was a high correlation between the two sets of answers, ranging from .70 to .98, between the "responses of 11 stores that were inadvertently interviewed twice." But this is disquieting on several grounds. For one thing, it shows that they were so confident about the accuracy of their data that they didn't even purposefully plan to test for the possibility of erroneous response or bias, or other sources of data incompatibility. For another, these correlations are merely between two sets of answers to the same questions. Even if they were both wildly inaccurate as measured against reality, they would still be expected not to diverge too much from each other. Third, the higher correlation, .98, was for the price of a meal, a mere control variable; the lower one of .7 was for the number of employees. But the latter variable is absolutely crucial to their analysis, while the former is of relatively less interest.[5] Furthermore, this correlation attests, at best, only to the accuracy of the first wave of interviews. Of far greater importance is the second wave, from which the actual employment experience after the legislated boost in compensation is derived. CK may have confidence in the accuracy of their data, but the data are suspect, if only on the grounds that they buttress a counterintuitive finding, namely, that minimum wage increases have positive employment effects.[6]

---

[4]Addison and Blackburn (1999. p. 393) reach a similar conclusion, albeit through a different "reduced form" methodology. Their analysis focuses not directly upon the employment effects of minimum wages, but rather on the implications of this law for poverty rates. They maintain that "a 25% increase in the minimum wage should lower the 1996 poverty rate ... by 9%..." For further support of this enactment see Quigley (1996, p. 513) who concludes: "Reforming the minimum wage by raising it and indexing it for inflation is a critical step toward attaining Franklin Delano Roosevelt's goal of assisting the nation's working poor by providing 'a fair day's pay for a fair day's work.'" For views critical of minimum wages, see Hutchison (1997), Berman (1995), and Deere, Murphy and Welch (1995), Block (1996), Almeida and Block (forthcoming), Ritchie, McGee and Block (forthcoming), McInerney (1997), and McCormick and Block (forthcoming).

[5]This does not mean of no interest at all. Surely, in the words of an anonymous referee, if "one or ... all producers (can) raise price to compensate for increased labor cost (this) would be of considerable interest." I agree. But if this occurs, then employers need not much reduce the size of their labor force in the face of a mandated wage increase. However, it is the latter phenomenon which is the major issue of debate; the former is only important, insofar as it impacts on the latter. Further, there is reason to believe that the ability of producers to raise price in reaction to an increase in the minimum wage level is rather limited. For, if they could increase final goods prices, now, without any sharp negative repercussions, why was this option not available to them before this change in the law?

[6]This, despite the recognized unreliability of survey data, particularly unpaid (that is, the respondents were not compensated for their labor and thus it may be expected to be of lower quality), and by telephone.

Choice of subjects. The second difficulty concerns their choice of subjects: fast-food employees. To be sure, there may be some advantages in this choice; such establishments, after all, are a leading employer of unskilled labor, and data are easy to collect. But while the "products of fast food restaurants are relatively homogeneous" (1994, p. 774), a relatively peripheral issue, this does not really apply to job requirements, an issue crucial to their findings. That is to say, while the typical consumer cannot readily tell one McDonalds' burger from another, this is not true of the ability of the manager to distinguish one employee of such an establishment from another.

How does heterogeneity enter into such low-skill occupations? In many ways: employees must come to work on time and stay on for their full tour of duty, otherwise staffing decisions are rendered more difficult; those who deal with the public must attain a "the customer is always right" outlook, and a demeanor consistent with this thought; even those behind the scenes (e.g., the cooks) must cooperate with one another. The last thing management wants is intra-staff fights or altercations between employees and customers. Employees, moreover, must take direction from management without resentment, even though this often occurs across racial, ethnic, and gender lines. In addition to reliability, cheerfulness, and docility, there is of course enthusiasm, a hard work ethic, cleanliness, etc. Workers can and do differ in all of these dimensions.

If CK were seeking homogeneity of job specifications, they might far better have looked at factory work.[7] There are, to be sure, heterogeneity "problems" there too. There is no field of human endeavor where heterogeneity is completely absent. But heterogeneity is likely to be of less importance on an assembly line.

Nor can we ignore the fact that heterogeneity is very important: Employers will fire their least productive workers in response to an artificially mandated wage increase and substitute better ones in their places. (The reason better ones were not originally hired is that poorer ones could be had at lower wages. But now, with the advent of the increase in the minimum wage, the latter option is no longer available, and the former seems preferable.)

Narrowness of sample. A third problem arises with the narrowness of CK's choice of subjects. Coerced rises in wage minima will have adverse employment effects on all those with productivity below stipulated levels. By excluding all but some 8,000+ workers in total (some 400 stores with an average of just above 20 employees each) from their purview, CK look at a small cohort of the total economies of New Jersey and Pennsylvania. This buy-one-ticket-in-a-lottery strategy is practically guaranteed to fail to take into account widespread unemployment effects. Suppose, for example, that CK had concentrated only on low-skilled gas station attendants before the advent of self service. The unemployment effects of a similar wage increase might have been equally nonidentifiable, spilling over to other industries more able to adjust quickly than this one.

Temporal adjustment. A fourth drawback to their analysis concerns timing; CK confine themselves to the relatively short run. And not unrelated to this, there is also a difficulty with the percentage change (18 percent) in the required wage rate. They gave this industry a scant seven months to respond to the new mandate.[8] Had they dealt in similar fashion with the operators of manual elevators when the minimum wage rose from $.40 to $.75 per hour when few were yet automated, they undoubtedly would have found almost zero negative employment impact. And yet within a few years of this experiment, virtually all of these jobs were obliterated by minimum-wage-inspired automation. What, then, is the relevant lag effect in this particular case? I expect that this is far longer than a mere seven months, since technological innovations in the fast-food industry will probably take far longer than in vertical transportation.

It is exceedingly speculative to try to anticipate the possible course of future inventiveness in this field that will be engendered by this minimum wage increase. It is an entrepreneurial, not an economic, task. But as CK do not vouchsafe us with even a discussion of lags, this oversight must be made good somehow.

How, then, will entrepreneurs substitute out the suddenly more expensive factors of production, unskilled labor, which has now been priced out of the market? One possibility, of course, is robotics or artificial intelligence. But this, unlike automatic elevators in an earlier epoch, seems decades away. Another insight might be garnered by borrowing a leaf from grocers, pharmacists, and other retail merchants, who utilize universal product codes. Yet another unskilled-labor-saving device might be to copy the trend in filling stations and make more use of self service.

---

[7] Deere, Murphy and Welch (1995, p. 52) use the homogeneity assumption to shine a spotlight on the economics of minimum wages: "If the world were simple, then the implications of increases in the minimum wage would also be simple. For example, if wages were the only form of compensation, if there were no fringe benefits or job amenities, and if all workers were of uniform quality, then everyone would get the same wage. A minimum that attempted to raise the wage would reduce employment."

[8] Comparing 1 April with 3 December the midpoint of their second wave interviews yields 7.1 months, not the 8 they report on page 775.

This has already been done, in some fast food cases, with regard to drinks.

How long will it take for this phenomenon to significantly reduce the demand for unskilled labor? This is extremely difficult to say. But judging from the results reported by CK, seven months would appear to be too short a time span to capture all of the forthcoming unemployment effects of this law.

CK do not attribute their findings to flawed statistics, an insufficient time lag, a relatively modest increase in the minimum wage level, or an exceedingly narrow coverage of industries, but rather to the presence of monopsony. On the face of it, this is not compelling. Say what you will about New Jersey, no one has yet seen fit to characterize it as a one company state. Even on the heroic assumption that monopsony is itself a logically coherent analytic construct, it strains all credulity to apply it to this case.[9]

What are the problems? Outsiders will enter the market to take advantage of the profits earned by the monopsonist; in the absence of entry barriers, monopsony, even if it could become established in the first instance, cannot long endure. Workers, too, are mobile, even if not perfectly so. Were their wages pushed below levels established by marginal revenue product, they would seek higher wages in areas less dominated by monopsony. Most important, even in the absence of these other considerations, the neoclassical monopsony analysis admits of only a very narrow window of opportunity for "ameliorative" legislation such as the minimum wage. This is because each monopsonist faces slightly different cost and demand schedules. And yet minimum wages can only have positive employment effects if their rise is between points Wm and Wc in Figure 1.[10] But if every monopsonist confronts different cost and demand schedules, in order not to generate negative employment effects a different minimum wage range must be mandated for each.[11] But this is clearly a manifest impossibility, something never even contemplated by advocates of government intervention into labor markets, such as CK.

Figure 1.

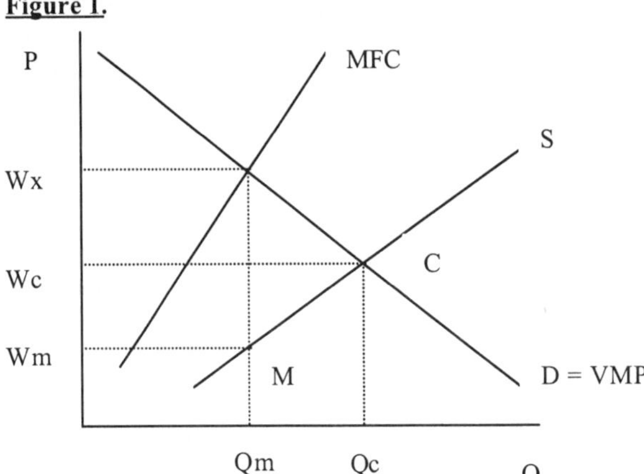

Then, too, there is the issue of general vs. specific training (Becker, 1964). In the former case, marginal revenue product is the same whether working for the present employer or for any other. In the latter case, if the employee migrates to another firm, his productivity is lower. This, in effect, "ties" the worker to his present employer, since skills are in part firm specific.

It is readily apparent that if there is any scope for monopsony, it is with regard to specific, not general, training. But this is very damaging for the CK analysis. For specific training is positively correlated with high skill;

---

[9]For a critical view, see Armentano (1972, 1982), Armstrong (1982), Block (1977, 1994) Rothbard (1962).

[10]S is the supply curve of labor, D is the demand curve (equal to the Value of the Marginal Product of Labor), and MFC is Marginal Factor cost to the monopsonist. The perfectly competitive industry in the factor market would locate at C, paying wages of Wc and hiring Qc of labor. The monopsonist would locate at M, and hire Qm workers at a wage of Wm.

[11] The range is depicted in Figure 1 by the distance Wm-Wx. This is because at a minimum wage lower than Wm the monopsonist will not be forced to increase his pay offer, and if it is higher than Wx, there will be unemployment effects even for the monopsonist.

those with general training tend to be less skilled. Thus, CK are attempting to apply the monopsonistic model to a sector of the labor market where it patently does not apply: to unskilled, generally-trained workers, who are free to take on other jobs, at least as far as the fear of diminution of productivity elsewhere is concerned. CK would have had a better case for linking highly-skilled workers to monopsony, but their wages usually far exceed mandated minimum wages.[12]

A further critique of the CK thesis has been articulated by Gary Becker. In his view, ceteris was not paribus in the research undertaken by CK. States Becker (1995, p. 10):

> "The higher federal minimum in 1990 and 1991 caused a much larger drop in New Jersey's teenage employment than Pennsylvania's, which could explain why employment did not fall more in New Jersey when that state increased its own minimum in 1992. New Jersey employers presumably anticipated the increase in their state's minimum when they sharply cut employment in responding to the earlier wage hike."

As well, there is a wealth of pertinent empirical evidence on this topic amassed by the economics profession over many decades. It shows the deleterious effects on employment for the unskilled; this applies to the minimum wage law in general, and to raising it to any given level in particular. This empirical record, alone, should have given CK pause for thought.

Even more daunting is the fact that their findings are contrary to economic law. It is a matter of basic theory that in equilibrium, wages tend to equal marginal revenue product.[13] If wages are above this level, losses and eventual bankruptcy may result.[14] If wages are below marginal revenue product, there is a proclivity for the employer to lose his labor force: either the quit rate will rise as employees seek better opportunities, or other employers will bid for the underpaid workers in an attempt to profit from them.[15] It is only when wages equal productivity that there is no internal impetus for further change.

This being the case, an elevation in the minimum wage level is bound to exceed the marginal revenue product of at least some additional workers. These people, then, are primary candidates for unemployment. A finding that the minimum wage level rose, and that employment did not fall but actually increased, is therefore highly anomalous. It cries out for explanation. On the level of pure theory, then, it must count against CK that -- apart from the economically dubious monopsony argument -- they felt no need to account for their anomalous findings.

One last issue overlooked by CK concerns non-monetary compensation: it is possible that the negative impact of minimum wages on unemployment may be masked by the phenomenon of fringe benefits and working conditions (see Wessel and McKenzie). In response to the mandate of a higher minimum wage, the employer can reduce the fringe benefits enjoyed by his workers and the amount of money he invests in their working conditions (e.g., air

---

[12] There are, perhaps, no more highly skilled workers than professional athletes. Their feats of derring-do, the required natural ability and years of intensive practice serve as a natural entry barrier. If anyone ought to be victimized by monopsony, according to neo-classical theory, it would be these workers. But even here, as shown by the continual rise of foreign and other competing leagues, these laborers, supposedly helpless in the face of the monopsonistic threat, are not at all victimized.

[13] I have reservations about the neoclassical model of industrial organization, which makes much of the distinction between monopoly (or monopsony) and perfect competition, based, mainly, on the number of firms in an industry and the concentration ratios of the leading four or eight firms, e.g., the Herfindahl index. On this see Armentano (1972, 1982, 1991), Armstrong (1982), Block (1977, 1982, 1994), DiLorenzo (1997), Boudreaux and DiLorenzo (1992), High (1984-1985), McChesney (1991), Rothbard (1970a), Shugart (1987), Smith (1983). But for purposes of the present paper, I adopt this model, and assume that the firm depicted in Figure 1 is either perfectly competitive or monopsonistic (I am comparing the two) in the labor market, while perfectly competitive in the product market it faces.

[14] Labor, of course, is only one factor of production. Where it comprises a small proportion of total costs, the firm may well survive by substituting now relatively cheaper factors for the now more expensive labor. However, in (service) industries where labor comprises a very high proportion of the total, losses and eventual bankruptcy are rendered more probable.

[15] We assume, here, that information is readily available to either the employer or employee or both.

conditioning, level of cleanliness, decor, music, etc.) If so, then the total money expenditure made in behalf of employees need not rise, even in the face of a legal requirement than money wages increase. Total wages equal money wages plus non-pay packet expenditures, and if the money wages rise while the non-monetary spending in behalf of employees falls, then total compensation need not rise. If so, there will be a minimum wage unaccompanied by additional unemployment in a context where traditional theory need not be jettisoned to any degree, neither that implied by CK (1994) nor even the more extreme version of this offered by Galbraith (2000).

But let it not be thought that the failure of the minimum wage raise to increase unemployment due to this phenomenon is an argument in behalf of this public policy. For there will be losses in utility, even if unemployment does not increase. Why? This is because we must assume that the previous allocation of total compensation over money wages, fringe benefits, and working conditions tended to be optimal. Certainly, the employer has every incentive to provide equally expensive non-monetary compensation to his employees when they prefer this, and not, when they do not. Specifically, better fringes and improved working conditions are supplied to workers only when they cost less for the firm to provide than they are worth to the worker. But when the minimum wage law forces the business to increase money compensation at the cost of more highly valued fringe benefits and better working conditions, this renders the total pay less attractive. These jobs may pay more, due to the minimum wage law, but since the higher monetary compensation comes at the expense of the even more highly valued non monetary aspects of compensation, it cannot be said that even the worker not unemployed by this law is a beneficiary of it.

In conclusion, we ask a question. Which is more reasonable? That economic law somehow has been repealed in New Jersey, or that CK's methodology is wanting in one or more of the ways indicated above?[16] The answer, at least from this quarter, is clear.

---

[16] Two public opinion surveys of economists asked 27 different questions about public policy issues, in an attempt to gauge the level consensus within the profession (Frey, et. al. 1984; Block and Walker, 1988). A statement which garnered amongst the highest degree of agreement of all these questions read as follows: "A minimum wage increases unemployment among young and unskilled workers."

Bibliography

Addison, John T., and McKinley L. Blackburn, "Minimum Wages and Poverty," 52 Industrial and Labor Relations Review, 393 (1999).

Almeida, Michael and Walter Block, "Higher Minimum Wages Advance Poverty," Commentaries on Law and Economics, forthcoming.

Armentano, Dominick T., The Myths of Antitrust, New Rochelle, N.Y.: Arlington House, 1972.

Armentano, Dominick T., Antitrust and Monopoly: Anatomy of a Policy Failure, New York: Wiley, 1982.

Armentano, Dominick T., Antitrust Policy: The Case for Repeal, Washington, D.C.: The Cato Institute, 1991.

Armstrong, Donald, Competition vs. Monopoly, Vancouver: The Fraser Institute, 1982.

Becker, Gary, Human Capital, New York: The National Bureau of Economic Research, 1964.

Becker, Gary, "It's Simple: Hike the Minimum Wage, and You Put People Out of Work," Business Week Magazine, 6 March 1995.

Berman, Richard B., "New Evidence on the Minimum Wage: The Crippling Flaws in the New Jersey Fast Food Study," Washington D.C.: Employment Policies Institute, 1995.

Block, Walter, "Austrian Monopoly Theory -- a Critique," The Journal of Libertarian Studies, Vol. I, No. 4, Fall 1977, pp. 271-279.

Block, Walter, Amending the Combines Investigation Act, Vancouver: The Fraser Institute, 1982.

Block, Walter, "Total Repeal of Anti-trust Legislation: A Critique of Bork, Brozen and Posner, Review of Austrian Economics, Vol. 8, No. 1, 1994, pp. 35-70.

Block, Walter, "Labor Market Disputes: A Comment on Albert Rees' 'Fairness in Wage Distribution,'" Journal of Interdisciplinary Economics, 1996, Vol. 7, No. 3, pp. 217-230.

Block, Walter, and Walker, Michael, "Entropy in the Canadian Economics Profession: Sampling Consensus on the Major Issues," Canadian Public Policy, Vol. XIV. No. 2, June 1988, pp. 137-150.

Boudreaux, Donald J., and DiLorenzo, Thomas J., "The Protectionist Roots of Antitrust," Review of Austrian Economics, Vol. 6, No. 2, 1992, pp. 81-96.

Card, David, and Krueger, Alan B., "Minimum Wages and Employment: A Case Study of the Fast-Food Industry in New Jersey and Pennsylvania," American Economic Review, Vol 84, No. 4, September 1994, pp.772-793.

Card, David, and Krueger, Alan B., 1995. Myth and Measurement: The New Economics of the Minimum Wage, Princeton, NJ: Princeton University Press.

Card, David, and Krueger, Alan B., "A Reanalysis of the Effect of the New Jersey Minimum Wage Increase on the Fast-Food Industry with Representative Payroll Data," working paper #393, Princeton University Industrial Relations Section, December 1997, Revised January 1999.

Deere, Donald, Kevin M. Murphy and Finis Welch, "Sense and Nonsense on the Minimum Wage," Regulation, No. 1, 1995.

DiLorenzo, Thomas J., 1997, "The Myth of Natural Monopoly," Review of Austrian Economics, Vol. 9, No. 2, pp. 43-58.

Frey, Bruno S., Werner W. Pommerehne, Friedrich Schneider and Guy Gilbert (1984) "Consensus and Dissension Among Economists: An Empirical Inquiry, American Economic Review, December, 74:5:986-94.

Galbraith, James K., "How the Economists Got It Wrong" on the web at: http://www.prospect.org/archives/V11-7/galbraith-j.html (2/17/00).

High, Jack, "Bork's Paradox: Static vs Dynamic Efficiency in Antitrust Analysis," Contemporary Policy Issues, Vol. 3, 1984-1985, pp. 21-34.

Hutchison, Harry, "Toward a Critical Race Reformist Conception of Minimum Wage Regimes: Exploding the Power of Myth, Fantasy and Hierarchy," 34 Harvard Journal on Legislation 93, 1997.

McChesney, Fred, "Antitrust and Regulation: Chicago's Contradictory Views," Cato Journal, Vol. 10, 1991.

McCormick, Paul, and Block, Walter, "The Minimum Wage: Does it Really Help Workers," Southern Connecticut State University Business Journal, forthcoming.

McInerney, Matthew, (1997), "Why the 'Living Wage' is Hurting Those Trying to Make a Living: A Critique of the Minimum Wage Law," Discourse No. 15, pp. 36-38.

Quigley, William P., "A Fair Day's Pay for a Fair Day's Work: Time to Raise and Index the Minimum Wage," 27 St. Mary's Law Journal 513, 1996.

Ritchie, R. J., McGee, Robert W., and Block, Walter, "The Minimum Wage: Misguided Altruism,"Commentaries on Law and Economics, forthcoming.

Rothbard, Murray N., Man, Economy and State, Los Angeles, Nash, 1962.

Shugart II, William F., "Don't Revise the Clayton Act, Scrap It!," 6 Cato Journal, 925, 1987.

Smith, Jr., Fred L., "Why not Abolish Antitrust?," Regulation, Jan-Feb 1983, 23.

Wessel and McKenzie, "to be supplied".

Heritage Stumbles on Minimum Wage

Walter Block

The Heritage Foundation is no flaming libertarian organization. Not for them the radical privatization of such things as bodies of water, roads, even social security, much less courts, armies, and police.

But they are, after all, a conservative organization, so a person would think he could rely on them, at least, to be sound on the issue of minimum-wage legislation. Even liberal economists, for goodness sake, know that this law prices low productive workers (read: some minorities, teens, school dropouts, the handicapped, and other clients of the welfare state) out of the employment market, consigning them to the forced idleness of unemployment.

It appears, however, that Heritage cannot be wholly relied upon, fully, even on this question.

They start out, reasonably enough, it might be conceded by those more charitable than myself, with their headline: "Why Congress Should Consider Alternatives to Raising the Minimum Wage," (The Heritage Foundation Executive Memorandum #627, September 30, 1999). Good; certainly it would be a move in the wrong direction to raise the level at which the minimum is pegged.

But even here, a quibble: why worry about raising the minimum, when the only just and economic policy would be to eliminate it entirely?

But perhaps this criticism is unwarranted. I am not a member of the inside-the-beltway crowd. I therefore plead a lack of sophistication. I am, admittedly, not current with the need of the cognoscenti to move the debate a centimeter or so to the right, and never, never, never, ever to challenge basic postulates. Radical maniac that I am, I thought it reasonable to tell the truth on this matter, allowing the chips to fall where they may.

Mea culpa

But what, then, are we to make of the following statement by the authors of the piece, Stuart M. Butler, Ph.D, and Angela Antonelli: "In a strong economy with a tight labor market ... raising the minimum wage from \$5.15 to \$6.15 per hour ... may seem a harmless way to boost incomes. But in the next economic downturn, this extra cost of living could mean pink slips for many of those same Americans a rise in the minimum wage is intended to help."

To say this is to make not one but several fatal concessions to this pernicious law. First of all, what is this "intended to help" business? Only terminally ignorant people think that the minimum wage law arises from benevolence. All those who have so much as taken a basic course in economics 101 know full well that the motivation behind these deleterious enactments is to artificially boost the wages of the low-productivity workers who compete with big labor, thus clearing the way for increases in their pay packets.

This law is a sop to a large constituency of the Democratic party, labor unions. Why else would these highly-paid workers, earning far in excess of either \$5.15 or \$6.15, be so intent upon "raising" the remuneration of those so far below them in the labor pecking order? As every freshman economics student knows, if an artificial, government-mandated raise in wages can actually help its intended "beneficiaries," why not raise it to \$1 million per hour, and be done with the economic problem entirely?

A second mistake, and perhaps an even more serious one because of its presumed complexity, is this distinction between the boom and the bust phase of the business cycle. It is a completely artificial one. The minimum wage effectively unemploys all those whose productivity lies below the level pegged. In other words, at a minimum wage level of \$5.15, all those with productivities of less than that amount are unemployed, assuming profit maximization behavior on the part of employers.

With a minimum of \$6.15, there will be an additional cohort of jobless people, those with marginal productivity of anywhere between these two amounts. True, this holds precisely true only in equilibrium, but there is no reason to think that the economy is more or less out of equilibrium in the boom-or-bust phase of the cycle. Therefore, the distinction offered by Butler and Antonelli is a spurious one.

An implication of the Heritage Foundation Executive Memorandum is that if, somehow, we could be assured of continual economic upturn, then the minimum wage law would be a welcome one. Nothing could be further from the truth.

The Living Wage: What's Wrong?

Walter Block and William Barnett II

The latest on the minimum-wage front, brought to us by the academic minions of "social justice," is a private, not a public, effort to raise the pay of low-wage workers. Emanating first from prestigious institutions of higher learning such as Harvard and Yale, this initiative has spread like wildfire to colleges all around the country.

The gist of the program is to raise the wages of janitors and others at the lower end of the pay distribution to $10 or $12 an hour, and to boycott suppliers who do not undertake a similar program. A minimum wage of $5.15, it would appear, might be all well and good, but something twice that amount is necessary if it is to be a "living wage."

It is entirely legitimate for a private university to offer whatever pay scale it wishes and to boycott any businesses whatsoever, for any reason it chooses. However, institutions of higher learning are supposedly distinguished by rational dialogue, and it is in this vein that we wish to register an objection to this unwise policy.

Let us consider several reasons for declining to pay labor more than is necessary to attract a sufficient number of job applicants and for ending discrimination against firms that pay market wages.

1. Universities attempt to raise funds from the entire business community (among many other constituents). Making invidious comparisons between firms — singling out those that operate under market conditions for implicit condemnation — can hardly be conducive to this end. But this is mere pragmatism, unworthy perhaps of even being considered.

2. The program will likely not have its intended effect of boosting the wages of low-skilled workers. Suppose the typical university subcontractor pays its unskilled employees $6 an hour and the "social justice" wage is $10. People in this stratum of the labor force would give their eyeteeth for such a position, since it pays 40 percent more than the market says the job is worth. Would not everyone and his uncle making under $10 gladly take up such a job? How will the limited number of spaces be allotted to the vast hordes of people? Would it unduly challenge credulity to think that some of the few selected would be willing to make a side payment to the hiring staff? Or that this might be demanded of applicants? Or that nepotism, favoritism, and other

forms of discrimination might arise? After all, if prices are not allowed to allocate labor resources, other criteria will be used.

3. If you want to give money to poor people, why not just go ahead and do it? Why tie it to their jobs of all things? That is, why conflate charity with an attempt to disrupt the labor market? Universities, at least private ones, are part of the market. Therefore they cannot disrupt it with any voluntary act on their part, even of this sort. But why even try? Why offer extra money to the unskilled in the form of higher salaries when you can use these funds for education or training or anything else under the sun?

People are Different

4. The notion of justice underlying the "living wage" is predicated on the philosophy that income differences are unfair. It is patently obvious that in a market-based society, the primary reason some people are wealthy and others very much not so is that they have different initial endowments of intelligence, work ethic, ambition, talents, and entrepreneurial skills, as well as inherited material wealth and even luck. But despite these differences, the market tends to diminish income differences that would otherwise exist. This is because to become rich under free enterprise, you must enrich the lives of many other people; at the apex of the economic pyramid you gain a great deal, but you also drag onto a higher economic plane practically an entire society. (Think of Bill Gates or Henry Ford.) If we all lived on tiny islands as hermits, without economic interaction, some of us would be far wealthier than others. If material differences in wealth are unfair, then are not the very causes thereof, different endowments of human capital, also unfair?

Egalitarianism as a philosophy is dead from the neck up, insofar as even its adherents do not and, indeed, cannot take it seriously. Yet suppose there were a magical machine that could transfer IQ points (or beauty or health or hair follicles or musical ability) from those that have "a lot" to those who have "too little." It would be the rare egalitarian who would follow through on this pernicious philosophy. In contrast, the freedom philosophy requires only that people keep their hands off other people and their property, something far more peaceful and just, and also more readily attainable.

Critique of Minimum Wage Petition

Walter Block

Hundreds of Economists Say: Raise the Minimum Wage

Suppose almost 700 leading physicists signed a petition stating that gravity was a myth; or an equal number of physicians publicly rejected the germ theory of disease; or several hundred geologists came to the conclusion that the earth was flat; or almost 1,000 famous astronomers maintained that the sun revolves around the earth instead of the other way around. Posit, further, that the signatories in each of these cases included over a dozen leaders in the fields, winners of Nobel Prizes in their respective disciplines, and/or members of the faculties of some of our most prestigious universities.

Most people would conclude that the world has turned upside down. After they have gotten over their initial shock at these claims, however, they might begin to have darker thoughts. They might reason, "They recall malfunctioning tires, television sets, toasters and automobiles, don't they? Maybe they ought to reconsider the credentials of professionals who reject the germ theory, the roundness of our planet, and so on." And, if word got out that these scholars took up such astounding positions not for scientific reasons, but rather for essentially political ones, they might well become greatly disappointed and even more distrustful of the discipline of economics. Why? Because these claims serve as litmus tests for minimal knowledge in these fields. If you held any of the opinions expressed above, you would expose yourself as an ignoramus in, respectively, medicine, geology, and astronomy. And, as a quack, you would be quickly drummed out of these disciplines.

Hard as it is to believe, something of this sort has recently occurred with regard to economics. Some 665 economists have signed a statement claiming that a raise in the minimum wage level will not increase unemployment, and can play a role in alleviating poverty. Several of them have won Nobel Prizes in economics. Virtually all of them boast PhD degrees in the dismal science. All of them are affiliated with prestigious organizations, many of them very much so.

Now, just as the germ theory in medicine, the roundness of the earth in geology and so on serve as a sine qua non in these physical sciences, so does the deleterious effects of minimum wage laws on low-skilled workers in economics. Every Basic Economics 101 textbook, even those written, paradoxically, by some of the signatories of this document, make this basic elementary point. If there is anything drummed into students taking introductory courses in economics, this is it. For a bunch of economists to reject this is highly problematic.

Why, then, do I include this statement in this book, with all the signatories to it, each and every one of them? To make it that much less likely that their behavior will never be forgotten. One of these days justice will prevail, and the eminent reputations of all those who signed this document will be called into question. In the meantime, we must content ourselves with exposing its errors.

Let us now consider their missive in some detail. It appears below, in quotation marks. My interspersed comments are in *italics*.

"The minimum wage has been an important part of our nation's economy for 68 years."

True enough. One of the only very few correct statements in the entire document.

"It is based on the principle of valuing work by establishing an hourly wage floor beneath which employers cannot pay their workers."

The minimum wage law does nothing of the kind. It does not at all establish anything, let alone a "wage floor." Rather, it states that it shall be illegal for an employer to pay, and employee to receive (this much is usually a dead letter law insofar as the government will not prosecute the supposed victim of the piece, the low-wage laborer) any amount less than that stipulated by law. But there is no requirement, none at all, that a job, any job, be offered at this wage.

A "wage floor," moreover, implies that when the minimum wage level rises, it pulls wages up with it. If we must use mechanical analogies to economic phenomena, a far better one would be that of the barrier over which one has to jump in order to land a job; and, the higher is this hurdle, the fewer those who can catapult over it.

"In so doing, the minimum wage helps to equalize the imbalance in bargaining power that low-wage workers face in the labor market."

When wages are below their equilibrium (productivity) levels, market forces push them up, as the demand for labor exceeds the supply. When wages are above their equilibrium (productivity) levels, market forces push them down, as the supply of labor exceeds the demand. It matters not that there are more employees than employers. What determines wages (assuming that worker productivity is the same for many potential employers, which is typically true for low-skilled workers) is thus productivity, not "bargaining power." (DiLorenzo, 2004)

"The minimum wage is also an important tool in fighting poverty."

It impossible to "fight" poverty by preventing unskilled workers from landing jobs.

"The value of the 1997 increase in the federal minimum wage has been fully eroded. The real value of today's federal minimum wage is less than it has been since 1951. Moreover, the ratio of the minimum wage to the average hourly wage of non-supervisory workers is 31%, its lowest level since World War II. This decline is causing hardship for low-wage workers and their families."

The facts cited here are all correct. And, thanks to the erosion of the real value of the minimum wage, unemployment for the poorest laborers is lower than it otherwise would have been. This decline, then, reduces *hardship.*

"We believe that a modest increase in the minimum wage would improve the well-being of low-wage workers and would not have the adverse effects that critics have claimed. In particular, we share the view the Council of Economic Advisors expressed in the 1999 Economic Report of the President that 'the weight of the evidence suggests that modest increases in the minimum wage have had very little or no effect on employment.' While controversy about the precise employment effects of the minimum wage continues, research has shown that most of the beneficiaries are adults, most are female, and the vast majority are members of low-income working families."

Why a "modest" increase? If a rise will help the poor, why be so niggardly? Why not raise it substantially? Because it is clear, even to the anti-market economists who signed the petition, that a boost to, say, $100 per hour would devastate employment opportunities for the unskilled; indeed, for all *of those with a productivity level lower than that amount. Nor is it a matter of "dosage is all": A little bit of arsenic is good, more is bad. The "model" utilized by the petitioners is that a rising floor of minimum wages will carry everyone upward along with it. If so, why not raise it to $100, and cure poverty in one fell swoop?*

As for the "weight of the evidence," if the increase is modest enough it is no wonder few effects can be found. But the overwhelming majority of econometric regressions suggest significant unemployment effects. Is it for this reason that surveys of many *economists, not just petitions from a few hundred, demonstrate a strong consensus against this law (Frey, 1984; Block & Walker, 1988)?*

"As economists who are concerned about the problems facing low-wage workers, we believe the Fair Minimum Wage Act of 2005's proposed phased-in increase in the federal minimum wage to $7.25 falls well within the range of options where the benefits to the labor market, workers, and the overall economy would be positive."

There is no "range" of a minimum wage law increase within which unskilled workers will not become unemployed. Why would a firm continue to employ a laborer who produces less than what he must be paid by law?

"Twenty-two states and the District of Columbia have set their minimum wages above the federal level. Arizona, Colorado, Missouri, Montana, Nevada and Ohio, are considering similar measures. As with a federal increase, modest increases in state minimum wages in the range of $1.00 to $2.50 and indexing to protect against inflation can significantly improve the lives of low-income workers and their families, without the adverse effects that critics have claimed."

At least without indexing, the real value of the minimum wage level atrophies over time, due to inflation. In this way, some low-skilled workers may find their way back into the labor force. Indexing precludes this respite from this pernicious legislation.

Leading economists who endorse the statement on increase in minimum wage:

Henry Aaron The Brookings Institution
Kenneth Arrow*+ Stanford University
William Baumol+ Princeton University and New York University
Rebecca Blank University of Michigan
Alan Blinder Princeton University
Peter Diamond+ Massachusetts Institute of Technology
Ronald Ehrenberg Cornell University
Clive Granger* University of California, San Diego
Lawrence Katz Harvard University (AEA Executive Committee)
Lawrence Klein*+ University of Pennsylvania
Frank Levy Massachusetts Institute of Technology
Lawrence Mishel Economic Policy Institute
Alice Rivlin+ The Brookings Institution (former Vice Chair of the Federal Reserve and Director of the Office of Management and Budget)
Robert Solow*+ Massachusetts Institute of Technology
Joseph Stiglitz* Columbia University

*Note: * Nobel Laureate*
+ Past president, American Economics Association

Affiliations are for identification only and should not be construed as official endorsement by the listed institutions. 650 of their fellow economists agree.

For more information, visit epi.org/minwage or contact the Economic Policy Institute at 202/775-8810.

Economists supporting increase in minimum wage:

Katherine G. Abraham University of Maryland • **Frank Ackerman** Tufts University • **F. Gerard Adams** Northeastern University • **Randy Albelda** University of Massachusetts — Boston • **James Albrecht** Georgetown University • **Jennifer Alix-Garcia** University of Montana • **Sylvia A. Allegretto** Economic Policy Institute • **Beth Almeida** International Association of Machinists and Aerospace Workers • **Abbas Alnasrawi** University of Vermont • **Gar Alperovitz** University of Maryland — College Park • **Joseph Altonji** Yale University • **Nurul Aman** University of Massachusetts — Boston • **Teresa L. Amott** Hobart and William Smith Colleges • **Alice Amsden** Massachusetts Institute of Technology • **Bernard E. Anderson** University of Pennsylvania • **Robert M. Anderson** University of California — Berkeley • **Bahreinian Aniss** California State University — Sacramento • **Kate Antonovics** University of California — San Diego • **Eileen Appelbaum** Rutgers University • **David D. Arsen** Michigan State University • **Michael Ash** University of Massachusetts — Amherst • **Glen Atkinson** University of Nevada — Reno • **Rose-Marie Avin** University of Wisconsin — Eau Claire • **M.V. Lee Badgett** University of Massachusetts — Amherst • **Aniss Bahreinian** Sacramento City College • **Ron Baiman** Loyola University Chicago • **Asatar Bair** City College of San Francisco • **Katie Baird** University of Washington — Tacoma • **Dean Baker** Center for Economic and Policy Research • **Radhika Balakrishnan** Marymount Manhattan College • **Stephen E. Baldwin** KRA Corporation • **Erol Balkan** Hamilton College • **Jennifer Ball** Washburn University • **Brad Barham** University of Wisconsin — Madison • **Drucilla K. Barker** Hollins College • **David Barkin** Universidad Autonoma Metropolitana • **James N. Baron** Yale University • **Chuck Barone** Dickinson College • **Christopher B. Barrett** Cornell University • **Richard Barrett** University of Montana • **Laurie J. Bassi** McBassi & Company • **Francis M. Bator** Harvard University • **Rosemary Batt** Cornell University • **Sandy Baum** Skidmore College • **Amanda Bayer** Swarthmore College • **Sohrab Behdad** Denison University • **Peter F. Bell** State University of New York — Purchase • **Dale L. Belman** Michigan State University • **Michael Belzer** Wayne State University • **Lourdes Beneria** Cornell University • **Barbara R. Bergmann** American University and University of Maryland • **Eli Berman** University of California — San Diego • **Alexandra Bernasek** Colorado State University • **Jared Bernstein** Economic Policy Institute • **Michael Bernstein** University of California — San Diego • **Charles L. Betsey** Howard University • **David M. Betson** University of Notre Dame • **Carole Biewener** Simmons College • **Sherrilyn Billger** Illinois State University • **Richard E. Bilsborrow** University of North Carolina — Chapel Hil • **Cyrus Bina** University of Minnesota — Morris • **Melissa Binder** University of New Mexico • **L. Josh Bivens** Economic Policy Institute • **Stanley Black** University of North Carolina — Chapel Hill • **Ron Blackwell** AFL — CIO • **Margaret Blair** Vanderbilt University Law School • **Gail Blattenberger** University of Utah • **Robert A. Blecker** American University • **Barry

Bluestone Northeastern University • **Peter Bohmer** Evergreen State College • **David Boldt** State University of West Georgia • **Roger E. Bolton** Williams College • **James F. Booker** Siena College • **Jeff Bookwalter** University of Montana • **Barry Bosworth** The Brookings Institution • **Heather Boushey** Center for Economic and Policy Research • **Roger Even Bove** West Chester University • **Samuel Bowles** Santa Fe Institute • **James K. Boyce** University of Massachusetts — Amherst • **Ralph Bradburd** Williams College • **Michael E. Bradley** University of Maryland — Baltimore County • **Elissa Braunstein** Colorado State University • **David Breneman** University of Virginia • **Mark Brenner** Labor Notes Magazine • **Vernon M. Briggs** Cornell University • **Byron W. Brown** Michigan State University • **Christopher Brown** Arkansas State University • **Clair Brown** University of California — Berkeley • **Philip H. Brown** Colby College • **Michael Brun** Illinois State University • **Neil H. Buchanan** Rutgers School of Law and New York University School of Law • **Robert Buchele** Smith College • **Stephen Buckles** Vanderbilt University • **Stephen V. Burks** University Of Minnesota — Morris • **Joyce Burnette** Wabash College • **Paul D. Bush** California State University — Fresno • **Alison Butler** Wilamette University • **Antonio G. Callari** Franklin and Marshall College • **Al Campbell** University of Utah • **James Campen** University of Massachusetts — Boston • **Maria Cancian** University of Wisconsin — Madison • **Paul Cantor** Norwalk Community College • **Anthony Carnevale** National Center on Education and the Economy • **Jeffrey P. Carpenter** Middlebury College • **Francoise Carre** University of Massachusetts — Boston • **Michael J. Carter** University of Massachusetts — Lowell • **Susan B. Carter** University of California — Riverside • **Karl E. Case** Wellesley College • **J. Dennis Chasse** State University of New York — Brockport • **Howard Chernick** Hunter College, City University of New York • **Robert Cherry** Brooklyn College — City University of New York • **Graciela Chichilnisky** Columbia University • **Lawrence Chimerine** Radnor International Consulting, Inc. • **Menzie D. Chinn** University of Wisconsin — Madison • **Charles R. Chittle** Bowling Green State University • **Kimberly Christensen** State University of New York — Purchase • **Richard D. Coe** New College of Florida • **Robert M. Coen** Northwestern University • **Steve Cohn** Knox College • **Rachel Connelly** Bowdoin College • **Karen Smith Conway** University of New Hampshire • **Patrick Conway** University of North Carolina — Chapel Hill • **David R. Cormier** West Virginia University • **James V. Cornehls** University of Texas — Arlington • **Richard R. Cornwall** Middlebury College • **Paul N. Courant** University of Michigan — Ann Arbor • **James R. Crotty** University of Massachusetts — Amherst • **James M. Cypher** California State University — Fresno • **Douglas Dalenberg** University of Montana • **Herman E. Daly** University of Maryland • **Anita Dancs** National Priorities Project • **Nasser Daneshvary** University of Nevada — Las Vegas • **David Danning** University of Massachusetts — Boston • **Sheldon Danziger** University of Michigan — Ann Arbor • **Jane D'Arista** Financial Markets Center • **Paul Davidson** The New School for Social Research • **Jayne Dean** Wagner College • **Gregory E. DeFreitas** Hofstra University • **Bradford DeLong** University of California — Berkeley

• **James G. Devine** Loyola Marymount College • **Ranjit S. Dighe** State University of New York — Oswego • **John DiNardo** University of Michigan — Ann Arbor • **Randall Dodd** Financial Policy Forum • **Peter B. Doeringer** Boston University • **Peter Dorman** Evergreen State College • **Robert Drago** Pennsylvania State University • **Laura Dresser** University of Wisconsin • **Richard B. Du Boff** Bryn Mawr College • **Arindrajit Dube** University of California — Berkeley • **Marie Duggan** Keene State College • **Lloyd J. Dumas** University of Texas — Dallas • **Christopher Dunn** Earth and Its People Foundation • **Steven N. Durlauf** University of Wisconsin — Madison • **Amitava K. Dutt** University of Notre Dame • **Jan Dutta** Rutgers University • **Gary A. Dymski** University of California — Riverside • **Peter J. Eaton** University of Missouri — Kansas City • **Fritz Efaw** University of Tennessee — Chattanooga • **Catherine S. Elliott** New College of Florida • **Richard W. England** University of New Hampshire • **Ernie Englander** George Washington University • **Gerald Epstein** University of Massachusetts — Amherst • **Sharon J. Erenburg** Eastern Michigan University • **Susan L. Ettner** University of California — Los Angeles • **Linda Ewing** United Auto Workers • **Colleen A. Fahy** Assumption College • **Loretta Fairchild** Nebraska Wesleyan University • **David Fairris** University of California — Riverside • **Warren E. Farb** International Capital Mobility Domestic Investment • **Martin Farnham** University of Victoria • **Jeff Faux** Economic Policy Institute • **Sasan Fayazmanesh** California State University — Fresno • **Rashi Fein** Harvard Medical School • **Robert M. Feinberg** American University • **Susan F. Feiner** University of Southern Maine • **Marshall Feldman** University of Rhode Island • **Marianne A. Ferber** University of Illinois — Urbana — Champaign • **William D. Ferguson** Grinnell College • **Rudy Fichtenbaum** Wright State University • **Deborah M. Figart** Richard Stockton College • **Bart D. Finzel** University of Minnesota — Morris • **Lydia Fischer** United Auto Workers, retired • **Peter Fisher** University of Iowa • **John Fitzgerald** Bowdoin College • **Sean Flaherty** Franklin and Marshall College • **Kenneth Flamm** University of Texas — Austin • **Maria S. Floro** American University • **Nancy Folbre** University of Massachusetts — Amherst • **Christina M. Fong** Carnegie Mellon University • **Catherine Forman** Quinnipiac University • **Harold A. Forman** United Food and Commercial Workers • **Mathew Forstater** University of Missouri — Kansas City • **Liana Fox** Economic Policy Institute • **Donald G. Freeman** Sam Houston State University • **Gerald Friedman** University of Massachusetts — Amherst • **Sheldon Friedman** AFL — CIO • **Alan Frishman** Hobart and William Smith Colleges • **Scott T. Fullwiler** Wartburg College • **Kevin Furey** Chemeketa Community College • **Jason Furman** New York University • **David Gabel** Queens College • **James K. Galbraith** University of Texas — Austin • **Monica Galizzi** University of Massachusetts — Lowell • **David E. Gallo** California State University — Chico • **Byron Gangnes** University of Hawaii — Manoa • **Irwin Garfinkel** Columbia University • **Rob Garnett** Texas Christian University • **Garance Genicot** Georgetown University • **Christophre Georges** Hamilton College • **Malcolm Getz** Vanderbilt University • **Teresa Ghilarducci** University of Notre Dame • **Karen J.**

Gibson Portland State University • **Richard J. Gilbert** University of California — Berkeley • **Helen Lachs Ginsburg** Brooklyn College — City University of New York • **Herbert Gintis** University of Massachusetts — Amherst • **Neil Gladstein** International Association of Machinists and Aerospace Workers • **Amy Glasmeier** Penn State University • **Norman J. Glickman** Rutgers University • **Robert Glover** University of Texas — Austin • **Arthur S. Goldberger** University of Wisconsin — Madison • **Lonnie Golden** Penn State University — Abington College • **Dan Goldhaber** University of Washington • **Marshall I. Goldman** Wellesley College • **Steven M. Goldman** University of California — Berkeley • **William W. Goldsmith** Cornell University • **Donald Goldstein** Allegheny College • **Nance Goldstein** University of Southern Maine • **Nick Gomersall** Luther College • **Eban S. Goodstein** Lewis and Clark College • **Neva Goodwin** Tufts University • **Roger Gordon** University of California — San Diego • **Peter Gottschalk** Boston College • **Elise Gould** Economic Policy Institute • **Harvey Gram** Queens College, City University of New York • **Jim Grant** Lewis & Clark College • **Ulla Grapard** Colgate University • **Daphne Greenwood** University of Colorado — Colorado Springs • **Karl Gregory** Oakland University • **Christopher Gunn** Hobart and William Smith Colleges • **Steven C. Hackett** Humboldt State University • **Joseph E. Harrington** Johns Hopkins University • **Douglas N. Harris** Florida State University • **Jonathan M. Harris** Tufts University • **Martin Hart-Landsberg** Lewis & Clark College • **Robert Haveman** University of Wisconsin — Madison • **Sue Headlee** American University • **Carol E. Heim** University of Massachusetts — Amherst • **James Heintz** University of Massachusetts — Amherst • **Paul A. Heise** Lebanon Valley College • **Susan Helper** Case Western Reserve University • **John F. Henry** University of Missouri — Kansas City • **Barry Herman** The New School • **Edward S. Herman** University of Pennsylvania • **Guillermo E. Herrera** Bowdoin College • **Joni Hersch** Vanderbilt University Law School • **Thomas Hertel** Purdue University • **Steven Herzenberg** Keystone Research Center • **Donald D. Hester** University of Wisconsin — Madison • **Gillian Hewitson** Franklin and Marshall College • **Bert G. Hickman** Stanford University • **Marianne T. Hill** Center for Policy Research and Planning • **Martha S. Hill** University of Michigan — Ann Arbor • **Michael G. Hillard** University of Southern Maine • **Rod Hissong** University of Texas — Arlington • **P. Sai-Wing Ho** University of Denver • **Emily P. Hoffman** Western Michigan University • **Harry J. Holzer** Georgetown University and Urban Institute • **Marjorie Honig** Hunter College, City University of New York • **Barbara E. Hopkins** Wright State University • **Mark R. Hopkins** Gettysburg College • **Ann Horowitz** University of Florida • **Ismael Hossein-Zadeh** Drake University • **Charles W. Howe** University of Colorado — Boulder • **Candace Howes** Connecticut College • **Frank M. Howland** Wabash College • **David C. Huffman** Bridgewater College • **Saul H. Hymans** University of Michigan — Ann Arbor • **Frederick S. Inaba** Washington State University • **Alan G. Isaac** American University • **Doreen Isenberg** University of Redlands • **Jonathan Isham** Middlebury College • **Sanford M. Jacoby** University of California — Los Angeles •

Robert G. James California State University — Chico • **Kenneth P. Jameson** University of Utah • **Russell A. Janis** University of Massachusetts — Amherst • **Elizabeth J. Jensen** Hamilton College • **Pascale Joassart** University of Massachusetts — Boston • **Jerome Joffe** St. John's University • **Laurie Johnson** University of Denver • **William Johnson** Arizona State University • **Lawrence D. Jones** University of British Columbia • **Alexander J. Julius** New York University • **Bernard Jump** Syracuse University • **Fadhel Kaboub** Drew University • **Shulamit Kahn** Boston University • **Linda Kamas** Santa Clara University • **Sheila B. Kamerman** Columbia University • **John Kane** State University of New York — Oswego • **Billie Kanter** California State University — Chico • **J.K. Kapler** University of Massachusetts — Boston • **Roger T. Kaufman** Smith College • **David E. Kaun** University of California — Santa Cruz • **Thomas A. Kemp** University of Wisconsin — Eau Claire • **Peter B. Kenen** Princeton University • **Farida C. Khan** University of Wisconsin — Parkside • **Kwan S. Kim** University of Notre Dame • **Marlene Kim** University of Massachusetts — Boston • **Christopher T. King** University of Texas — Austin • **Mary C. King** Portland State University • **Lori G. Kletzer** University of California — Santa Cruz • **Janet T. Knoedler** Bucknell University • **Tim Koechlin** Vassar College • **Andrew I. Kohen** James Madison University • **Denise Eby Konan** University of Hawaii — Manoa • **Ebru Kongar** Dickinson Colleg • **James Konow** Loyola Marymount University • **Krishna Kool** University of Rio Grande • **Douglas Koritz** Buffalo State College • **Daniel J. Kovenock** Purdue University • **Kate Krause** University of New Mexico • **Vadaken N. Krishnan** Bowling Green State University • **Douglas Kruse** Rutgers University • **David Laibman** Brooklyn College — City University of New York • **Robert M. La-Jeunesse** University of Newcastle • **Kevin Lang** Boston University • **Catherine Langlois** Georgetown University • **Mehrene Larudee** DePaul University • **Gary A. Latanich** Arkansas State University • **Robert Z. Lawrence** Harvard University — Kennedy School of Government • **Daniel Lawson** Drew University • **William Lazonick** University of Massachusetts — Lowell • **Joelle J. Leclaire** Buffalo State College • **Frederic S. Lee** University of Missouri — Kansas City • **Marvin Lee** San Jose State University • **Sang-Hyop Lee** University of Hawaii — Manoa • **Woojin Lee** University of Massachusetts — Amherst • **Thomas D. Legg** University of Minnesota • **J. Paul Leigh** University of California — Davis • **Charles Levenstein** University of Massachusetts — Lowell • **Margaret C. Levenstein** University of Michigan — Ann Arbor • **Henry M. Levin** Columbia University • **Herbert S. Levine** University of Pennsylvania • **Mark Levinson** Economic Policy Institute • **Oren M. Levin-Waldman** Metropolitan College of New York • **Mark K. Levitan** Community Service Society of New York • **Stephen Levy** Center for Continuing Study of California Economy • **Arthur Lewbel** Boston College • **Lynne Y. Lewis** Bates College • **David L. Lindauer** Wellesley College • **Victor D. Lippit** University of California — Riverside • **Pamela J. Loprest** Urban Institute • **Richard Lotspeich** Indiana State University • **Michael C. Lovell** Wesleyan University • **Milton Lower** Retired Senior Economist, US House of Representatives • **Stephanie**

Luce University of Massachusetts — Amherst • **Robert Lucore** United American Nurses • **Jens Otto Ludwig** Georgetown University • **Dan Luria** Michigan Manufacturing Technology Center • **Devon Lynch** University of Denver • **Lisa M. Lynch** Tufts University • **Robert G. Lynch** Washington College • **Catherine Lynde** University of Massachusetts — Boston • **Arthur MacEwan** University of Massachusetts — Boston • **Hasan MacNeil** California State University — Chico • **Allan MacNeill** Webster University • **Craig R. MacPhee** University of Nebraska — Lincoln • **Diane J. Macunovich** University of Redlands • **Janice F. Madden** University of Pennsylvania • **Mark H. Maier** Glendale Community College • **Thomas N. Maloney** University of Utah • **Jay R. Mandle** Colgate University • **Andrea Maneschi** Vanderbilt University • **Mar** San Francisco State University • **Dave E. Marcotte** University of Maryland — Baltimore County • **Robert A. Margo** Boston University • **Ann R. Markusen** University of Minnesota — Twin Cities • **Ray Marshall** University of Texas LBJ School of Public Affairs • **Stephen Martin** Purdue University • **Patrick L. Mason** Florida State University • **Thomas Masterson** Westfield State College • **Julie A. Matthaei** Wellesley College • **Peter Hans Matthews** Middlebury College • **Anne Mayhew** University of Tennessee — Knoxville • **Alan K. McAdams** Cornell University • **Timothy D. McBride** St. Louis University School of Public Health • **Elaine McCrate** University of Vermont • **Kate McGovern** Springfield College • **Richard D. McGrath** Armstrong Atlantic State University • **Richard McIntyre** University of Rhode Island • **Hannah McKinney** Kalamazoo College • **Judith Record McKinney** Hobart and William Smith Colleges • **Andrew McLennan** University of Sydney • **Charles W. McMillion** MBG Information Services • **Ellen Meara** Harvard Medical School • **Martin Melkonian** Hofstra University • **Jo Beth Mertens** Hobart and William Smith Colleges • **Peter B. Meyer** University of Louisville and Northern Kentucky University • **Thomas R. Michl** Colgate University • **Edward Miguel** University of California — Berkeley • **William Milberg** The New School • **John A. Miller** Wheaton College • **S.M. Miller** Cambridge Institute and Boston University • **Jerry Miner** Syracuse University • **Daniel J.B. Mitchel** University of California — Los Angeles • **Edward B. Montgomery** University of Maryland • **Sarah Montgomery** Mount Holyoke College • **Robert E. Moore** Georgia State University • **Barbara A. Morgan** Johns Hopkins University • **John R. Morris** University of Colorado — Denver • **Monique Morrissey** Economic Policy Institute • **Lawrence B. Morse** North Carolina A&T State University • **Saeed Mortazavi** Humboldt State University • **Fred Moseley** Mount Holyoke College • **Philip I. Moss** University of Massachusetts — Lowell • **Tracy Mott** University of Denver • **Steven D. Mullins** Drury University • **Alicia H. Munnell** Boston College • **Richard J. Murnane** Harvard University • **Matthew D. Murphy** Gainesville State College • **Michael Murray** Bates College • **Peggy B. Musgrave** University of California — Santa Cruz • **Richard A. Musgrave** Harvard University • **Ellen Mutari** Richard Stockton College • **Sirisha Naidu** Wright State University • **Michele Naples** The College of New Jersey • **Tara Natarajan** St. Michael's College • **Julie A. Nelson** Tufts University • **Reynold F.**

Nesiba Augustana College • **Donald A. Nichols** University of Wisconsin — Madison • **Eric Nilsson** California State University — San Bernardino • **Laurie Nisonoff** Hampshire College • **Emily Northrop** Southwestern University • **Bruce Norton** San Antonio College • **Stephen A. O'Connell** Swarthmore College • **Mehmet Odekon** Skidmore College • **Paulette Olson** Wright State University • **Paul Ong** University of California — Los Angeles • **Van Doorn Ooms** Committee for Economic Development • **Jonathan M. Orszag** Competition Policy Associates, Inc. • **Paul Osterman** Massachusetts Institute of Technology • **Shaianne T. Osterreich** Ithaca College • **Rudolph A. Oswald** George Meany Labor Studies Center • **Spencer J. Pack** Connecticut College • **Arnold Packer** Johns Hopkins University • **Dimitri B. Papadimitriou** The Levy Economic Institute of Bard College ✸ **James A. Parrott** Fiscal Policy Institute • **Manuel Pastor** University of California — Santa Cruz • **Eva A. Paus** Mount Holyoke College • **Jim Peach** New Mexico State University • **M. Stephen Pendleton** Buffalo State College • **Michael Perelman** California State University — Chico • **Kenneth Peres** Communications Workers of America • **George L. Perry** The Brookings Institution • **Joseph Persky** University of Illinois — Chicago • **Karen A. Pfeifer** Smith College • **Bruce Pietrykowski** University of Michigan — Dearborn • **Michael J. Piore** Massachusetts Institute of Technology • **Karen R. Polenske** Massachusetts Institute of Technology • **Robert Pollin** University of Massachusetts — Amherst • **Marshall Pomer** Macroeconomic Policy Institute • **Tod Porter** Youngstown State University • **Shirley L. Porterfield** University of Missouri — St. Louis • **Michael J. Potepan** San Francisco State University • **Marilyn Power** Sarah Lawrence College • **Thomas Power** University of Montana • **Robert E. Prasch** Middlebury College • **Mark A. Price** Keystone Research Center • **Jean L. Pyle** University of Massachusetts — Lowell • **Paddy Quick** St. Francis College • **John M. Quigley** University of California — Berkeley • **Willard W. Radell, Jr.** Indiana University of Pennsylvania • **Fredric Raines** Washington University in St. Louis • **Steven Raphael** University of California — Berkeley • **Salim Rashid** University of Illinois — Urbana — Champaign • **Wendy L. Rayack** Wesleyan University • **Randall Reback** Barnard College, Columbia University • **Robert Rebelein** Vassar College • **James B. Rebitzer** Case Western Reserve University • **Daniel I. Rees** University of Colorado — Denver • **Michael Reich** University of California — Berkeley • **Robert B. Reich** University of California — Berkeley • **Cordelia Reimers** Hunter College and The Graduate Center — City University of New York • **Donald Renner** Minnesota State University — Mankato • **Trudi Renwick** Fiscal Policy Institute • **Andrew Reschovsky** University of Wisconsin — Madison • **Lee A. Reynis** University of New Mexico • **Daniel Richards** Tufts University • **Bruce Roberts** University of Southern Maine • **Barbara J. Robles** Arizona State University • **John Roche** St. John Fisher College • **Charles P. Rock** Rollins College • **William M. Rodgers III** Rutgers University • **Dani Rodrik** Harvard University • **John E. Roemer** Yale University • **William O. Rohlf** Drury University • **Gerard Roland** University of California — Berkeley • **Frank Roosevelt** Sarah

Lawrence College • **Jaime Ros** University of Notre Dame • **Nancy E. Rose** California State University — San Bernardino • **Howard F. Rosen** Trade Adjustment Assistance Coalition • **Joshua L. Rosenbloom** University of Kansas • **William W. Ross** Fu Associates, Ltd. • **Roy J. Rotheim** Skidmore College • **Jesse Rothstein** Princeton University • **Geoffrey Rothwell** Stanford University • **Joydeep Roy** Economic Policy Institute • **David Runsten** Community Alliance with Family Farmers • **Lynda Rush** California State Polytechnic University — Pomona • **Gregory M. Saltzman** Albion College and the University of Michigan • **Sydney Saltzman** Cornell University • **Dominick Salvatore** Fordham University • **Blair Sandler** San Francisco, California • **Daniel E. Saros** Valparaiso University • **Michael Sattinger** University at Albany • **Dawn Saunders** Castleton State College • **Larry Sawers** American University • **Max Sawicky** Economic Policy Institute • **Peter V. Schaeffer** West Virginia University • **William C. Schaniel** University of West Georgia • **A. Allan Schmid** Michigan State University • **Stephen J. Schmidt** Union College • **John Schmitt** Center for Economic and Policy Research • **Juliet B. Schor** Boston College • **C. Heike Schotten** University of Massachusetts — Boston • **Eric A. Schutz** Rollins College • **Elliot Sclar** Columbia University • **Allen J. Scott** University of California — Los Angeles • **Bruce R. Scott** Harvard Business School • **Robert Scott** Economic Policy Institute • **Stephanie Seguino** University of Vermont • **Laurence Seidman** University of Delaware • **Janet Seiz** Grinnell College • **Willi Semmler** The New School • **Mina Zeynep Senses** Johns Hopkins University • **Jean Shackelford** Bucknell University • **Harry G. Shaffer** University of Kansas • **Sumitra Shah** St. John's University • **Robert J. Shapiro** Sonecon LLC • **Mohammed Sharif** University of Rhode Island • **Lois B. Shaw** Institute for Women's Policy Research • **Heidi Shierholz** University of Toronto • **Deep Shikha** College of St. Catherine • **Richard L. Shirey** Siena College • **Steven Shulman** Colorado State University • **Laurence Shute** California State Polytechnic University — Pomona • **Stephen J. Silvia** American University • **Michael E. Simmons** North Carolina A&T State University • **Margaret C. Simms** Joint Center for Political and Economic Studies • **Chris Skelley** Rollins College • **Max J. Skidmore** University of Missouri — Kansas City • **Peter Skott** University of Massachusetts — Amherst • **Courtenay M. Slater** Arlington, Virginia • **Timothy M. Smeeding** Syracuse University • **Janet Spitz** College of Saint Rose • **William Spriggs** Howard University • **James L. Starkey** University of Rhode Island • **Martha A. Starr** American University • **Howard Stein** University of Michigan — Ann Arbor • **Mary Huff Stevenson** University of Massachusetts — Boston • **James B. Stewart** Pennsylvania State University • **Jeffrey Stewart** Northern Kentucky University • **Robert J. Stonebraker** Winthrop University • **Michael Storper** University of California — Los Angeles • **Diana Strassmann** Rice University • **Cornelia J. Strawser** Consultant • **Frederick R. Strobel** New College of Florida • **James I. Sturgeon** University of Missouri — Kansas City • **David M. Sturges** Colgate University • **William A. Sundstrom** Santa Clara University • **Jonathan Sunshine** Reston, Virginia • **Paul Swaim** Organisation for Economic Co-operation and Development • **Craig Swan**

University of Minnesota — Twin Cities • **Paul A. Swanson** William Paterson University • **William K. Tabb** Queens College • **Peter Temin** Massachusetts Institute of Technology • **Judith Tendler** Massachusetts Instittue of Technology • **David Terkla** University of Massachusetts — Boston • **Kenneth Thomas** University of Missouri — St. Louis • **Frank Thompson** University of Michigan — Ann Arbor • **Ross D. Thomson** University of Vermont • **Emanuel D. Thorne** Brooklyn College — City University of New York • **Jill Tiefenthaler** Colgate University • **Thomas H. Tietenberg** Colby College • **Chris Tilly** University of Massachusetts — Lowell • **Renee Toback** Empire State College • **Mayo C. Toruño** California State University — San Bernardino • **W. Scott Trees** Siena College • **A. Dale Tussing** Syracuse University • **James Tybout** Penn State University • **Christopher Udry** Yale University • **Daniel A. Underwood** Peninsula College • **Lynn Unruh** University of Central Florida • **Leanne Ussher** Queens College, City University of New York • **David Vail** Bowdoin College • **Vivian Grace Valdmanis** University of the Sciences in Philadelphia • **William Van Lear** Belmont Abbey College • **Lane Vanderslice** Hunger Notes • **Lise Vesterlund** University of Pittsburgh • **Michael G. Vogt** Eastern Michigan University • **Paula B. Voos** Rutgers University • **Mark Votruba** Case Western Reserve University • **Susan Vroman** Georgetown University • **Howard M. Wachtel** American University • **Jeffrey Waddoups** University of Nevada — Las Vegas • **Norman Waitzman** University of Utah • **Lawrence A. Waldman** University of New Mexico • **John F. Walker** Portland State University • **William Waller** Hobart and William Smith Colleges • **Jennifer Warlick** University of Notre Dame • **Matthew Warning** University of Puget Sound • **Bernard Wasow** The Century Foundation • **Robert W. Wassmer** California State University — Sacramento • **Sidney Weintraub** Center for Strategic and International Studies • **Mark Weisbrot** Center for Economic and Policy Research • **Charles L. Weise** Gettysburg College • **Thomas E. Weisskopf** University of Michigan — Ann Arbor • **Christian E. Weller** Center for American Progress • **Fred M. Westfield** Vanderbilt University • **Charles J. Whalen** Perspectives on Work • **Cathleen L. Whiting** Williamette University • **Howard Wial** The Brookings Institution • **Linda Wilcox Young** Southern Oregon University • **Arthur R. Williams** Rochester — Minnesota • **Robert G. Williams** Guilford College • **John Willoughby** American University • **Valerie Rawlston Wilson** National Urban League • **Jon D. Wisman** American University • **Barbara L. Wolfe** University of Wisconsin — Madison • **Edward Wolff** New York University • **Martin Wolfson** University of Notre Dame • **Brenda Wyss** Wheaton College • **Yavuz Yasar** University of Denver • **Anne Yeagle** University of Utah • **Erinc Yelden** University of Massachusetts — Amherst • **Ben E. Young** University of Missouri — Kansas City • **Edward G. Young** University of Wisconsin — Eau Claire • **June Zaccone** National Jobs for All Coalition and Hofstra University • **Ajit Zacharias** Levy Economics Institute of Bard College • **David A. Zalewski** Providence College • **Henry W. Zaretsky** Henry W. Zaretsky & Associates, Inc. • **Jim Zelenski** Regis University • **Andrew Zimbalist** Smith College • **John Zysman** University of California — Berkeley

Source: http://www.epi.org/minwage/epi_minimum_wage_2006.pdf

References

Block, W and MA Walker (1988). Entropy in the Canadian economics profession: Sampling consensus on the major issues, *Canadian Public Policy*, XIV(2), 137–150.

DiLorenzo, T (September 6, 2004). Do capitalist have superior bargaining power? *Mises.org Daily Articles*. http://www.mises.org/story/1602 [December 6, 2007].

Economic Policy Institute (2006). Hundreds of Economists Says: Raise the Minimum Wage. Washington, DC: EPI. http://www.epi.org/content.cfm/minwagestmt2006

Frey, BS, WW Pommerehne, F Schneider and Guy Gilbert (1984). Consensus and dissension among economists: An empirical inquiry, *American Economic Review*, 74(5), 986–994.

IV. Immigration

A LIBERTARIAN CASE
FOR FREE IMMIGRATION

Walter Block[*]

"None are too many."—Reply of an anonymous senior official in the government of Canadian Prime Minister McKenzie King to the question, "How many Jews fleeing Nazi Germany should be allowed into this country?"[1]

All merchants shall have safe and secure exit from England and entry to England, with the right to tarry there and to move about as well by land as by water, for buying and selling by the ancient and right customs, quite from all evil tolls, except (in time of war) such merchants as are of the land at war with us. And if such are found in our land at the beginning of the war, they shall be detained, without injury to their bodies or goods, until information be received by us, or by our chief justiciar, how the merchants of our land found in the land at war with us are treated; and if our men are safe there, the others shall be safe in our land.[2]

In this paper I will attempt to analyze laws limiting emigration, migration, and immigration from the libertarian perspective. I will defend the view that the totally free movement of goods, factors of production, money, and, most important of all, people, is part and parcel of this traditional libertarian

---

[*]Walter Block is chairman of the economics and finance department at the University of Central Arkansas, and is a senior fellow of the Ludwig von Mises Institute. He is co-editor of *The Journal of Libertarian Studies* and the *Quarterly Journal of Austrian Economics*, and has served as editor of other scholarly journals. Among the books he has written are *Defending the Undefendable, Amending the Combines Investigation Act*, and *The U.S. Bishops and their Critics: An Economic and Ethical Perspective*. He is the co-author of *Lexicon of Economic Thought*, and *Economic Freedom of the World, 1975–1995*, among other works.

I would like to thank David Kennedy and Anthony Sullivan of the Earhart Foundation, and their colleagues on the Board of Directors, for financial support during the summer of 1997, which was instrumental in the writing of this article. The views expressed herein are of course mine, not theirs. I would like to thank Professors Larry White and Arnold Aberman for several suggestions which greatly improved this paper.

[1]Irving Abella and Harold Troper, *None is Too Many: Canada and the Jews of Europe, 1933–1948* (Toronto: Lester and Orpen Dennys, 1982), p. ix. I owe this citation to Phil Bryden and Jenny Forbes.

[2]From chapter 41 of the Magna Carta, cited in Samuel E. Thorne, et al., *The Great Charter: Four Essays on Magna Carta and the History of Our Liberty* (New York: Pantheon, 1965), p. 133. I wish to thank Ralph Raico for bringing this quotation to my attention.

Journal of Libertarian Studies 13:2 (Summer 1998): 167–186

philosophy. Like tariffs and exchange controls, migration barriers of whatever type are egregious violations of laissez-faire capitalism.

I begin by briefly reviewing the libertarian philosophical perspective. Next, I appraise each of these elements of the movement of peoples from this vantage point. Then, I consider—and reject—a series of possible objections. I conclude with an overview.

LIBERTARIANISM

Libertarianism is a political philosophy; as such, it is a theory of the just use of violence. Here, the legitimate utilization of force is only defensive: one may employ arms only to repel an invasion, i.e., to protect one's person and his property from external physical threat, and for no other reason. According to Murray N. Rothbard:

> The libertarian creed rests upon one central axiom: that no man or group of men may aggress against the person or property of anyone else. This may be called the "non-aggression axiom." "Aggression" is defined as the initiation of the use or threat of physical violence against the person or property of anyone else. Aggression is therefore synonymous with invasion.
>
> If no man may aggress against another, if, in short, everyone has the absolute right to be "free" from aggression, then this at once implies that the libertarian stands foursquare for what are generally known as "civil liberties": the freedom to speak, publish, assemble, and to engage in . . . "victimless crimes."[3]

I shall contend that emigration, migration, and immigration all fall under the rubric of "victimless crime." That is, not a one of these three *per se* violates the non-aggression axiom.[4] Therefore, at least for the libertarian, no restrictions or prohibitions whatsoever should be placed in the path of these essentially peaceful activities.

Before considering the specifics, let us clear the decks of one possible misconception: that the libertarian can be a "moderate"

---

[3]Murray N. Rothbard, *For a New Liberty* (New York: Macmillan, 1978), p. 23. For another definitive vision of libertarianism, see Hans-Hermann Hoppe, *The Economics and Ethics of Private Property: Studies in Political Economy and Philosophy* (Boston: Kluwer, 1993).

[4]For a listing of dozens of other archetypes, none of which necessarily violate the libertarian non-aggression axiom, and all of which are reviled by many, see Walter Block, *Defending the Undefendable* (New York: Fox and Wilkes, [1976] 1985).

on this question, advocating fully opening the borders at some times, completely closing them on other occasions, and leaving them slightly ajar if it seems warranted. Typically, such a policy is advocated based on considerations of assimilation, as in the following statement of "plain-spoken reasoning" by William F. Buckley:

> At various points in history we have opened, and then gently closed, our borders, pending economic and social assimilation. If there is dogged unemployment, there is no manifest need for more labor. If pockets of immigrants are resisting the assimilation that over generations has been the solvent of American citizenship, then energies should go to accosting multiculturalism, rather than encouraging its increase.[5]

Such a position, whatever its merits on other grounds, is simply not available to the libertarian, who requires consistency with Rothbard's non-aggression axiom. Pragmatic matters such as assimilation can form no part of the libertarian world view. The only issue is: do emigration, migration, and immigration constitute, *per se*, a physical trespass against person and property or a threat thereof? If so, then libertarians must oppose them totally; if not, they must oppose any and all limits to them. There does not appear to be any middle ground or compromise position consistent with libertarianism.[6] That is, if the transfer of peoples does indeed constitute a violation of the libertarian axiom, as does murder, rape, theft, etc., then it must be completely prohibited. There can be no countenance for *partially* restricted immigration,[7] any more than for *partially* restricted murder. Buckley-type pragmatism applied to murder would mean that in some decades there should be no law at all opposing this heinous act, in other epochs we should very strictly prohibit it, and that in still other time periods we should adopt a more moderate position, perhaps allowing only a certain number of murders. Perhaps our choice should be dictated by life expectancy, or numbers of elderly people in the population.[8] Say what you will about the

---

[5]William F. Buckley, "Immigration Advocates Resist Reasoning," *Conservative Chronicle* (February 12, 1997): 20.

[6]For the view that at least on some issues the libertarian position occupies a middle ground, or compromise, see Walter Block, "Compromising the Uncompromisable: the Austrian Golden Mean," *Cultural Dynamics* 9, no. 2 (July 1997): 211–38.

[7]We are here implicitly assuming that the migrant will find a private property owner who is willing to take him in. Below, we subject this assumption to intensive examination.

[8]This would constitute a "modest proposal" for solving the Ponzi scheme elements of social security bankruptcy. We could hold "open season" on retirees, while protecting the lives of those still in the labor force.

pragmatic benefits of this idea, it clearly falls outside the purview of libertarians.

Rothbard makes much the same point in another context:

> "Economic power," then, is simply the right under freedom to refuse to make an exchange. Every man has this power. Every man has the same right to refuse to make a preferred exchange.
>
> Now, it should become evident that the "middle-of-the-road" statist, who concedes the evil of violence but adds that the violence of government is sometimes necessary to counteract the "private coercion of economic power," is caught in an impossible contradiction. A refuses to make an exchange with B. What are we to say, or what is the government to do, if B brandishes a gun and orders A to make the exchange? This is the crucial question. There are only two positions we may take on the matter: *either* that B is committing violence and should be stopped at once, *or* that B is perfectly justified in taking this step because he is simply "counteracting the subtle coercion" of economic power wielded by A. Either the defense agency must rush to the defense of A, or it deliberately refuses to do so, perhaps aiding B (or doing B's work for him). *There is no middle ground!*[9]

The identical situation exists with regard to migration. Here, A, the migrant, is peacefully coming to visit his friend or relative in another land.[10] Whereupon B pounces on him, and forces him at the point of a gun to return to his place of origin. What should the libertarian defense agency do? Again, *there is no middle ground!* It must either support A or B. It cannot possibly do both.

The legality of migration is an all-or-none matter: either migration is *per se* legitimate, in which case it would be improper to interfere with it in any way, or it is *per se* invasive, in which case it should be prohibited, totally and comprehensively, just as in the case of murder and rape.

EMIGRATION

Ponder the barriers to emigration which long existed behind the Iron Curtain, and still do for countries such as North Korea and Cuba. Civilized people of all ideological dispositions regard

---

[9]Murray N. Rothbard, *Power and Market: Government and the Economy* (Menlo Park, Calif.: Institute for Humane Studies, 1977), p. 229, emphasis in original.

[10]For how long? Who knows? Whose business is it anyway?

these as barbarous relics from the past—harking back to a time of serfdom, or actual slavery. A country which will not allow its citizens to leave is nothing better than a vast jail, no matter how many Olympic medals the prisoners may have won, no matter how many Sputniks the inmates may have launched.

Such a stark statement can be made on the basis of the libertarian philosophy. For here, people own themselves absolutely. It is a moral outrage for them to be enslaved by the state. Restrictions are sometimes justified on the grounds that would-be emigrants have benefitted from public education, provided free of charge by the government. They are compelled to pay exit fees, or are prohibited from leaving outright, on the ground that they will take with them information given to them by the state, which continues to be "its" property. Since there is no way to leave without taking this education with them, the emigrants are prohibited from departing. We in the west, for the most part, see this merely as an excuse for a quasi-slave system—as an attempt to cover unlawful imprisonment with a thin veneer of legitimacy and property rights. But no state provides education to the populace "for free." On the contrary, schooling is financed from funds taken from the people in the first place, though taxes.

Even if, somehow, the government gave education to the citizens for free, it would still not follow that governments are entitled to enslave them on this ground. No, the only slavery even arguably compatible with libertarianism would be that agreed to in advance by freely contracting parties—a sort of "indentured servitude" for life. But no such contracts have ever been signed. Thus, there is no warrant to assume that the hapless people suffering from Communism or Nazism were treated appropriately,[11] even under our heroic assumption of "free" education.

As a matter of fact, one may interpret the curious historical institution of slavery[12] along these lines. That is, chattel slavery is but a special instance of the lack of freedom to emigrate. What makes it slavery is that the slave cannot quit, or emigrate from the situation, any time he feels like picking up and leaving. If he could, it would not be slavery but merely a peculiar voluntary employment contract. In other words, the right of emigration is so important that its absence implies outright slavery.

---

[11]For another analysis of the view that state actions can be justified on the basis of a "contract" which was never signed, see Lysander Spooner, *No Treason: The Constitution of No Authority* (Larkspur, Colo.: Ralph Myles, [1870] 1966).

[12]See Jeffrey Rogers Hummel, *Emancipating Slaves, Enslaving Free Men: A History of the American Civil War* (Chicago: Open Court, 1996), for a thorough-going analysis of slavery.

There is a further connection between emigration and immigration. Suppose the world contained one totalitarian country, while the rest of them were "free." If all other nations enact immigration prohibitions, this is tantamount to the imposition of emigration restrictions by the government of the one country from which people wish to flee. While it is a basic implication of the libertarian non-aggression axiom that people have a right to emigrate, at least one other nation must allow them to immigrate, or the exercise of this right will become impossible as a practical matter.

MIGRATION

If there is to be a third category, migration, distinct from immigration and emigration, then it must be confined to that aspect of travel during which a person is neither under the control of the host country (emigration) nor the receiving one (immigration). It would apply to the ocean, after the migrant has vacated the country of origin, e.g., Cuba, or traveled to that small no-man's land or demilitarized zone between such places as North and South Korea.

There is no real difficulty for the libertarian in such a case. Shooting down a fleeing family in cold blood, no matter which nation is doing the killing, is murder. Should this have to be said, that murder is contrary to the libertarian axiom of non-aggression? Within limits, it matters not one whit why the persons involved are escaping—whether for economic reasons, or to have a freer life, or because they are tired of totalitarianism.[13]

IMMIGRATION

A moment's reflection will convince any disinterested party that immigration is not necessarily invasive. Immigration consists of no more than moving to a foreign country. For the purist libertarian, national boundaries are only lines on a map, demarcating one "country" from another; there is no such thing as a legitimate nation-state. According to Rothbard:

> [T]here can be no such thing as an "international trade"
> problem. For nations might then possibly continue as

---

[13]Of course, if they are themselves murderers, and are escaping to another country in order to avoid paying the just penalties for their foul deeds, or are escaping with private property stolen from its rightful owners, this is an entirely different matter. No longer do we have here innocent people merely attempting to better their own lives. Now, the "migrants" are themselves the criminals.

Walter Block – A Libertarian Case for Free Immigration   173

cultural expressions, but not as economically meaningful units. Since there would be neither trade nor other barriers between nations nor currency differences, "international trade" would become a mere appendage to a general study of interspatial trade. It would not matter whether the trade was within or outside a nation.[14]

Therefore, immigration across national boundaries should be analyzed in an identical manner to that migration which takes place within a country. If it is non-invasive for Jones to change his locale from one place in Misesania to another in that country, then it cannot be invasive for him to move from Rothbardania to Misesania. Alternatively, if migration across international borders is somehow illegitimate, this should apply to the domestic variety as well.

As long as the immigrant moves to a piece of private property whose owner is willing to take him in (maybe for a fee), there can be nothing untoward about such a transaction. This, along with all other capitalist acts between consenting adults, must be considered valid in the libertarian world. Note that there is no freedom of movement of the person *per se*. This is always subject to the willingness of property owners in the host nation to accept the immigrant onto their land. Rothbard explains:

> [T]he private ownership of all streets would resolve the problem of the "human right" to freedom of immigration. There is no question about the fact the current immigration barriers restrict, not so much a "human right" to immigrate, but the right of property owners to rent or sell property to immigrants. There can be no human right to immigrate, for on *whose* property does someone else have the right to trample? In short, if "Primus" wishes to migrate now from some other country to the United States, we cannot say that he has the absolute right to immigrate to this land area; for what of those property owners who don't *want* him on their property? On the other hand, there may be, and undoubtedly are, other property owners who would jump at the chance to rent or sell property to Primus, and the current laws now invade their property rights by preventing them from doing so.[15]

It is almost a certainty that there will in fact always be "other property owners who would jump at the chance to rent or

---

[14]Murray N. Rothbard, *Man, Economy, and State: A Treatise on Economic Principles* (Auburn, Ala.: Ludwig von Mises Institute, [1962] 1993), p. 550. See also Rothbard, *For A New Liberty;* and Ludwig von Mises, *Nation, State, and Economy*, Leland Yeager, trans. (New York: New York University Press, 1983).

[15]Murray N. Rothbard, *The Ethics of Liberty* (Atlantic Highlands, N.J.: Humanities Press, 1982), p. 119.

sell property to" immigrants. If this is not obvious based on common sense experience, the economics of discrimination suggests no other possible conclusion.[16] If there are many owners who refuse to rent or sell to immigrants, the price the latter will have to pay will be high. But this will tend to induce those landowners on the margin to agree to accept immigrants. It must be the rare case indeed where in a country of millions of property owners there is not a single one willing to accept newcomers, even at the very highest prices they are willing to pay. In such a rare case, all those who adhere to libertarianism must indeed unite in opposing immigration,[17] for, with Rothbard, there is no one "on whose property . . . someone else ha[s] the right to trample."

But this is a theoretical curiosity, not something relevant to reality, or to public policy analysis. In real world countries, certainly including the U.S., there can be found thousands, if not millions, of landowners willing to sell or rent space to people from all parts of the globe, no matter how obscure. For example, restaurateurs specializing in the foods common to foreign lands may wish to hire authentic foreign-born cooks. As a practical matter, it is inconceivable that some citizen property owners, whose families themselves immigrated in the past, would not be interested in taking in their countrymen, particularly at the very high remuneration available if most landlords do not wish to deal with the immigrants.

The case is equally clear for allowing immigrants to settle on unowned land. When there is virgin territory, there is no legitimate reason for immigrants (or domestic citizens) to be prevented from bringing it into fruitful production. States Rothbard: "*Everyone* should have the right to appropriate as his property previously unowned land or other resources."[18] "Everyone," presumably, includes immigrants as well as citizens or residents of the home country.

Mises, from a utilitarian rather than a natural-rights libertarian position, considered immigration an important element of freedom and progress:

---

[16]On this topic, see Gary Becker, *The Economics of Discrimination* (Chicago: University of Chicago Press, 1957); Thomas Sowell, *Race and Economics* (New York: Longman, 1975); idem, *The Economics and Politics of Race: An International Perspective* (New York: Morrow, 1983).

[17]That is, opposing it totally, as private property rights violations. However, even in this case there would be no need for a law prohibiting immigration, only one banning trespass in general.

[18]Rothbard, *Ethics of Liberty*, p. 240, emphasis added. See also Hoppe, *Economics and Ethics of Private Property*.

> The principles of freedom, which have gradually been
> gaining ground everywhere since the eighteenth century,
> gave people freedom of movement. The growing security
> of law facilitates capital movements, improvement of
> transportation facilities, and the location of production
> away from the points of consumption. That coincides—
> not by chance—with a great revolution in the entire
> technique of production and with drawing the entire
> earth's surface into world trade. The world is gradually
> approaching a condition of free movement of persons
> and capital goods.[19]

One last point under this topic. If immigration were *per se* invasive, then, perhaps with the exception of Indians,[20] as Americans are all either immigrants or descended from them, our occupancy of this country would be legally questionable. Since no advocate of immigration restrictions has ever expressed any such reservations, there is a problem of logical consistency here.

OBJECTIONS

Let us now deal with several possible objections to the foregoing.

Allowing unrestricted immigration is equivalent to allowing the invasion of a foreign army

States Mises:

> Under present conditions, the adoption of a policy of
> outright laissez faire and laissez passer on the part of
> the civilized nations of the West would be equivalent to
> an unconditional surrender to the totalitarian nations.
> Take, for instance, the case of migration barriers. Unre-
> strictedly opening the doors of the Americas, of Austral-
> ia, and of Western Europe to immigrants would today be
> equivalent to opening the doors to the vanguards of the
> armies of Germany, Italy, and Japan.[21]

It must be remembered that these words were first published in 1944, and written some time before that; hence, perhaps, the fear of military invasion. But this appears to be an idiosyncratic use of language. No advocate of laissez-faire capitalism ever conceived of this position as anything akin to total pacifism. Unrestricted immigration, in this perspective, does not at all include

---

[19]Mises, *Nation, State, and Economy*, p. 58.

[20]But the ancestors of native peoples, too, had to come from somewhere. If so, then they, too, are at least indirectly "guilty" of the crime of immigration.

[21]Ludwig von Mises, *Omnipotent Government: The Rise of the Total State and Total War* (New Rochelle, N.Y.: Arlington House, 1969), p. 10.

allowing invading armies *carte blanche* access to the home country.[22] On the contrary, it refers to peaceful settlement therein. It is perfectly consistent with the libertarian philosophy to oppose with the utmost determination an invading army, while throwing completely open the doors to peaceful settlers.

Unrestricted immigration will create or exacerbate unemployment

This objection illustrates nothing so much as economic illiteracy. It assumes that there is only so much work in a nation to be done, and that if immigrants do more of it, there will be just that much left for present occupants. If it were true, any and every technological advance would prove a dire threat to our economy.[23] For example, the pick and shovel, to say nothing of the truck, can do the work of thousands of people, compared to tea spoons, or, better yet, bare fingernails. Are we to rid ourselves of these technological advances in order to improve our economy, and combat unemployment? Hardly.

Unrestricted immigration will reduce the real wages of the workers already in residence

This contention, more perhaps than any other, explains the vicious opposition to immigration traditionally displayed by union leaders such as Cesar Chavez.[24] This charge, however, cannot be denied; it is true that under some circumstances, workers in the receiving country (and capital and land in the country or origin), will lose out.[25] Conversely, capitalists and land owners in the receiving country gain from the cooperation of a larger supply of labor, and workers remaining in the country of origin gain from the increased local scarcity of their services.

The owner of any resource, labor or any other, tends to be subject to a loss in wealth, at least relatively, when confronted by increasing supplies of a substitute factor of production. It is possible that these losses as a producer will be more than offset by gains as a consumer (due to the lower prices of final goods), but this need not at all be the case. It is also possible for an individual domestic worker's loss in wages to be more than offset by

---

[22]This would apply, also, to carriers of communicable diseases. They are in effect, if not by intention, an "invading army" in that if they are allowed in the recipient country, they will spread their germs to innocent people.

[23]See Henry Hazlitt, *Economics in One Lesson* (New York: Arlington House, 1979); Mises, *Omnipotent Government*, p. 105; and Julian Simon, *The Economic Consequences of Immigration* (Oxford: Basil Blackwell, 1989).

[24]See Ira Mehlman, "Funding Fraud," *National Review* (March 24, 1997): 30.

[25]See Ludwig von Mises, *Human Action*, 3rd rev. ed. (Chicago: Regnery, 1966), pp. 377, 627.

gains in his invested wealth (e.g., he may have pension funds invested in stock ownership), but again, this need not be the case.

But as Hoppe has shown, people have the right only to the physical aspects of their property, not to its value.[26] For the latter is determined on the market by the human actions of thousands of people, exercising their demand for and providing supplies of commodities. To say that X has a right to the value of his property is thus to say that he has a right to make economic decisions for these thousands of other people whose choices determine the value of his property, a manifest absurdity.

Unrestricted immigration will increase crime

There is no doubt that were the U.S. to open its doors to all and sundry, some number of criminals would take advantage of this opportunity. There are good "pickings" to be had here, after all.

But this is really an indictment of our criminal justice system,[27] not of open immigration. Nowadays, liberals wax eloquent about the cost of crime. The bill for incarcerating a criminal exceeds that of tuition at our most prestigious universities. When one imagines hordes of immigrants coming to this country, committing crimes, and then putting additional strain on our very limited supply of jails, it is easy to contemplate the closing of the borders.

In actuality, a libertarian society serious about crime would not experience so much of it in the first place. For one thing, it would legalize drugs. It is the prohibition, not the use of drugs, that leads to criminal behavior. The very high prices of illegal drugs which are due to prohibition, and are not intrinsic to addictive substances themselves, serve as a magnet for the underworld. When alcohol was prohibited, it was associated with criminal gang activity; when it was legalized, this connection was cut asunder.[28]

---

[26]Hoppe, *Economics and Ethics of Private Property.*

[27]See Rothbard, *Power and Market;* also idem, *For A New Liberty,* pp. 215–41.

[28]For more on this point, see Walter Block, "Drug Prohibition: A Legal and Economic Analysis," *Journal of Business Ethics* 12 (1993): 107–18; idem, "Drug Prohibition, Individual Virtue, and Positive Economics," *Review of Political Economy* 8, no. 4 (October 1996): 433–36; David Boaz, ed., *The Crisis in Drug Prohibition* (Washington, D.C.: Cato Institute, 1990); Milton Friedman, "An Open Letter to Bill Bennett," *Wall Street Journal* (September 7, 1989); Ronald Hamowy, ed., *Dealing With Drugs: Consequences of Government Control* (San Francisco: Pacific Institute, 1987); Thomas Szasz, *Ceremonial Chemistry: The Ritual Persecution of Drugs, Addicts, and Pushers,* rev. ed. (Holmes Beach Fl.: Learning Publications, 1985); and Mark Thornton, *The Economics of Prohibition* (Salt Lake City: University of Utah Press, 1991).

A libertarian society, moreover, would get tougher on genuine criminals. There would be no more cozy jails with color TVs, air conditioning, or recreation rooms. If indentured servitude for convicts were brought back, prisons could be run by private enterprise. Instead of draining taxpayers of vast amounts of money to house inmates, they could turn a profit.

Under such a system, apart from the undoubted harm they would perpetrate on their victims, immigrants who become criminals would not cost "society" a dime. On the contrary, through their sweat and tears they could be forced to make a positive contribution.

Unrestricted immigration will promote welfarism

The argument here is that immigrants come to our shores not to breathe the heady wine of economic freedom, but to avail themselves of our stupendously generous welfare system. This is not so much a quarrel with immigration as it is with welfare. Says Hoppe in this context:

> It would also be wrongheaded to attack the case for free immigration by pointing out that because of the existence of a welfare state, immigration has become, to a significant extent, the immigration of welfare bums, who, even if the United States is below the optimal population point, do not increase but rather decrease average living standards. For this is not an argument against immigration but rather against the welfare state. To be sure, the welfare state should be destroyed, root and branch. However, the problems of immigration and welfare are analytically distinct, and they must be treated accordingly.[29]

Let it be said loudly and clearly: end welfare for all people, but at the very least for immigrants and their descendants, and by definition immigrants will no longer be attracted to our shores in order to receive such funds.

But there is another problem with this line of argument: it proves far too much. For if immigrants are to be prohibited from entry into this country on the ground that they *might in the future* go on the welfare rolls, and thus in effect steal from the long-suffering taxpayer, Pandora's box will be flung wide open. If we can physically invade people for what they *might* do in the future,[30] the sky is the limit. Surely we can engage in preventive

---

[29]Hans-Hermann Hoppe, "Free Immigration or Forced Integration," *Chronicles* 19, no. 7 (July 1995): 25.

[30]Make no mistake about it: an immigration barrier is a physical invasion of innocent people. Here comes Mr. X, say, from Turkey, peaceably going about his

Walter Block – A Libertarian Case for Free Immigration   179

detention of all teenaged males—the guilty along with the innocent—on the ground that this cohort commits more than its proportionate share of crimes. But surely this would be a great injustice.

And what about child bearing for the present occupants of
this great country of ours? It cannot be denied that any children
born today might, some years into the future, avail themselves of
our welfare program. But if we can preclude the entry of immigrants on this ground, this goes as well for having babies. Becoming pregnant ought to be a crime, on these grounds. At least the
Chinese Communists limited people to one child per couple. If
opponents of totally open immigration on the ground that they
might become welfare recipients are logically consistent, they
would have to oppose any childbearing, whatever.[31]

***Legally unrestricted immigration is indeed the libertarian
position, the only possible libertarian position, but it should
not be implemented until the every other plank in this
program is first put into effect***

This is a very powerful objection to the argument being presented here. For suppose unlimited immigration is made the order of the day while minimum wages, unions, welfare, and a law
code soft on criminals are still in place in the host country. Then,
it might well be maintained, the host nation would be subjected
to increased crime, welfarism, and unemployment. An open-door
policy would imply not economic freedom, but forced integration
with all the dregs of the world with enough money to reach our
shores.

However strong this objection may be, Rothbard, albeit arguing in another context, provides us with the definitive rebuttal.[32]
Rothbard noted that Alan Greenspan, in his youth, was a strong
advocate of a gold standard,[33] but as head of the Fed, never from

---

business of settling on the land of his cousin in Arkansas, for example. Yet, before
he can get there, the minions of the government interfere with his peaceful right
of passage, and either jail him or forcibly return him to his country of origin.

[31]One might argue that the rich could escape this implication, by, say, posting a
bond so that their children never need become welfare recipients. This might or
might not work, depending upon such matters as future inflation, productivity,
and precisely how these bonds are financed. In any case, however, it would also be
possible to eliminate this argument by merely requiring that all *immigrants* sign an
agreement never to go on welfare, and/or by posting a bond so that *their* children
never need become welfare recipients.

[32]Murray N. Rothbard, "Alan Greenspan: A Minority Report on the New Fed
Chairman," *The Free Market* (August 1987).

[33]Alan Greenspan, "Gold and Economic Freedom," in *Capitalism: The Unknown
Ideal*, Ayn Rand, ed., (New York: Signet: 1967), pp. 96–101 (reprinted from *The
Objectivist*, 1966).

him a word of this has been heard.[34] Has Greenspan changed his mind? Or is he a total hypocrite? In Rothbard's view, neither is true. On the contrary, Greenspan *does* favor laissez-faire capitalism and gold, but only on a high philosophical level where he doesn't have to *do* anything about it. In contrast, he does not champion it as a practical matter, for then he would be called upon to show some evidence of his beliefs. Says Rothbard:

> There is one thing, however, that makes Greenspan unique, and that sets him off from (the) Establishment. . . . And that is that he is a follower of Ayn Rand, and therefore "philosophically" believes in laissez faire and even the gold standard. But as *The New York Times* and other important media hastened to assure us, Alan only believes in laissez faire "on the high philosophical level." In *practice* in the policies he advocates, he is a centrist like everyone else because he is a "pragmatist." . . .[35]

> Thus, Greenspan is only in favor of the gold standard if all conditions are right: if the budget is balanced, trade is free, inflation is licked, everyone has the right philosophy, etc. In the same way, he might say he only favors free trade if all conditions are right: if the budget is balanced, unions are weak, we have a gold standard, the right philosophy, etc. In short, *never* are one's "high philosophical principles" applied to one's actions. It becomes almost piquant for the Establishment to have this man in its camp.

This, it must be acknowledged, is a devastating critique of the Greenspan position. And, if this be so, we cannot avoid the conclusion that the same argument constitutes a knock-out blow against the defense of immigration restrictions on the ground that every other aspect of full free enterprise must be reached.

There is a certain pattern underlying the position of these "postponement libertarians," the paleo-libertarians who favor full, free, open, and unrestricted immigration—but only after the entire libertarian vision has been attained. The underlying coherence of this perspective is that we should, whenever possible, attempt to achieve the same results now, under statism, as would ensue were we to be living in the fully free society.

Take the case of the bum in the library. What, if anything, should be done about him? If this is a private library, then the plumb-line or pure libertarian would agree fully with his paleo

---

[34]Although see William R. Bradford, "Greenspan: Deep-Cover Radical for Capitalism?" *Liberty* 11, no. 2, (November 1997), p. 40.
[35]Rothbard, "Alan Greenspan," p. 3.

cousin: throw the bum out! More specifically, the law should *allow* the owner of the library to forcibly evict such a person, if need be, at his own discretion. Cognizance would be taken of the fact that if the proprietor allowed this smelly person to occupy his premises, he would soon be forced into bankruptcy, as normal paying customers would avoid his establishment like the plague.

But what if it is a public library? Here, the paleos and their libertarian colleagues part company. The latter would argue that the public libraries are *per se* illegitimate. As such, they are akin to an unowned good. Any occupant has as much right to them as any other. If we are in a revolutionary state of war, then the first homesteader may seize control. But if not, as at present, then, given "just war" considerations, any reasonable interference with public property would be legitimate.[36]

The paleos or postponement libertarians take a sharply divergent view: one should treat these libraries in as close an approximation as possible to how they would be used in the fully free society. Since, on that happy day, the overwhelmingly likely scenario is that they will be owned by a profit maximizer who will have a "no bums" policy,[37] this is exactly how the public library should be treated right now. Namely, what we should do to the bum in the public library today is exactly what would be done to him by the private owner: kick him out.

There are difficulties with this stance. First, as we have already seen, it is extremely likely that in the fully free society, virtually all immigrants would be taken in by a landowner in the host country. Therefore, if the paleos are to remain consistent with their own position, they should eschew all legislated immigration barriers.

Secondly, and even apart from this consideration, the postponement libertarian perspective is vulnerable to rebuttal by *reductio ad absurdum*. If we should not allow unrestricted immigration until we have achieved the free society, but instead should curtail immigration in an effort to approximate what would take place under a fully libertarian society, let us apply this insight to other realms of controversy.

---

[36]One could "stink up" the library with unwashed body odor, or leave litter around in it, or "liberate" some books, but one could not plant land mines on the premises to blow up innocent library users.

[37]Consider the Body Shop, or Ben and Jerry's Ice Cream, or any other "ethically oriented" company. Even they operate in this manner. That is, they may donate a part of their profits to unsavory enterprises (from a libertarian point of view), but they presumably do not employ "bum" types of people in the manufacture of their products.

Public schooling is a disaster. Certainly, in the present journal, there is no need to document such a claim.[38] That being the case, the libertarian position is clear: get rid of public education, forthwith, even if we have not attained complete liberty in other sectors of society.

But those who would be true to the paleo-libertarian position on immigration cannot avail themselves of this conclusion. Instead, they would have to ask: what would education be like in the free society? They would then have to endeavor to treat public schools as much like that as possible.[39] But if there is one thing that is clear, it is that in the free society the educational industry, like all others, would allow competition. How, then, to apply this principle? Simple. Embrace educational vouchers. Get in harness with those such as Milton Friedman who have long advocated this form of competition for the public schools.[40]

Here is a second example. The U.S. welfare policy is a disaster. The libertarian position is once again crystal clear: abolish welfare forthwith, no matter what the status of the remainder of the economy.[41] But the paleo or postponement libertarians are once again precluded from embracing so clear, just, and simple a solution. If they are to remain true to their immigration position, they will have to reason as follows: the problem with welfare as presently constituted is that it has a built-in marginal tax rate of 100%. If the dole is now $500 per week and the recipient earns a salary of $100, this payment will be reduced to $400, leaving the welfare "client" no better off financially. But thanks to Milton

---

[38]But for a curious and very limited defense of public education, see Michael Levin, *Why Race Matters* (Westport, Conn.: Praeger, 1977).

[39]The postponement libertarians could not advocate privatizing all public schools since the remainder of the economy is not yet fully free. They are limited to treating public property in manner as similar as possible to how it would be used in the free society.

[40]Milton Friedman, *Capitalism and Freedom* (Chicago: University of Chicago Press, 1962).

[41]True, it is harsh on the poor to totally eliminate welfare while the minimum wage, anti-peddler laws, etc., are still in effect. But two wrongs do not make a right. Just because the state victimizes the poor by making it illegal, and thus difficult, to earn money, does not make it right for government to injure a second group of people, taxpayers, and demand money from them at the point of a gun so as to transfer some of their funds to the first set of victims. In any case, were welfare to be totally eliminated right now, this would set up irresistible forces to end such employment barriers. This is analogous to the case for eliminating immigration restrictions on behalf of breaking up welfare. If hordes of poor foreigners poured onto our shores in order to take advantage of generous welfare provisions, that would immeasurably hasten the day that they were eliminated.

Friedman's negative income tax plan,[42] this problem can be overcome.[43]

Unrestricted immigration will assault the institutions which make a free society possible in the first place

This, too, is a very powerful objection, for it cannot be denied that many of the people who might enter the U.S. under an open-door policy come from parts of the world where freedom is non-existent, unheard of, or denigrated.

Nevertheless, the case for free immigration is not without a response. First of all, the U.S. is no longer the freest country in the world, if it ever was; there are several others which beat us out for this honorific.[44] Thus, not all immigrants are likely to be less conducive to freedom than are we.[45] Second, there have been immigrants in our history who have improved our freedom immeasurably. The names Ludwig von Mises, Friedrich A. Hayek, Israel Kirzner, William Hutt, Ludwig Lachmann, Hans Hoppe, Yuri Maltsev, Kurt Leube, James Ahiakpor, George Ayittey, Nathaniel Branden, Barbara Branden, Sam Konkin, Harry Watson, David Henderson, and Ayn Rand leap immediately to mind in this context. A closed-door policy in the past might well have made it impossible for these people to contribute to our society. And this is to say nothing of all the children and grandchildren of immigrants who have made significant contributions. How could it be otherwise, given that virtually all of us are "the children and grandchildren of immigrants"?

Third, just how, precisely, is it contemplated that the new immigrants will bring in to disrepute the mores, habits, and institutions which undergird our liberties? The most likely method is through voting. That is, hordes of people from other continents will come to our shores, settle down, and then vote for Nazism,

---

[42]Friedman, *Capitalism and Freedom.*

[43]Let it be remembered that each of these examples is a *reductio* of the paleo position on immigration. I certainly do not favor school vouchers or the negative income tax. My claim is only that if the postponement libertarians remain true to their views on immigration, logic will force them to embrace these latter positions as well.

[44]See James Gwartney, Robert Lawson, and Walter Block, *Economic Freedom of the World, 1975–1995* (Vancouver, B.C.: Fraser Institute, 1996).

[45]The restrictionist might reply: let us limit immigration to those countries which are actually freer than our own. But people who come to the U.S. from totalitarian countries are likely to do so because they *dislike* such regimes. Many of the strongest supporters of freedom in the U.S. are first and second generation Polish-Americans, Lithuanian-Americans, Cuban-Americans, etc. In any case, this is a mere empirical issue, unworthy, perhaps, of even noting. Underlying it, at least for the libertarian, is that immigration is a victimless crime, and should no more be legally banned than should prostitution or drug use.

Communism, welfare statism, or some such. It cannot be rationally denied that this is a plausible scenario. The only problem
with it is that it, again, assumes a real world situation (one with
a welfare state, a pro-criminal penal system, etc.) instead of the
ideal libertarian one. It is crucial that this not be done, however.
For if it is, we conflate these other issues with that of immigration; we are seemingly arguing against an open-door policy, when
actually, our real problem is with welfarism, criminal coddling,
etc. If we are to generate a *libertarian* theory of immigration, we
must argue in a *ceteris paribus* manner.[46]

It is the same in this case.[47] The real difficulty here concerns
promiscuous voting, not immigrants who might vote "incorrect-
ly." The problem, even apart from new entrants to our country, is
that those who are already citizens now have the "right" to
vote on, not whether or not, but how much of other people's property they can legally steal through the ballot box. *This* is the
real threat to liberty. In a free society, all the wrong-thinking
immigrants in the world would be powerless to overturn (what is
left of) our free institutions, for there would be no possibility of
voting to seize other people's property.

But suppose these foreign hordes enter our pure libertarian
society where no such decisions are even allowed to be politically contemplated, let alone enacted, and then proceed to do just
that. After all, at one time in our history we were far more free
than we are now. It was people—many of them, it must be conceded, immigrants—who undermined our free institutions.

One answer is that we never had a fully libertarian society.
Had we, the courts would have ruled against any property-grab-
bing initiative or referendum. The police would have dealt firm-
ly with any property-destroying or denigrating riots engaged in
by Communist or Nazi or welfarist immigrants. On the assumption that these foreigners were civilians, not an actual invading
army, there seems little reason to believe they would have succeeded in their nefarious "foreign" schemes.

But suppose they did, somehow, overturn us, in a fully peaceful manner (perhaps through the sheer eloquence of their orato-
ry), so that no physical sanctions against them would be compatible with libertarianism. Then what? Then we have a division

---

[46]Rothbard, in *For A New Liberty,* pp. 238–39, argues in this way when he refutes
the objection of the Russian menace to a stateless U.S. society.

[47]This objection is but a variation of the first one considered, above. Only now instead of bearing rifles, the invading "army" will be issued votes, as soon as they
have been naturalized.

between the libertarian axiom of non-aggression and what might considered, from the pragmatic or utilitarian point of view, to be the good society.

But this should occasion no surprise or any embarrassment for the libertarian position. If you pack enough into your assumptions, you can overturn any principle, even an entirely appropriate and valid one such as the libertarian non-aggression axiom. For example, suppose that the all-powerful "Martians" threaten that unless we kill innocent person A, they will blow up the world. Surely, then, we would be presented with a stark choice indeed: violate the libertarian basic premise, or bring forth the end to all human life. One response might be "Justice though the heavens fall!" Another might be to say that the libertarian axiom is pro-life in all but such contrived situations. Or, to treat more realistic scenarios where utilitarianism and libertarianism might diverge, there is the fact that if we outlawed homosexuality, or engaged in preventive detention of male teenagers, we would undoubtedly reduce the incidence of AIDS and crime, respectively. Happily, we shrink back from such perversions of justice, because of elemental decency, e.g., adherence to the libertarian non-aggression axiom. Should we do any less in the case of immigration? Certainly not.

CONCLUSION

If one is against immigration, there are ways to reduce it which are fully compatible with libertarianism. For one thing, unilaterally declare full free trade with all nations. Trade in goods, services, and capital is an economic substitute for immigration. That is, there are two ways to right any imbalance between capital and labor: bring labor to the areas where population is below its optimum size (immigration), and bring capital and goods to the areas where they are below their optimum sizes (free trade in capital and goods). As the latter are typically far cheaper than the former, a regime of full free trade would eliminate much of the economic incentive toward migration.

Are libertarians moderates or extremists on the issues of emigration, migration, and immigration? The libertarian position on migration does *not* constitute a compromise in that it is indubitably an all-or-none proposition: either migration is totally legitimate, in which case there should be no interferences with it whatsoever, or it is a violation of the non-aggression axiom, in which case it should be banned, fully. I have argued in this paper that the former position is the only correct one.

But libertarianism constitutes a compromise position on this issue in two other senses. First, immigration is allowed if and only if there are property owners willing to sponsor (presumably for a fee, but not necessarily so) the new entrants, and not otherwise. Second, there are people on both right and left who oppose borders totally open to peaceful settlement (Chavez, Buckley), and libertarians find themselves safely on the other side of this unholy alliance.

For example, states Buckley: "The idea of totally open borders—anybody who wants to can come on in—is the stuff of libertarian fancy, nice for tone poems by such as Ayn Rand, but not very good national policy."[48]

It is not often that viewpoints are so starkly contrasted. We have, at least in this case, achieved real disagreement. It is clear that whatever the merits of this conservative perspective, it is not a libertarian one. Buckley is absolutely correct in labeling this Ayn Rand viewpoint as libertarian—and no one who dissents from it can to that extent call himself a libertarian.

---

[48]Buckely, "Immigration Advocates Resist Reasoning," p. 20.

Is There a Right to Immigration?: A Libertarian Perspective

Walter Block and Gene Callahan

Introduction

The question of whether there is a right to immigrate, and, if so, what limits may exist on that right, is a controversial one for rights theorists. Even thinkers sharing many fundamental normative principles can come to radically different policy conclusions on immigration. To illustrate some of the difficulties with the issue, we will examine the controversy over immigration among those who adhere to the classic liberal view on the primacy of property rights for normative politics—what is often, today, called libertarianism.

There are some who take the position that any compromise whatsoever with free and unrestricted immigration must perforce be ruled incompatible with libertarianism. After all, the immigrant, merely by appearing at our shores, particularly at the invitation of a citizen and property owner, cannot be said by that fact alone to have initiated violence against an innocent person.[1] Not being guilty of a violation of the basic libertarian principle of not initiating aggression, there is no justification for visiting any violence upon him. Since forceful removal from our shores would indeed constitute an initiation of force against him, this would be improper.

A number of libertarians argue to the contrary.[2] We will employ Hoppe (1995, 1998, 2001) as our paradigmatic example, both because of the clarity of his arguments and his prominence among those making this case. Hoppe maintains that there can no more be a libertarian defense for unrestricted immigration than there can be one for unrestricted trespass, or for forced integration, or for the violation of the law of free association, or for the elimination of property rights. Just as trespass and coerced integration violate private property rights, so does "free" immigration.

Hoppe's Case for Restricted Immigration

Let us consider the specifics. Hoppe (1999) begins his analysis by considering what the situation would be in regards to "immigration" in a stateless society based on private property:

All land is privately owned, including all streets, rivers, airports, harbors, etc. With respect to some pieces of land, the property title may be unrestricted; that is, the owner is permitted to do with his property whatever he pleases as long as he does not physically damage the property owned by others. With respect to other territories, the property title may be more or less severely restricted. As is currently the case in some housing developments, the owner may be bound by contractual limitations on what he can do with his property (voluntary zoning), which might include residential vs. commercial use, no buildings more than four stories high, no sale or rent to Jews, Germans, Catholics, homosexuals, Haitians, families with or without children, or smokers, for example.

Clearly, under this scenario there exists no such thing as freedom of immigration. Rather, there exists the freedom of many independent private property owners to admit or exclude others from their own property in accordance with their own unrestricted or restricted property titles. Admission to some territories might be easy, while to others it might be nearly impossible. In any case, however, admission to the property of the admitting person does not imply a "freedom to move around," unless other property owners consent to such movements. There will be as much immigration or non-immigration, inclusivity or exclusivity, desegregation or segregation, non-discrimination or discrimination based on racial, ethnic, linguistic, religious, cultural or whatever other grounds as individual owners or associations of individual owners allow.

Hoppe then examines our current situation, where various states assert control over all of the land on earth:

In an anarcho-capitalist society there is no government and, accordingly, no clear-cut distinction between inlanders (domestic citizens) and foreigners. This distinction comes into existence only with the establishment of a government, i.e., an institution which possesses a territorial monopoly of aggression (taxation). The territory over which a government's taxing power extends becomes "inland," and everyone residing outside of this territory becomes a foreigner. [The existence of s]tate borders . . . implies a two-fold distortion with respect to peoples' natural inclination to associate with others. First, inlanders cannot exclude the government (the taxman) from their own property, but are subject to what one might call "forced integration" by government agents. Second, in order to be able to intrude on its subjects' private property so as to tax them, a government must invariably take control of existing roads, and it will employ its tax revenue to produce even more roads to gain even better access to all private property, as a potential tax source. Thus, this over-production of roads . . . involves forced domestic integration (artificial desegregation of separate localities).

Moreover, with the establishment of a government and state borders, immigration takes on an entirely new meaning. Immigration becomes immigration by foreigners across state borders, and the decision as to whether or not a person should be admitted no longer rests with private property owners or associations of such owners but with the government as the ultimate sovereign of all domestic residents and the ultimate super-owner of all their properties. Now, if the government excludes a person while even one domestic resident wants to admit this very person onto his property, the result is forced exclusion (a phenomenon that does not exist under private property anarchism). Furthermore, if the government admits a person while there is not even one domestic resident who wants to have this person on his property, the result is forced integration (also non-existent under private property anarchism).

And Hoppe contends that democratic governments may have especially perverse incentives regarding immigration:

For a democratic ruler, it also matters little whether bums or geniuses, below or above-average civilized and productive people immigrate into the country. Nor is he much concerned about the distinction between temporary workers (owners of work permits) and permanent, property owning immigrants (naturalized citizens). In fact, bums and unproductive people may well be preferable as residents and citizens, because they cause more so-called "social problems," and democratic rulers thrive on the existence of such problems. Moreover, bums and inferior people will likely support his egalitarian policies, whereas geniuses and superior people will not.

So what, for Hoppe, does this imply in regards to a preferred immigration policy in a democratic state?

What should one hope for and advocate as the relatively correct immigration policy, however, as long as the democratic central state is still in place and successfully arrogates the power to determine a uniform national immigration policy? The best one may hope for, even if it goes against the "nature" of a democracy and thus is not very likely to happen, is that the democratic rulers act as if they were the personal owners of the country and as if they had to decide who to include and who to exclude from their own personal property (into their very own houses). This means following a policy of utmost discrimination: of strict discrimination in favor of the human qualities of skill, character, and cultural compatibility.

More specifically, it means distinguishing strictly between "citizens" (naturalized immigrants) and "resident aliens" and excluding the latter from all welfare entitlements. It means requiring as necessary, for resident alien status as well as for citizenship, the personal sponsorship by a resident citizen and his assumption of liability for all property damage caused by the immigrant. It implies requiring an existing employment contract with a resident citizen; moreover, for both categories but especially that of citizenship, it implies that all immigrants must demonstrate through tests not only (English) language proficiency, but all-around superior (above-average) intellectual performance and character structure as well as a compatible system of values—with the predictable result of a systematic pro-European immigration bias.

As we see it, while Hoppe voices some valid concerns about immigration under the democratic welfare states that dominate North America and Europe today, his solution is at odds with the libertarian view of human rights, and contains errors in its analysis of the entrepreneurial function of the ruler of a private domain. Let us explore our difficulties with Hoppe's view.

The Monarchical Ruler as Entrepreneur

As we have seen, Hoppe hopes that a democratic government, for instance that of the United States, will act like an entrepreneur running the firm "USA Inc." when deciding immigration policy. There is a major problem with the case he attempts to build from that postulate: his analysis of how entrepreneurs behave in their efforts to assemble the factors of production is flawed.

For one thing, entrepreneurs do not attempt to acquire *superior* or *above-average* factors of production—they attempt to acquire the *most profitable* factors of production. Only an extremely foolish entrepreneur would, upon determining his business will need 1000 computers, tell his purchasing manager to acquire the 1000 best computers, or even 1000 above-average computers. A wise entrepreneur will attempt to acquire just those 1000 computers such that the difference between the cost of the computer and the projected value of the product it can produce is the greatest.

Similarly, no sensible entrepreneur, on determining he needs 1000 people to staff his company, tells his HR person to hire 1000 superior people. In order to maximize profits, HR should hire the 1000 people for whom there is the greatest difference between the cost of employing them and the projected value of their output. No company will hire an MIT graduate as its janitor, or a person with an IQ of 150 to answer the phone. It is true that some firms, such as a small, creative programming shop, might indeed want to hire *mostly* people who are, for instance, very intelligent. But a company supplying maintenance services to area office buildings would probably have very little interest in the SAT scores of applicants. Similarly, a hereditary monarch who is "staffing" his country, would not want the "best" people, he would want the people who will profit the nation i.e., himself, the most.

At many times and places in history we find illustrations of these facts. The citizens of Athens did not try to hire other Athenian citizens to clean their homes and harvest their olives. The people most fit for such jobs, so they thought, were "barbarians," in other words, less able non-Greeks.

Medieval nobles did not attempt to persuade other dukes and earls to come live on their lands. What they wanted were peasants and craftsmen. The head of an upper class, nineteenth-century British household did not try to find other members of the gentry to serve as scullery maids and gardeners; he hired lower class folk who would do the work he needed done at the lowest possible cost. Southern slave owners did not attempt to buy the slaves most culturally compatible with themselves; they sought slaves who they felt could best endure the harsh conditions of agricultural workers in the Deep South in the summer.

Hoppe seems to ignore the law of comparative advantage when it comes to analyzing immigration. He contends that kings would like to keep "people of inferior productive capabilities" out of their kingdoms. This implies that, if the whole world were privatized, such people would have to leave the planet! But as Mises and many previous economists noted, the great binding force holding human society together is that all people who are able to produce at all, whatever their capabilities may be, always can find a comparative advantage that enables profitable trade with others. Mises (1998: 159–164, 168, 175) found the principle so important that he preferred to call it "the law of association." It is true that anti-social people, whom we might refer to as the "counter-productive," are unwelcome in all societies. That is why we have law and law enforcement. Otherwise, all people cooperate in creating the "Great Society" based on trade and the division of labor.

Here, Hoppe (1999) might protest that he recognizes quite well the advantages offered by free trade. But, he asserts, such trade can take place at a distance:

"Note that none of this, not even the most exclusive form of segregationism, has anything to do with a rejection of free trade and the adoption of protectionism. From the fact that one does not want to associate with or live in the neighborhood of Blacks, Turks, Catholics or Hindus, etc., it does not follow that one does not want to trade with them from a distance."

Certainly, at times the law of comparative advantage will operate such that people are best off trading at a distance, or, at least, that they can successfully do so in this manner. But if a Vietnamese lady would very much prefer doing our wives' nails to working in a rice paddy, and each of our wives' would very much prefer that Vietnamese lady, rather than some gum-chewing teenager from the local high school, perform this service, they will have a rather difficult time realizing these gains from trade if the Vietnamese lady cannot enter the United States.

Hoppe might attempt to answer our objection with his assertion that people with "inferior productive capabilities" would be "be admitted temporarily, if at all, as seasonal workers" by a proprietary sovereign. Perhaps the Vietnamese lady could do our wives' nails six months a year.

But this is merely an arbitrary assumption on his part, without foundation in economics or history. *Sometimes*, it might appear most profitable to the ruler to have the workers come and go. Other times, having them continuously close at hand might seem to be the best strategy. Southern U.S. slave owners did not admit Africans only as migratory workers, shipping them back to Africa each winter; rather, they *bought* them and had them live on their own property, indeed, often in their own houses.

When we examine Hoppe's criterion of "English language proficiency," again we find that it *may* be a consideration for an entrepreneur who is building his workforce, or it may not. If a businessman is hiring people to staff his customer support lines, he certainly will hope they can speak English fairly well. On the other hand, if he needs carpenters, he might be quite happy with an entire crew that speaks only Lithuanian, as long as he can find a bilingual foreman. The criterion of a "compatible system of values" falls to similar analysis: such a qualification is only as important as an entrepreneur deems it to be for his particular project.

The Bum in the Library

We also differ from Hoppe on the relationship of public property to the citizens who ostensibly own such property. We can illustrate our differences by examining the case of "the bum in the library." While Hoppe finds that the state has a right, indeed an obligation, to "throw the bum out," we hold that the bum can be interpreted as homesteading property that is now under illegitimate control, in other words, that is essentially unowned.

Hoppe (2001, 159–160, fn10), citing and criticizing Block (1998, 180–181), says:

"What, if anything, should be done about [the bum in the library]? If this is a private library, . . . the law should *allow* the owner of the library to forcibly evict such a person, if need be, at his own discretion. . . . But what if it is a public library? . . . [Block holds that public libraries] are akin to an unowned good. Any occupant has a much right to them as any other. If we are in a revolutionary state of war, then the first homesteader may seize control. But if not, as at present, then, given 'just war' considerations, any reasonable interference with public property would be legitimate. . . . One could 'stink up' the library with unwashed body odor, or leave litter around in it, or 'liberate' some books, but one could not plant land mines on the premises to blow up innocent library users."

"The fundamental error in this argument, according to which everyone, foreign immigrants no less than domestic bums, has an equal right to domestic public property, is Block's claim that public property 'is akin to an unowned good.' In fact, there exists a fundamental difference between unowned goods and public property. The latter is *de facto* owned by the taxpaying members of the domestic public. They have financed this property; hence, they, in accordance with the amount of taxes paid by individual members, must be regarded as its legitimate owners. Neither the bum, who has presumably paid no taxes, nor any foreigner, who has most definitely not paid any domestic taxes, can thus be assumed to have any rights regarding public property whatsoever."

Our analysis of the bum in the library is very different. Hoppe (2001: 160) avers that the library is *de facto* "owned by the taxpaying members of the domestic public." We believe that this is an error. These premises are, indeed, owned *de jure*[3] by the taxpaying members of the domestic public. But as far as *de facto* is concerned, the real owners are state officials.

The distinction we are making is offered in other words by Rothbard (1990: 241), who makes it in terms of punishment and defense: "In current law, the victim is in even worse straits when it comes to defending the integrity of his own land or movable property. There, he is not allowed to use deadly force in defending his own home, much less other land or properties. The reasoning seems to be that since a victim would not be allowed to kill a thief who steals his watch, he should not be permitted to shoot the thief in the process of stealing the watch or in pursuing him. But punishment and defense of person or property are not the same, and must be treated differently. Punishment is an act of retribution after the crime has been committed and the criminal apprehended, tried, and convicted. Defence [*sic*], while the crime has been committed, or until property is recovered and the criminal apprehended, is a very different story."

In these terms, Hoppe is in effect speaking of punishment, while we are speaking on the basis of defense. That is, our criticism of him is that, in effect, he is confusing the two concepts in this situation.

In order to see the import of this point more clearly, imagine the following scenario: We are partisans fighting the Soviets. We break into a garage of theirs, and are about ready to throw a Molotov cocktail at one of their trucks. Along comes a Russian Hoppean[4] who says to us, "Stop, that truck was paid for with my taxes (among those of many other innocent people); it is, really, in effect, private property, owned by me and my fellow long suffering tax payers. If you destroy that vehicle you are stealing from me." Our answer to this fellow is that he may well indeed be the *de jure* rightful (part) owner of the truck, but as far as the *de facto* situation is concerned, the conveyance is now owned by the Communists, they are using it for altogether nefarious purposes, and we are thus entirely justified in blowing it up. If he persists in his demands that we cease and desist from our altogether righteous behavior, we begin to no longer consider him an innocent victim of taxation, but rather a supporter of that very Stalinist system he purportedly opposes.

Yes to Hoppe, if we are now deciding upon whom, in justice, should be the owners of the library: the bum and the foreigner are way down on the list. But no to Hoppe, a thousand times no, if the bum or the foreigner is *the only one* now attacking this public property. Hoppe is in effect calling upon libertarians to resist *not the state*, but those very people who are now busily attacking it.[5]

The point is, Hoppe is confusing a real life *process* of privatization with the ivory tower libertarian *theory* of how it can best be attained. He in effect conflates a flow and a stock. Yes, under libertarian judicial supervision the library would be turned over to the taxpayers, exactly as articulated by Hoppe. But what are we to make of attempts on the part of other people to seize control over what we consider to be unjustified public property? Are we to reject them, because they do not accord with the theory? Not a bit of it. Very much to the contrary, we as libertarians must *applaud* the transfer of such property from public to private hands, no matter *who* is the owner of the appendages.[6] We can always worry about getting this property into the exact right hands, later. But right now, we are faced with a stark choice between two and only two alternatives: *either* the bum gets to ruin the library (and the partisan blows up the Soviet truck) *or* the status quo ante prevails. When put in these terms, it is not too difficult to discern the proper libertarian answer.

In Rand (1957), the fictional hero John Galt "liberated" money from the government, and gave it to Hank Reardon, a deserving businessman. But suppose a bum, or a foreigner, had acted with regard to the state in exactly the same manner as did Galt, but kept the proceeds for himself. Hoppe, presumably, would oppose such an action; he would do so, presumably, because he thinks that the rightful owners are the taxpayers, and that this would amount to a theft from them. We would, in sharp contrast, *support* this wealth transfer, because based on our reading of libertarianism, while the best outcome would indeed be the one depicted by Rand and implicitly supported by Hoppe, the *second* best would be the scenario where the thieves were deprived of their ill-gotten booty; that is, where the "bum" or the foreigner, relieved the illicit government of this money. The worst alternative of the three, from this perspective, would be the status quo, where the crooks keep the swag.

Our disquiet with Hoppe's analysis is that it makes the best the enemy of the good. To be sure, all of us in this debate favor the first (e.g., Randian) situation, where someone returns the stolen property to the long-suffering taxpayers. But given that this option is unavailable (often the reality), we are forced to choose between leaving the money with the thief or seeing it in the hands of another (e.g., the bum or the foreigner in this context). Which is more consonant with libertarianism? Clearly, it is the *latter*. For an illegitimate government is a thief; it has taken money through the use of force from the proper owners, the taxpayers. In very sharp contrast, neither the bum or the foreigner nor the John Galt character is a *robber*. *They* have not taken property from its rightful holders. Rather, they have *liberated* it from crooks.[7]

Hoppe is undoubtedly correct if we are in the context of a trial run by libertarians, where property is to be allocated to its rightful owners. However, we contend, that at present, we are not in any such situation. Rather, the position is that government now controls these properties, and the libertarian solution is for them to be privatized. We agree with Hoppe, fully, on the goal: complete privatization of all property. But our intellectual opponent acts as if we have, in some sense, *already* attained this objective. Therefore, he opposes the bum[8] who acts so as to ruin the library or the rebel who attempts to blow up the Communist government truck as contrary to this goal. We, on the other hand, realize full well that we have not at all yet attained a situation of complete liberty. We thus welcome, and not only pragmatically, acts which are either intended to undermine the present unjust system, or which have that effect, whether they are intended to do so or not.

What conclusions for immigration policy can we draw from this analysis of the bum in the library case? Simply, contrary to Hoppe, that the foreigner is *not* guilty of a trespass if he seizes or liberates public property. Thus, there is no apodictic argument to be made against his mere presence on our shores (we assume he does not trespass on *private* property.) Immigration is *not* logically equivalent to trespass or forced integration.

There is another difficulty with Hoppe's position. The argument that since the private library owner would throw the bum out, that therefore it is justified for the statist owner to do so too, is incompatible with Austrian insights into the socialist calculation debate. It is not a given that a private library owner would throw out the bum; it is due to Austrian analysis we know that the economist has no particular insight into what is essentially an entrepreneurial decision.

Unowned land

If the bum is justified in taking over the library, it is even easier to see a role for the foreigner with regard to totally unowned property. In the United States there are vast stretches of land west of the Mississippi and in Alaska that are "owned"[9] by government, which have never been taken away from the people by force as have tax revenues (although the government did prevent citizens from homesteading these territories in the first place). Suppose a foreigner locates himself on some of this acreage, and homesteads it, in the teeth of governmental proscriptions to the contrary. That is, the foreigner violates the enactment against homesteading this unowned property, while domestic citizens sit idly by and obey this law. We ask how Hoppe can regard this as trespass on private property, since by stipulation this is *not* private property.

Hoppe (2001: 121–122, fn. 1) states in a different context: "Now, if a man used his body ('labor') in order to appropriate, i.e., bring under his control, some other nature-given things (unowned 'land'), this action demonstrates that he values these things. Hence, he must have gained utility in appropriating them. At the same time, his action does not make anyone else worse off, for in appropriating previously unowned resources nothing is taken away from others. Others could have appropriated these resources, too, if they had considered them valuable. Yet, they demonstrably did not do so. Indeed, their failure to appropriate them demonstrates their preference for *not* appropriating them. Thus, they cannot possibly be said to have lost any utility as a result of another's appropriation."

The Hoppe (2001: 160) who likens immigration to trespass on private property stands condemned by the Hoppe (2001: 121–122) who states that homesteaders do not violate any rights of non-homesteaders, nor do they even harm them economically speaking. The homesteaders in the scenario we are offering for consideration are the foreigners, and the non-homesteaders the U.S. residents who obey the law against homesteading this land which is claimed by government, but which is actually unowned, at least according to libertarian principles.[10] If it is indeed the case that foreigners undertake the homesteading domestics might have (perhaps even, in retrospect, *should* have undertaken) but refrained from doing, then the former can take just title to the lands involved. They can do so without violating any rights of the latter. If so, then surely immigration is justified at least when the lands they enter are truly unowned.

We will illustrate this point by considering Canada, a relatively uninhabited country. As far as its actual settlement is concerned, it most resembles Chile, only stretching east to west, not north to south. That is, the overwhelming majority of its population resides with 200 miles or so of the border it shares with the United States. As for the rest of this vast terrain claimed by the government of Canada (the country is *not* presently run under anarcho-capitalist law) it is mostly frozen tundra, empty woodlands, icebergs, etc.

Now suppose there are a billion Chinese, or Martians for that matter, who live in inhumane (or un-Martian) overcrowded conditions. They are eying the empty parts of Canada with grave interest. These lands spell life or death for them. According to Hoppe, they can only settle there with the consent of the Canadian government, which would be justified in imposing rather strict conditions on their entry. To make this claim based on private property rights is highly problematic. In our view, if the Canadians want to preclude from entry these new inhabitants, let them first homestead these presently unowned areas. If they do not themselves first do so, their right to prevent others from doing this cannot be justified on libertarian grounds.

Nor can it sensibly be argued that these lands are sub marginal, uninhabitable, and thus not even worth discussing in the present context. This may have been true in the past, for Canadians in our example, but for the Chinese and Martians they are anything but. To argue in this way would be to overlook the subjectivist insights of Austrian economists, who correctly note that objective valuation applies to neither land nor anything else. Rather, goods take on values from the subjective evaluations of human (well, Martian too) actors (Mises, 1998; Buchanan, 1969; DiLorenzo, 1990). Yes, if a country such as Switzerland were to allow immigrants in the millions, let alone billions, that country would very likely suffer egregiously, apart from the civil strife it would unleash. But the same hardly applies to countries with vast unsettled wilderness, such as Brazil, Russia, Australia, and the United States.

Children

Children, as all parents know, present special problems; nowhere is this more so than in the case of Hoppe's views on immigration.

Here, the same arguments that apply to immigrants from other nations (they come to the United States for welfare,[11] they will vote Communist, they will cause unemployment, etc.)[12] also apply to *new babies*, at an 18-year or so remove. For purposes of this analysis, new babies may be regarded as immigrants to this country from somewhere else: the country of Storkovia, from Mars, from heaven, wherever. Just as we have no right to limit new births on the grounds that when they grow up the new children will go on welfare, be criminals, vote badly, etc., so too do we have no right to limit immigration now on this basis. *Any* such argument against immigration applies equally well to bearing children.

The case of children can be employed as a reductio ad absurdum of the Hoppean system. Immigration is forced integration? Then so is childbirth. Immigration is per se a trespass against private property rights? Then so is bringing a new child into the world. Let us put forward a hypothetical, but, we contend, typical argument:

> Many libertarians have been far too complacent in the face of a growing threat to our cultural cohesion, our way of life, and our liberty. We're talking, of course, about the thousands of people who arrive in our country everyday, hoping to make it their new home.
>
> Those arrivals present us with a myriad of social problems. They do not speak our language. They are unfamiliar with our culture. It will take time to assimilate them all, and the government's effort to promote multi-culturalism through the public schools and other government institutions can only lengthen that assimilation time.
>
> Few of these strangers arrive in America with job offers in hand. The odds are high that many of them will rely, at some point in their lives, on government handouts. And studies show that the longer new arrivals reside in the country, the more likely they are to receive welfare.
>
> They will make use of public transportation, public roads, public utilities, public schools, and so on, further straining resources that are already stretched thin in many cases. Their arrival results in a "dumbing down" of the public education system, prompting politicians to throw even more money at it.

All of the above means an increased tax burden on the productive members of society, many of whom already work over half their day to pay their federal, state, and local income taxes, sales taxes, excise taxes, tariffs, and fees.

Because of the lure of government largesse dangled before them, the new arrivals represent a ready-made voting block for a bigger state. Unfamiliar with the American tradition of limited government, the arguments against expanded social programs seem remote and abstract to them, while the benefits appear immediate and tangible. The resultant swelling of the class of tax consumers portends an ominous increase in the scope of the welfare state.

As we have time to watch them adapting to our country, we find their customs strange. Whether it is their music, dress, dating, or manners, their distinct cultures present what appears to be an unbridgeable gulf between them and traditional American life. Americans find themselves longing, as Peter Brimelow (1995) put it, "for some degree of ethnic and cultural coherence."

Libertarians are correctly suspicious of any increases in government power. In the case of these new Americans, however, it should be clear that the cause of liberty is advanced, not retarded, by limiting their influx. So great are their numbers, and so enormous is the difficulty in assimilating them, that the current situation amounts to little less than a foreign invasion of our shores. Libertarians should at least be able to agree that as long as we have any government, its most essential role is to protect the nation from foreign invasion!

Of course, in a purely libertarian society, it would be property owners who would have the right to accept or reject anyone wishing to live on or otherwise use their property. But we don't live in that society. Property owners today are limited by law from excluding individuals from their place of employment due to affirmative-action and other anti-discrimination laws, and from their neighborhoods by similar "civil-rights" legislation.

Simply reducing the number of arrivals allowed in the country each year would be a step forward. But given the vast numbers who have already arrived in the past two decades, it would be wiser to place a several-year moratorium on all new. . . .

What's that you say? Immigration?! You think we've been talking about immigrants?

We've been talking about babies. What we need is a several-year moratorium on births. Our battle cry should be, "Outlaw babies, for the sake of our liberty!"

Clearly, the above is not a libertarian position.

The Irish Problem[13]

When the above argument appeared elsewhere (Callahan, 2002a), the question was raised as to whether there are actually deep similarities between babies and immigrants, or if the argument is merely a rhetorical trick. We will show that there are, and it was not.

Let us imagine we are in the United States in 1854. Irish immigrants have been pouring into the country. In the eyes of the bulk of the population of the United States, they are of sub-standard intelligence, indigent, inebriate, disorderly, possessed of a bizarre culture (Celtic-Catholicism) that is at odds with the predominant United States culture (Anglo-Protestantism), and just generally undesirable. The anti-immigrant American Party is a dynamic populist force in United States politics.

Let us further imagine that, by some miracle, the current American apparatus of anti-discrimination laws and "social" benefits suddenly springs full-grown into existence. American residents are able to collect unemployment; receive AFDC, Social Security, Medicare, and Medicaid payments; and sue for discrimination in the workplace, in housing, and at places they shop.

Suddenly, all of those "No Irish Need Apply" signs are useless. The Irish can sue their way into places that do not want them. Furthermore, their drunkenness and idleness no longer concern just them and their neighbors—they can now get on the dole. What was once a severe annoyance has now become a terrible burden.

We contend that, in the alternative science fiction-ish 1854 United States described above, every argument that could be put forward against free Irish immigration would apply equally well against free Irish procreation.

We should first note that the Irish who had come in under more libertarian arrangements would be justly staying wherever they were at the time the new welfare state came into being. Therefore, there is no libertarian rationale for violating their right to remain where they are.

Let us proceed to consider two further situations: each Irish family decides to have ten children, or each family decides to have no children, but invite ten relatives from Ireland to come live with them. (If you want to include rental contracts forbidding non-immediate-family from living in an apartment in the picture, you can just suitably increase the number of invitees by the Irish-Americans who do not have such contracts.)

Now, it is clear that both sets of newcomers are justly arriving in the country, in that they are invited to live where they will be staying. The problem, from the point of those who are quite understandably worried about the "Irish question" under the new regime, is that they will not stay there. Whether children or immigrants, at some point they will tend to wander off the property.

Not only will they get off the property, but with public roads and anti-discrimination laws in existence, they cannot be kept out of many places. People who detest the Irish will be forced to rent to them, sell to them, hire them, and so on. It's true that Irish babies will not be ready to head out and violate the property rights of "Gaelophobes" quite as quickly as will immigrants. But can a substantive issue of human rights turn on a few years difference in when a potential rights violation will occur?

Both babies and immigrants will tend to be more numerous than they would have been without government social programs in existence, although, of course, the incentives apply to the parents in the first case but to the immigrants themselves in the second.

While the children might be slightly more Americanized than the immigrants, the difference would only be of degree, not of kind. The children would, most likely, remain Catholic, be raised in Irish neighborhoods, attend Irish Catholic schools, and know mostly first- or second-generation Irish immigrants during their life. They will be scarcely less foreign than newly arrived immigrants.

An Attempted Answer by Hoppe

Hoppe is not without a reply to these arguments. He (2001: 167) states: "the receiving party (the mentor of the immigrant) must assume legal responsibility for the actions of his invitee for the duration of his stay. The invitor is held liable to the full extent of his property for any crimes by the invitee committed against the person or property of any third party (as parents are held accountable for crimes committed by their offspring as long as these are members of the parental household.) This obligation, which implies that invitors will have to carry liability insurance for all of their guests, ends once the invitee has left the country, or once another domestic property owner has assumed liability for the person in question by admitting him onto his property."

But this simply will not do. It opens up a Pandora's Box of objections and difficulties.

One implication, by analogy, is that people ought to be held responsible for the crimes of their child as long as the child lives on their property. One could, conceivably, make a case for this for a very young child, completely under the control of the parents. But what of a youngster aged 14, 17, or even 20? The older in age we go along this succession, the further away we remove ourselves from the libertarian doctrine of *individual* responsibility.

And then when this child moves away from the parental abode into a rental apartment, it would appear that the landlord would become responsible for him, according to Hoppe. Surely that is a travesty of justice. Nor would this appear to apply only to children moving away from home. Rather, as a general principle, Hoppe would hold *all* landlords responsible for the crimes of *all* of their tenants. This would pretty much spell the death knell for renting. A more counterintuitive non-libertarian scenario could hardly be imagined.

Moreover, suppose that A allows B and C into his restaurant, as customers, whereupon B attacks C. Then, according to this Hoppean logic, it would not be B who is responsible for this attack on C, but rather A, the property owner. If B murders C, the presumption would appear to be that A would be made to pay for this crime; B would presumably get off scot-free, since the real criminal of the piece, A, has already been caught and punished.

It is one thing to hold entrepreneurs responsible for roughhousing on their property in the *economic* sense: those who do not provide suitable protection for customers will lose revenue. But it is quite another thing to hold restaurateurs accountable for such malfeasance in the *legal* sense; that is, to punish them, instead of the actual malefactors.

Nor will resorting to "insurance" protect Hoppe (2001: 239–265) from these implications. Insurance, in this context, is merely a cloaking device, obfuscating matters. The bottom line is that only the perpetrators of crimes, not the owners of property upon which the crime is committed, are guilty of criminal behavior under a libertarian theory of law.

States Rothbard (1990: 245–246) in this regard: "Under strict liability theory, it might be assumed that if 'A hit B,' then A is the aggressor, and that A—and only A—is liable to B. And yet the legal doctrine has arisen and triumphed, approved even by Professor Epstein, in which sometimes C, innocent and not the aggressor, is also held liable. This is the notorious theory of 'vicarious liability.'" Rothbard (1990: 246) is "properly scornful of the tortured reasoning by which the courts have tried to justify [this] legal concept so at war with libertarianism, individualism and capitalism. . . . " It cannot be denied that Rothbard is discussing employer responsibility for the acts of employees, while we are debating Hoppe over land-lord responsibility for the acts of tenants. Yet the analogy is quite close.

In Hoppe's interpretation, parents offer security for their children, while those who invite immigrants do not do so for their invitees. That is why mothers and fathers are justified in bringing children into the world, while immigration runs contrary to the libertarian legal code, and our analogy fails. If so, then, he should at least allow, right now, all immigration for those who can find mentors in the United States willing to support them. But if Hoppe acquiesced in this practice, his opposition to immigration would vanish in one fell swoop, on the reasonable assumption that millions of property owners would be willing to undertake such a risk.

Hoppe might object on the ground that these mentors might renege and declare bankruptcy, rendering their promises unreliable, and free immigration unjustified. But so can parents go back on the responsibility that Hoppe assigns to them as guarantors in this regard. The implication, here, is either that no one would be justified in giving birth to children, a manifest absurdity, or that only those who are sufficiently wealthy to post bonds sufficient to cover pretty much any damage their children might wreak would be entitled to start a family. The latter may be more acceptable, but it takes us quite a bit down the road away from the usual libertarian assumption that population control is illicit. Further, it leads onto the treacherous ground of preventive detention.[14] For example, it is currently the case that in the United States teenaged, black males commit a share of crimes disproportionate to their numbers in the general population. According to the logic we are attributing to Hoppe, he would be compelled to assent in locking them up at least until they grow to maturity, surely an act contrary to libertarianism.

Upon initially learning of our analogy between children and immigrants, Hoppe (2002) responded with a note to one of the present authors: "

[the analogy between immigrants and babies] just doesn't [work] as soon as you consider the time dimension in the process of property acquisition, and accordingly the establishment of easements, carefully enough. Certainly the babies of domestic tax payers have a right to domestic public goods (and their parents have an easement to have the kids), because their parents [were forced to finance] these goods. . . . The kids inherit ownership from the parents. Obviously, foreigners have no such inheritance claim to domestic public goods."

But there are a number of problems with Hoppe's answer. Firstly, it is a very curious sort of "inheritance" in which the person "passing on" the right to use public goods to his "heirs" hasn't yet died! And no matter how much in taxes he has paid and how many children he has, he can "leave" them all full rights to all public goods to which he has a right.

True, Hoppe could claim that this is just like genetic inheritance, or inheriting the parent's last name, so the parent need not be dead for this to occur. But there is a significant dissimilarity to these cases: names, and genetic inheritance are not at all scarce goods, while residence in a country with limited land certainly is. That is, a child can take on a parent's name without depriving the latter of that nomenclature in the least. Similarly, the parent gives the child a genetic code without in the slightest depriving himself of that benefit. In sharp contrast, however, there is a limited amount of land on the earth, let alone in any one country. Thus when parents have children, and remain alive to live alongside of them, there is that much less land for the parents (or anyone else) to enjoy.

Consider the example of an Irish homeowner in the fictional 1854 United States described above. If he can have his children "inherit" his right to the public road, then why can't he pass it on to the relatives, friends, or even strangers he invites from the old sod as well? He could just as easily leave his whiskey still to his cousin as to his children, so why not his right to use the road? We have seen that he need not be dead to "leave" this right to his heirs, nor does there seem to be any intrinsic limit to how many people to whom he can bequeath it.

Therefore, we contend that if Hoppe's point is true, then so is a rewritten version that runs like this:

"Certainly the *guests* of domestic tax payers have a right to domestic public goods (and their *hosts* have an easement to invite the *guests*), because their *hosts* were forced to finance these goods. The *guests* inherit ownership from the *hosts*."

Hoppe might attempt to limit the right to invite immigrants based on the amount in taxes the invitor has paid. However, the implication of that approach is that the number of children also should be commensurate with taxes paid. For example, such and such an amount of taxation would entitle a person to have one child; a little more, then two children. Those who have *not* paid taxes (or, rather, whose taxes have not exceeded their subsidies)[15] would not be able to have any children at all. This is not quite the program of "one child per family" practiced by the Mainland Chinese government, but it comes perilously close.

Another difficulty is that the age of initial childbearing would be unduly increased, perhaps even to biologically dangerous levels, at least in the case of the relatively impecunious. Typically, a couple might have their first child at, say, age 25. But if both husband and wife had just graduated from college a year or so before, it might well be the case that this would be too young to justify bringing in immigrants from the foreign country of "Storkovia." It might be that a couple with their earning power would not pass from net tax consumer to net tax provider (see Calhoun (1953: 16–18) until age 30, 35, or even 40.

Hoppe might be willing to accept the logical implication of his stance, to wit, that greater obstacles ought be placed in the path of the poor who want to bear children than of the rich. It is the word "placed" that is key to our disagreement. If someone wants merely to assert that, in so far as they have fewer resources than the rich, the style in which the poor can raise their children is quite justly more cramped, then we would have no problem with his argument. (For libertarians, that one has a right to have children in no way entails an obligation upon others to support one's children.) As a general principle, again for libertarians, it is unobjectionable that the rich have greater command over goods and services than the poor; after all, unless that were true, there would be no point in being rich.

It is one thing to insist that all people be able to use their money in any non-invasive purchases they wish. But it is entirely a different matter to coercively forbid non-invasive acts, such as child bearing. Here, to contend that the rich can have as many children as they wish, while the poor must be constrained from doing so, is to violate the rights of the latter. That is, while it is theft to transfer boats and cars forcibly from rich to poor, it is a rights violation to prevent the poor from buying any boat or car that they can afford—even a yacht or a limousine. To justify such a policy based on the fact that one suspects that some poor person "can't really afford" the luxury item is to substitute one's own judgment for that of the other person. The poor now suffer not merely from having fewer resources than the rich, but also by losing a measure of control over their own choices that the rich continue to enjoy. This runs directly contrary to the core principles of libertarianism.

In fact, a law restricting childbirth on the basis of taxes paid is analogous to forbidding anyone who is poor from *ever* owning a yacht. (After all, the poor man who buys and uses a boat will add to the crowding of the public waterways.) While we can embrace a view saying that there is no injustice in the poor finding it more difficult to own large boats than the rich, we cannot do so to a view that says the poor fellow who scrimps and saves to afford the yacht of his dreams should be legally denied the right to buy it.

There is yet another difficulty for libertarians with Hoppe's position on this matter: It is akin to the justification used to defend interferences with liberty such as socialized medicine. Per such reasoning, you can be forced to wear a motorcycle helmet or refrain from smoking, since otherwise you might impose costs on all others who have medical coverage. In a libertarian society, whether the individual wears protective headgear or eats in a healthy manner is entirely up to his own discretion. He alone suffers the consequences of foolhardy action. But under socialized medicine, everyone else is forced to bear the costs of dangerous behavior, and this fact is used to justify forcing all members of society to protect their health willy-nilly.

Hoppe's analysis of child bearing opens up a similar door. Because of prior state interventions, what was previously considered a non-invasive act is suddenly criminalized. What aspect of liberty is safe under such a principle? The drug war is okay, as users are more likely to go on the dole. High taxes on fatty foods are fine, since the obese are more likely to have health problems. Zoning laws can now be recommended, since they force landowners to hold large lots, driving up lot prices and keeping the poor out of town.

In fact, as pointed out by Mises (1988), Ikeda (1997), and others, every governmental intervention leads to undesirable results that call for another intervention as a "fix." The state expands in a vast pattern of such interlocking interventions. The removal of any of them might have unpleasant effects for any number of people, even if they do not directly gain from the intervention. If, in attempting to reduce the size of the state, we restrict ourselves to only eliminating interventions when we can show that no innocent third parties are economically harmed by that elimination, we will never start. Nor, for the various reasons highlighted in the socialist calculation debate (Boettke, 1991; Hoppe, 1989, 1996; Mises, 1981) can we determine which interventions are least costly and eliminate them first. No, we must start wherever we can.

A Practical Objection

Hoppe (2001: 161) makes much of the fact that free trade implies a willing buyer and a willing seller—voluntary actions on both side of the transaction—while for the case of immigration, in sharp contrast, this does not apply. That is, if all land is privately owned, then, in addition to a willing immigrant, there must also be a landowner willing to take in the new arrivals.

A practical problem with Hoppe's perspective is that it is exceedingly likely there will always be *someone* with sufficient land holdings in the domestic country who will provide a sanctuary for immigrants—perhaps even for quite a few of them. There is nothing in his preferred immigration policy that would prevent the owner of a vast ranch in Texas from inviting the entire Masai people to move to his property and take up their traditional way of life there. It might simply entertain him to do so. Or he might be acting out of charitable or benevolent motives. But by far the most common reason for such invitations would be financial considerations. If the productivity of say, Argentinian labor is sufficiently higher in the United States than, in Argentina, then this difference can be capitalized, and used to finance immigration. Thus vast numbers of immigrants can legitimately arrive in this country, Hoppe's trespass objections notwithstanding (i.e., we stipulate that there are sufficient lands, held privately, enough of whose owners welcome the newcomers and provide surety for them).[16]

Hoppe's second line of defense it that these newcomers still would not be able to get out onto the roads and other people's property. That is, he contends, not only must there be a sufficient number of land owners to welcome the new immigrants, but the society as a whole, or at least the nation's road owners, must also allow them onto their property, if the new arrivals are to become involved in the economy. As *everything* would be private in the present scenario, this would include streets, highways, avenues, indeed, *all* traffic thoroughfare and arteries.

There are several replies available to the pro-legalized immigration side of this debate. The first is to concede Hoppe's thrust and admit that immigrants will be confined to the lands of those who welcome them, while noting that this still varies greatly from the position Hoppe set out to defend.

The second is to deny his contention. An intensive discussion of this question would take us too far afield, but the conclusion reached by most analysts who have written about the subject from a private property libertarian perspective (e.g., Block, 1979) is, given that the landlord or the firm has a contractual right to use the road, this would extend to his tenants and employees. This is because a man would be unlikely to purchase land in the first place if he would not have access to the streets abutting his property; and also, because it is unlikely that the local street proprietor would not also have a contract with the owners of the other traffic arteries contiguous with his own holdings that each of them would allow motorist customers of the others to enter his own property.[17] Of course, certain groups might adopt a policy of self-imposed isolation, but the standard arguments for the benefits of the division of labor and free trade make it likely they would be few in number.

This being the case, we believe that the number of immigrants in a fully laissez-faire society would approximate the number who would arrive under a government policy of unfettered immigration. No one could be turned away as long as there was *either* unowned land, *or* landlords willing to take large numbers of immigrants onto their holdings. Nor would they be confined to the property of any one or even the many host firms or people who had specifically invited them to the domestic country. Similarly, no one would be barred from entry in a society such as ours, with its public property.

Gordon (1997) summarizes the Rothbard–Hoppe position: "The result of doing so (e.g., adopting this perspective) is apt to be carefully controlled immigration, not unrestricted entry." But, as we have shown, any holdout with large acreage can invite in *anyone he wants*. This may be "carefully controlled" in the sense that only private property owners can sponsor immigrants, but not in the usual immigration sense that the newcomers have to have property of their own, be intelligent, literate, be able to post a bond, etc.

A Reductio

If Hoppe is right that there is no right to international migration, since this would violate existing property rights, it implies that there is no right to intra-national migration either,[18] and for the same reason. For example, the migration of blacks from the southern states to northeastern cities in the 1940s, that of the Okies to California in the 1930s, and of the Jews from the lower east side of Manhattan to the surrounding suburbs of New York City would all be prohibited or restricted by law. Indeed, it is hard to see why any sorts of moves, even those of just a few miles, should not be regulated per his reasoning.

Hoppe's analysis is attractive to those who oppose immigration from foreign shores. It will be less so with regard to internal migration within one country, since it must be the rare person who opposes such movement of peoples. And yet if the argument implies in the one context it applies in the other, since national boundaries, while having political reality, are of no moment when it comes to the application of libertarian law.[19]

Culture, Value, and Private Property Rights

Consider these remarks of Rothbard (1994, p. 7):

"The question of open borders, or free immigration, has become an accelerating problem for classical liberals. This is first, because the welfare state increasingly subsidizes immigrants to enter and receive permanent assistance, and second, because cultural boundaries have become increasingly swamped. I began to rethink my position on immigration when, as the Soviet Union collapsed, it became clear that ethnic Russians had been encouraged to flood into Estonia and Latvia in order to destroy the cultures and languages of these peoples. Previously, it had been easy to dismiss as unrealistic Jean Raspail's anti-immigration novel *The Camp of the Saints*, in which virtually the entire population of India decides to move, in small boats, to France, and the French, infected by liberal ideology, cannot summon the will to prevent economic and cultural national destruction. As cultural and welfare-state problems have intensified, it became impossible to dismiss Raspail's concerns any longer."

We readily admit that we disapprove of a situation in which a state (the Soviet Union, in Rothbard's example) coercively relocates people. However, that is not the same thing as a state merely allowing people to enter its territory. First of all, no one has been more eloquent in making the point that what is legitimately owned is property itself, and *not* the value of that property, than Hoppe.[20] One can own a house itself but not its value, which depends upon the evaluations and actions of others (e.g., buyers and sellers). We suggest that the same considerations apply to the case of the value of culture. Here, too, *all* that one can legitimately own is one's physical property, *not* the value of it as impacted upon it by the culture of one's neighbors.

Take Raspail's scenario of the Indians and the French. Here, at least from our perspective, the Indians did nothing wrong[21] (at least in the barebones scenario as laid out by Rothbard.) They did not violate the libertarian proscription against initiatory violence. They were peaceable. They did not trespass. They purchased land and homes, or rented them, all on a voluntary basis (or homesteaded public or unowned property). Yes, they also perpetrated "economic and cultural national destruction" upon the French, but this is merely part and parcel of *values*, *not* private property rights. The French people, along with everyone else, have a right to their physical property. They have, in contrast, no right at all to its *value*; e.g., they have no right whatsoever, to resist "economic and cultural national destruction," at least not by violent means.

How, then, may they properly resist these incursions? In the same way that all of us may do so: by tying up neighbors in restrictive covenants, or joining gated communities, or housing cooperatives, or condominiums, or proprietary communities (see McCallum, 1970). These, based on the libertarian concept of free association, may licitly specify not only the type of fences and exterior color paint that can be utilized, but also the types of people who can live there.

But suppose there is but one (and there might well be, in any reasonable scenario, dozens, hundreds if not thousands of such people) Frenchman who is desirous of inviting Indian immigrants, hordes of them, to his own private property. If he owns several square miles of land, far from an insurmountable task at least in agricultural areas, he will be able to host literally *millions* of immigrants. We have already rejected Hoppe's contention that this holdout would be responsible for any crimes his invitees might commit, on the ground that people are responsible for their own legal transgressions, and cannot legitimately pass them off onto their landlords, employers, etc. As well, we resist his notion that the Indian immigrants would not be able to get out onto the (privatized) French roads.[22] Condos and covenants may provide some measure of protection against French "economic and cultural national destruction," but it is an empirical issue as to just how much.

Thus, it is not true that "the regime of open borders that exists de facto in the United States really amounts to a compulsory opening by the central state." Very much to the contrary, this state of affairs would exist even under anarcho-capitalism, provided, only, that there was at least *one* large-scale land-owning holdout.

The Pragmatic Aspects of Immigration Restriction for Libertarians

An anti-immigration libertarian might acknowledge the above reasoning, but contend that as a purely practical matter immigration must be restricted today, since the consequences of unrestricted immigration for liberty would be so pernicious under the existing political regimes in the United States and Western Europe. Cases in point include the welfare state, anti-discrimination (i.e., forced association) laws, and public property. In a different, more libertarian world, open borders might be practical, but not in ours.

Such a stance as a holding action, an attempt to keep the status quo from becoming more pro-immigrant, might work. But if a libertarian really means to reduce the number of immigrants from current levels, we must ask him if he has seriously contemplated the expansion of government power and intrusiveness necessary to fully control the borders of a nation like the United States?

Such an expansion would be needed to prevent the chief effect of new laws from decreasing legal immigration by increasing illegal immigration. After all, immigrants arrive here primarily because they perceive a demand for their labor. Prohibition in the face of demand simply drives the supply underground.

Since immigrants arrive in America by plane, boat, car, and foot, government surveillance of all parts of the country would have to be increased. The movements of tourists must be closely tracked to make sure they do not stay or work while they are "vacationing."

Illegal immigrants flood an area of several square miles around the house one of us lives in every morning in the spring, summer, and fall. The aftermath of 8:16 into the local train station is a Mexican diaspora. Are the neighbors upset that the State has sent an invading force into their community? No, the illegal aliens are going to work at their houses. Productive, private citizens and migrant workers are cooperating to evade the State's laws and peacefully conduct mutually beneficial private transactions.

Given private individuals' complicity in the "law-breaking," it is clear that a serious effort to reduce immigration would have to investigate and punish such "criminals" as well. Private homes would be subject to search to ensure they were not housing or employing any illegal aliens. Everyone's bank balance would be monitored and all suspicious payments traced.

That is not a pro-liberty scenario, unless one is of the "it can only get better by getting worse first" school of libertarian thought. We are not arguing, at least here, that this school of thought is wrong. But if that is the line of thinking to which an anti-immigrant libertarian is adhering, it should be made explicit, i.e.: "I favor an immigration crack-down because it will help to bring about a police state, hastening the day of full freedom." Hoppe, at least, does not make this argument.

So, what can be done, given that we do not dismiss the concerns of anti-immigration libertarians as baseless? We admit that the Anglo-American residents, in the imaginary 1854 United States we depict above, had many valid reasons to be worried about "the Irish problem." Cultural assimilation is not a difficulty to be sneered at. And we agree that the welfare state creates perverse incentives that result in a different kind of immigrants than would occur in a libertarian society. But we recommend addressing the problem in a libertarian fashion.

There are a number of reforms available to that would entail the government doing *less for* immigrants, rather than *more to* them (and to citizens):

- The United States could greatly extend the period after which citizenship can be granted. Since most libertarians do not believe that anyone should be able to vote away anyone else's property in any case, forbidding that power to immigrants alone does not take away anything from them that is justly theirs.
- Reserve automatic citizenship for the children of citizens, rather than for all children born in the United States. The argument follows that above.
- Make immigrants ineligible for government transfer programs. In the libertarian view, Social Security, Medicare, Medicaid, and so on, ought not exist. If immigrants are the group that we can wean off of such programs first, so be it. Again, for a libertar-

ian, this does not entail taking anything away from them that is justly theirs. Proposition 187 in California was an example of such an initiative, successful, at least, with the voters, if not the courts.

- Eliminate civil rights protection for immigrants. For libertarians, no one has the right to force himself onto another's property. Not applying anti-discrimination laws to immigrants might be a first step in ridding ourselves of such laws completely.

These measures go some ways toward handling what we feel are the valid concerns of anti-immigration libertarians. Furthermore, they reduce the risk that libertarians, ironically, will be the sponsors of a massive new Federal program, Operation Iron Borders, or whatever it would be named. And lest we face the complaint that we are being unrealistic in what we ask of the political system, we point out that the libertarian anti-immigration platform has not, so far, met with electoral success either, nor does it seem to be on the verge of doing so.

Conclusion

In one sense, there is no real debate between Hoppe and us. As libertarians, we all favor private property rights and oppose trespasses against them. If this implies "limiting" immigration, then we all favor doing just that; if not, not. We agree, further, that in a libertarian society, no foreigner would be allowed into any territory without the permission of at least one property owner.

However, we are not, "as ships passing in the night," failing to contradict each other's position; this is not a mere verbal dispute. For one thing, we appear to differ as regards an empirical issue: would an ideal libertarian society be one which could, in the main, heavily reduce the number of foreigners who enter and live in the territory now controlled, for example, by the U.S. government? Hoppe maintains that it will; we take the opposite position, based on our supposition that there will be large numbers of holdouts to any restrictive covenant amongst the hundreds of millions of people now residing in this area, and that without such specific contracts, the profit motive, if nothing else, will lead to the mass invitation of foreigners to our shores.

But it is when we come to the real world that we diverge from Hoppe even more sharply. Hoppe maintains that in the present context the U.S. government is in effect a manager[23] for the private property owners who live within the borders of the country. We maintain, in contrast, that the state cannot properly take on any such role.

States Higgs[24] in this regard: "Some of us . . . are disinclined to recognize that the United States or any other existing nation state has legitimate authority to establish any so-called borders within which it takes pleasure in exercising its coercive powers over the resident population. Such borders are nothing but artifacts of the interplay of the brute forces exercised by the various armed groups (that is, nation states) wielding established powers over the people who inhabit this planet. If the state cannot legitimately create borders in the first place, because its very existence is illegitimate, then it manifestly cannot promulgate just rules with regard to how open or closed any such borders will be."

We differ with Hoppe, moreover, as to the proper status of "public property" and whether any would-be homesteader may seize it; we further diverge as to the responsibility for crimes committed by children, tenants and employees, on the part of parents, landlords and employers. We say, as a matter of libertarian principle that there is no such legal responsibility; Hoppe takes the opposite stance.

Notes

1. The classical liberal position of open immigration always excepted criminals and those with communicable diseases, on the ground that persons of this sort would represent a physical threat against the citizenry.
2. For other critiques of open immigration from a libertarian point of view see Gordon (1995, 1997), Raico (1996) and Rothbard (1994). For other libertarian treatments of this issue see Simon (1998), Hospers (1998), de Soto (1998), Machan (1998), and North (1998).
3. Well, de libertarian jure, in any case.
4. It is more than passing curious that Hoppe, perhaps the foremost theoretician of anarcho-capitalism at present, nevertheless persists in defending the government. Hoppe (2001: 263) gives it as his view that: "If there were any aggression or provocation against the state at all, this would be the action of a particular person, and in this case the interest of the state and the insurance agencies would fully coincide. Both would want to see the attacker punished and held accountable for all damages." Here, the "insurance agencies" are Rothbard's (1978: 219) "Metropolitan Protection Companies," e.g., the very embodiment of anarcho-capitalism. How could it be to the interest of the latter to protect government property? Why, merely pragmatic considerations aside, should a libertarian applaud the capture and punishment of a person who commits aggression against an unjustified government? Surely, libertarian principle would incline us in the opposite direction.
5. Attacking it, that is, in the objective sense. The bum's understanding of the finer points of libertarian property theory might be altogether lacking, but at least he is acting so as to undermine unjustified public property. Hoppe is making the exact same mistake made by Rand (1967) when she opposed the student takeovers of the *public property* that was the University of California at Berkeley. See on this Block (2003).
6. This reminds us of the following joke. There was a flood, and a man was hanging for dear life from the top of a house, perched just above the raging water. A pious man, he prayed long and hard for God to rescue him. Along came a man in a rowboat who said to him, "Get in, I'll row you to safety." Replied the religious man, "Thanks, but I'm waiting for God to rescue me." Whereupon a helicopter pilot threw down a rope ladder, and offered him a ride to higher ground. Again said the religious man, "Thanks, but I'm waiting for God to rescue me." No sooner did the helicopter recede into the distance but the waters rose, and drowned the man. Appearing before St. Peter, the man remonstrated with God: "I was a pious man. I worshipped you all my life. I prayed for a rescue, but you abandoned me." Retorted God: "And who do you think sent the rowboat and the helicopter?"
7. We define a robber, crook, or thief as a person who unjustifiably takes property from its *rightful* owner. In contrast, if someone takes property from a criminal of this sort, he is not himself a thief but rather a liberator.
8. Support for our analysis is offered by Rothbard (1992: 119), who discusses not the bum in the library but rather, analogously, the "undesirable" on the street: "In New York City ... there are now hysterical pressures by residents of various neighborhoods to prevent McDonald's food stores from opening in their area, and in many cases they have been able to use the power of local government to prevent the stores from moving in. These, of course, are clear violations of the right of McDonald's to the property which they have purchased. But the residents *do* have a

point: the litter, and the attraction of 'undesirable' elements who would be 'attracted' to McDonald's and gather in front of it—on the *streets*. In short, what the residents are *really* complaining about is not so much the property right of McDonald's as what they consider the 'bad' use of the government streets. But as taxpayers and citizens, these 'undesirables' surely have the 'right' to walk on the streets, and of course they *could* gather on the spot, if they so desired, without the attraction of McDonald's." We must concede that there is one slight disanalogy: Hoppe assumes the bum has paid no taxes at all (a heroic assumption) while Rothbard more realistically stipulates that the 'undesirable' person has been victimized in this manner.

9. De facto but not de (libertarian) jure.

10. The government itself never homesteaded these territories; it only precluded others from doing so. For a further discussion of illicit pre-emption, see Block and Whitehead (forthcoming).

11. Milton Friedman has been quoted to the effect that (paraphrase) "Free immigration is incompatible with the welfare state, and thus we cannot have the former." The libertarian response, in contrast, is surely, "Free immigration is incompatible with the welfare state, and thus we cannot have the *latter*."

12. According to Hoppe (2001: 162, fn. 11): "Note, that even if immigrants were excluded from all tax-funded welfare entitlements as well as the democratic 'right' to vote, they would still be 'protected' and covered by all currently exiting anti-discrimination affirmative action laws, which would prevent domestic residents from 'arbitrarily' excluding them from employment, housing, and any other form of 'public' accommodation." This is disingenuous, since in the fully free society on the assumption of which we are contrasting present immigration law with Hoppe, all of these provisions would disappear. And in the present society, it is likely they would be swept away by the proverbial "hordes" of immigrants who would arrive on our shores. But suppose, for argument's sake, that this is not the case. That is, that these unjustified laws would remain on the books. If the immigrants have a right to enter the domestic country, anti-libertarian enactments passed by the legislature of the host nation cannot properly undermine it. To maintain that they can is to tread dangerously close to legal positivism, the doctrine that all man made law is per se rightful.

13. An earlier version of this section appeared in Callahan (2002b).

14. It also leads in another direction that neither Hoppe nor any libertarian could welcome: preemptive war, as in the U.S. attack on Iraq of 2003. For more on this see antiwar.com, lewrockwell.com, two libertarian opponents of military adventurism.

15. They are net tax consumers in Calhounian language.

16. There are also public lands, and territory never homesteaded by anyone, mentioned above.

17. An exception might be gated communities, where the contract would be "one way." That is, all owners of property in the gated community would be allowed out onto all roads and highways, but the reverse would not be true.

18. We owe this point to Michael Edelstein.

19. Although Hoppe does not explicitly discuss this reductio, we have no doubt he would embrace it as logically consistent with his overall view. That is, for him, this is no criticism at all, but merely a logical implication of his thesis. However, this depiction of traffic immobility within a country may not be acceptable to all (libertarian) opponents of open immigration.

20. Hoppe (1989, pp. 139ff; 1993, pp. 188ff, pp. 199ff; 206f). See also Hoppe and Block (2003).

21. In contrast, the Russians, presumably, engaged in violence vis-à-vis the Latvians and Estonians. If so, they would be booted out of these two countries under a libertarian regime.

22. However, there is a role that a privatized road industry can play in ameliorating this scare scenario of millions of Indians in France. As more people patronize the roadways, the price of so doing will tend to rise, which will reduce that tendency. Also, likely, many of those already located in France will be the owners of its highways. They may legitimately, under the libertarian code, choose to discriminate against foreigners. This phenomenon will likely play a far greater role in densely settled places such as France or Switzerland, and less so in relatively empty countries such as Canada, Russia, Australia, etc.

23. For a critique of putting government on a "business basis," see Rothbard (1956, 1970).

24. This is private e-mail correspondence to the mises@yahoogroups.com list; dated 1/27/02.

References

Block, Walter. 1969. "Against the Voluntary Military." *The Libertarian Forum* (August 15): p. 4.

Block, Walter. 1979. "Free Market Transportation: Denationalizing the Roads." *Journal of Libertarian Studies: An Interdisciplinary Review* 3, 2: 209–238.

Block, Walter. 1998. "A Libertarian Case for Free Immigration." *Journal of Libertarian Studies: An Interdisciplinary Review* 13, 2 (Summer 1998): 167–186.

Block, Walter. 2003. "Libertarianism vs Objectivism; A Response to Peter Schwartz." *Reason Papers*.

Block, Walter and Roy Whitehead, "Compromising the Uncompromisable: A Private Property Rights Approach to Resolving the Abortion Controversy," unpublished ms.

Boettke, Peter J. "The Austrian Critique and the Demise of Socialism: The Soviet Case," in *Austrian Economics: Perspectives on the Past and Prospects for the Future.* Vol. 17, Richard M. Ebeling, ed., Hillsdale, MI: Hillsdale College Press, 1991, pp. 181–232.

Brimelow, Peter. 1995. *Alien Nation: Common Sense about America's Immigration Disaster.* New York: Random House.

Buchanan, James M. 1969. *Cost and Choice: An Inquiry into Economic Theory,* Chicago: Markham.

Calhoun, John C. 1953. *A Disquisition on Government,* New York: Liberal Arts Press.

Callahan, Gene. 2002. "They're Coming to America: What Should We Do About It?" *Anti-State.com* (March): http://www.anti-state.com/callahan/callahan5.html.

Callahan, Gene. 2002. "The Irish Problem." *Anti-State.com*: http://www.anti-state.com/callahan/callahan6.html.

de Soto, Jesus Huerta. 1998. "A Libertarian Theory of Free Immigration." *The Journal of Libertarian Studies* 13, 2 (Summer 1998): 187–198.

DiLorenzo, Thomas J. 1990. "The Subjectivist Roots of James Buchanan's Economics," *The Review of Austrian Economics,* Vol. 4, 1990, pp. 180–195.

Gordon, David. 1997. "The Invisible Hoppe." *Mises Review* (Winter): *http://www.mises.org/misesreview_detail.asp?control=47&sortorder=issue.*

Gordon, David. 1995. "Come One, Come All?" Rev. of *Alien Nation: Common Sense about America's Immigration Disaster,* by Peter Brimelow. *Mises Review* (Summer).

Hoppe, Hans-Hermann. 1989. *A Theory of Socialism and Capitalism: Economics, Politics and Ethics.* Boston: Kluwer, 1989.

Hoppe, Hans-Hermann. 1993. *The Economics and Ethics of Private Property: Studies in Political Economy and Philosophy.* Boston: Kluwer, 1993.

Hoppe, Hans-Hermann. 1995. "Free Immigration or Forced Integration?" *Chronicles* (July): 25–27.

Hoppe, Hans-Hermann. 1996. "Socialism: A Property or Knowledge Problem?" *Review of Austrian Economics* 9, 1: 147–154.

Hoppe, Hans-Hermann. 1998. "The Case for Free Trade and Restricted Immigration." *The Journal of Libertarian Studies* 13, 2 (Summer 1998): 221–233.

Hoppe, Hans-Hermann. 1999. "On Free Immigration and Forced Integration." LewRockwell.com.http://www.lewrockwell.com/orig/hermann-hoppe1.html.

Hoppe, Hans-Hermann. Democracy—The God That Failed: The Economics and Politics of Monarchy, Democracy, and Natural Order. Rutgers University, N.J.: Transaction Publishers, 2001.

Hoppe, Hans-Hermann. 2002. E-mail to Walter Block. March 19.

Hoppe, Hans-Hermann and Walter Block. 2003. "Property and Exploitation," *International Journal of Value-Based Management,* 15, 3.

Hospers, John. 1998. "A Libertarian Argument Against Open Borders." *The Journal of Libertarian Studies* 13, 2 (Summer): 153–166.

Ikeda, Sanford. 1997. *Dynamics of the Mixed Economy: Toward a Theory of Interventionism.* London: Routledge, 1997.

McCallum, Spenser H. 1970. *The Art of Community,* Menlo Park, CA: Institute for Humane Studies, 1970.

Machan, Tibor. 1998. "Immigration into a Free Society," *The Journal of Libertarian Studies* 13, 2: 199–204.

Mises, Ludwig von. [1949] 1998. *Human Action: A Treatise on Economics.* Scholar's Edition. Auburn, Ala.: Ludwig von Mises Institute.

Mises, Ludwig von. 1981. *Socialism*, Indianapolis: Liberty Press / Liberty Classics.
North, Gary. 1998. "The Sanctuary Society and its Enemies," *The Journal of Libertarian Studies* 13, 2 (Summer): 205–220.
Raico, Ralph. 1996. "Mises on Fascism, Democracy, and Other Questions." *Journal of Libertarian Studies* 12, 1 (Spring): 1–27 (see especially 24–25).
Rand, Ayn. 1957. *Atlas Shrugged*, New York, Random House.
Rand, Ayn. 1967. "The Cashing-In: The Student Rebellion." *Capitalism: The Unknown Ideal*. New York: Signet Books, 1967: 236–269.
Rothbard, Murray, N. 1956. "Government in Business." *Freeman* (September): 39–41. Reprinted in *Essays on Liberty IV*. New York: Foundation for Economic Education, 1958: 183–187.
Rothbard, Murray N. 1970. *Power and Market: Government and the Economy*, Menlo Park Cal.: Institute for Humane Studies.
Rothbard, Murray N. 1978. *For a New Liberty*. Macmillan, New York.
Rothbard, Murray N. 1982. *The Ethics of Liberty*. Atlantic Highlands, N.J.: Humanities Press.
Rothbard, Murray, N. 1994. "Nations by Consent: Decomposing the Nation-State." *Journal of Libertarian Studies* 11, 1.
Simon, Julian. 1998. "Are There Grounds for Limiting Immigration?" *The Journal of Libertarian Studies* 13, 2 (Summer): 137–152.

On Immigration: Reply to Hoppe

Anthony Gregory and Walter Block

Hoppe takes the position that it would be justified for the state to limit the supposed unlimited libertarian right to free immigration. We criticize his analysis on the ground that it is not compatible with libertarian theory.

Keywords: Immigration; migration; integration; libertarianism; private property rights.

I. Introduction

Hans Hoppe is clearly one of the most creative, inventive and insightful libertarians now writing. This claim would be true if his only contribution was his "argument from argument" (1993, pp. 204–207), which placed the entire corpus of libertarian theory on an undeniable praxeological-like basis. But he has done more, far more. He has made sterling and original contributions to the theory of anarchism (2001), private property rights (1993), homesteading (1993), socialism and capitalism (1989) and insurance (2003), to single out just a few of his many contributions to this field.

Although the present paper is dedicated to a highly critical examination of Hoppe's contributions to the field of immigration, we readily acknowledge at the outset that here, too, even though we cannot see our way clear to agreeing with his conclusions, his splendid, imaginative and ingenious "footprints" can readily be seen. Who else but this scholar would think to model immigration along the lines of the importation of goods, pointing to disanalogies between them (2001)? No one other than he could analyze free immigration as a form of forced integration (2001) in violation of private property rights.

The early Rothbard took what may be called the traditional libertarian view. Here, libertarianism was ineradicably bound to the free movement of goods, investments *and* labor[1] across international boundaries. Indeed, the concept of "international boundaries" was itself a highly problematic one, and without it, the "problem" of free immigration did not even arise. Stated Rothbard (1993):

---

[1]The historian A.J.P. Taylor wrote: "In 1914 Europe was a single civilized community…. A man could travel across the length and breadth of the Continent without a passport until he reached … Russia and the Ottoman empire. He could settle in a foreign country for work or leisure without legal formalities…. Every currency was as good as gold…." (Cited in Stromberg, 1999).

"Tariffs and immigration barriers as a cause of war may be thought far afield from our study, but actually this relationship may be analyzed praxeologically. A tariff imposed by Government A prevents an exporter residing under Government B from making a sale. Furthermore, an immigration barrier imposed by Government A prevents a resident of B from migrating. Both of these impositions are effected by coercion. Tariffs as a prelude to war have often been discussed; less understood is the Lebensraum argument. "Overpopulation" of one particular country (insofar as it is not the result of a voluntary choice to remain in the homeland at the cost of a lower standard of living) is always the result of an immigration barrier imposed by another country. It may be thought that this barrier is purely a "domestic" one. But is it? By what right does the government of a territory proclaim the power to keep other people away? Under a purely free-market system, only individual property owners have the right to keep people off their property. The government's power rests on the implicit assumption that the government owns all the territory that it rules. Only then can the government keep people out of that territory."

"Caught in an insoluble contradiction are those believers in the free market and private property who still uphold immigration barriers. They can do so only if they concede that the State is the owner of all property, but in that case they cannot have true private property in their system at all. In a truly free-market system, such as we have outlined above, only first cultivators would have title to unowned property; property that has never been used would remain unowned until someone used it. At present, the State owns all unused property, but it is clear that this is conquest incompatible with the free market. In a truly free market, for example, it would be inconceivable that an Australian agency could arise, laying claim to "ownership" over the vast tracts of unused land on that continent and using force to prevent people from other areas from entering and cultivating that land. It would also be inconceivable that a State could keep people from other areas out of property that the "domestic" property owner wishes them to use. No one but the individual property owner himself would have sovereignty over a piece of property."

It is to Hoppe's credit that he saw an entirely different way of looking at this issue. In his analysis, the free movement of goods was fully justified, insofar as it was entirely a matter of voluntary interaction. That is, there was a willing seller abroad, and an agreeable buyer in the domestic country. Nothing could fit "plumb-line" libertarianism better than that. Then, too, this applied, completely, to investments. In this case also, there was a voluntary investor in the foreign nation, and an equally disposed recipient of the capital in the recipient one. Again, this is black-letter libertarianism, grounded, as are all justifiable acts, in mutual consent.

Matters were very, very different when it came to immigration, however. In this case there was a willing immigrant, it cannot be denied. No one would migrate from country A to B were he not doing this of his own accord.[2] But, according to Hoppe, there was no agreeable recipient at the other end of this "transaction." That is, it was not a voluntary commercial interaction at all. Rather, it was a unilateral move on the part of the immigrant onto a territory not all of whose owners accepted it. And, for those who did not welcome these people with open arms, the trip amounted to no less than a trespass. Governments that acquiesced in this were either initiating, or aiding and abetting a variant of forced integration. Brilliant. No other anti-immigration advocate[3] has ever come up with anything half as insightful.

The later Rothbard (1994, p. 7) took a position similar to the one supported by Hoppe. He reversed field, and supported governmental interference with the free movement of peoples:

"I began to rethink my views on immigration when, as the Soviet Union collapsed, it became clear that ethnic Russians had been encouraged to flood into Estonia and Latvia in order to destroy the cultures and languages of these people."

Let us allow Hoppe to state his position in his own words, lest any problems of misinterpretation cloud our analysis. Hoppe (2001, pp. 159–160) stated:

"A truly remarkable position is staked out by Walter Block, 'A Libertarian Case for Free Immigration,' (Journal of Libertarian Studies 13, no. 2 1998). Block does not deny the above predicted consequences of an 'open border policy.'"

To the contrary, he wrote:

"… suppose unlimited immigration is made the order of the day while minimum wages, unions, welfare and a law code soft on criminals are still in place in the host country. Then, it might well be maintained, the host country would be subjected to increased crime, welfarism, and unemployment. An open-door policy would imply not economic freedom, but forced integration with all the dregs of the world with enough money to reach our shores" (p. 179).

"Nonetheless, Block then goes on to advocate an open-door policy, *regardless* of these predictable consequences, and he claims that such a stand is required by the principles of libertarian political philosophy. Given Block's undeniable credentials as a leading contemporary theoretician of libertarianism, it is worthwhile explaining where his argument goes astray and why libertarianism requires *no* such thing as an open-door policy. Block's pro-immigration stand is based on an analogy. 'Take the case of the bum in the library,'" he states.

---

[2]If this were not the case, it would not be a matter of immigration but rather kidnapping.

[3]For another staunch opponent of open immigration see Brimelow (1995).

"What, if anything, should be done about him? If this is a private library, ... the law should *allow* the owner of the library to forcibly evict such a person, if need be, at his own discretion.... But what if it is a public library.... As such [libraries] are akin to an unowned good. Any occupant has a much right to them as any other. If we are in a revolutionary state of war, then the first homesteader may seize control. But if not, as at present, then, given 'just war' considerations, any reasonable interference with public property would be legitimate.... One could 'stink up' the library with unwashed body odor, or leave litter around in it, or 'liberate' some books, but one could not plant land mines on the premises to blow up innocent library users (p. 180–81)."

"The fundamental error in this argument, according to which everyone, foreign immigrants no less than domestic bums, has an equal right to domestic public property, is Block's claim that public property 'is akin to an unowned good.' In fact, there exists a fundamental difference between unowned goods and public property. The latter is *de facto* owned by the taxpaying members of the domestic public. They have financed this property; hence, they, in accordance with the amount of taxes paid by individual members, must be regarded as its legitimate owners. Neither the bum, who has presumably paid no taxes, nor any foreigner, who has most definitely not paid any domestic taxes, can thus be assumed to have any rights regarding public property whatsoever."

Notwithstanding the abovementioned accolades, in this paper we criticize Hoppe's theory of immigration as forced integration on the ground that it is vulnerable to a series of reductio ad absurdum. His views on immigration are inconsistent with his own (correct) perspectives on a myriad of other issues. We maintain that he cannot endorse, even tacitly or reluctantly, statist limits on immigration since such a stance is incompatible with his own views. We deal with free trade in Section II. Section III is devoted to other aspects of statism, including forced integration. In Section IV we take on N. Stephan Kinsella's majoritarian restitution argument, and deal with forestalling and the tragedy of the commons. In Sections V and VI we deal with homesteading and libertarian punishment theory, respectively. We conclude in Section VI with a discussion of the coercive and socialistic nature of immigration controls.

II. Free Trade

Although Hoppe is correct that the problems with the state's unjust socialization of resources can be compounded when numerous immigrants enter the country as part and parcel of "forced integration," no case can be made for the state restricting immigration, at least not on grounds compatible with libertarianism.

On way to see the flaw in his position is by use of the argument reductio ad absurdum. Hoppe says that free immigration means that unwilling taxpayers are forced to finance the living expenses of the new entrants. They use roads, for example.

But surely, goods that are imported into America under provisions of free trade are *also* driven around on roads, and otherwise move through socialized sectors of the economy. Many of the same folks who are forced to fund roads and consider their now-stolen private property to be "invaded" by immigrants, would also consider free-flowing goods from China and Mexico, trucked around on public roads, to be "invasive."

In accepting Hoppe's argument that once private property has been stolen, the state compounds the injustice when it allows immigrants to use the property, thus further "invading" the private property rights of the original owners, we are certainly entitled to draw similar conclusions about free trade.

The point is, what is sauce for the immigration goose is also sauce for the free trade gander. Hoppe cannot be allowed to have it both ways. He (correctly) favors complete free trade, but opposes equally open immigration. He takes this latter stance on the ground that the long-suffering taxpayer is in effect forced to subsidize the newcomers' use of highways and streets. Well and good. But then the same argument can be used against eliminating all tariffs. For the imported goods are *also* trucked around on taxpayer-financed thoroughfares. If he can object to immigrants using roadways, he is compelled by logical necessity to make the same objection to shipping these imported goods on streets and highways.

Another anomaly for the Hoppe position surfaces when we consider migration between cities and states within the US. If migration from, say, Norway or Brazil to the US constitutes an unwarranted "forced integration," then why does not movement of peoples from, say, Texas to Ohio fall under this rubric too? And if it does, then it also applies to labor mobility between cities, such as between New Orleans and Atlanta. And if this holds, then it also applies to migration *within* a city.

III. Statism and Forced Integration

Another logical problem with the Hoppe position is that keeping illegals off public property because of their supposed "invasiveness" could easily be extended to other matters, aside from free trade. Gun laws, drug laws, prostitution laws, drinking laws, smoking laws, laws against prayer — all of these things could be defended on the basis that many tax-paying property owners would not want such behavior on their own private property. Such examples are hardly without a real-world basis. Large numbers of Americans would not allow guests in their homes if those guests had machine guns or crack cocaine in their possession. Extending the principle of the freedom to exclude and set conditions for entry onto private property simply cannot be extended to the socialized public sphere, or else all sorts of unlibertarian, illiberal policies could be as easily justified as border controls. In other words, just because an individual, or many individuals, would not want Act X to occur on their property does not mean that, according to libertarian law, it can be prohibited as a general principle, even on so-called "public-property."

The question then, comes down to allowing the state to determine what and whom to allow and forbid on public land, based on what the taxpayers would decide as it concerns their own private property. This becomes impossible, for states cannot make such economic calculations. In the end, immigration controls empower the state and further the misconception that they can emulate market decisions.

Another good analogy is campaign finance legislation. Certainly, mass democracy and the central state are sins, and the buying of favors by special interest groups is all too much of a reality. And yet, the answer is not to further empower the state so as to limit the invasiveness caused by the socialized sphere of society! The answer is not to use legislative force to keep lobbyists out of the democratic process, so long as the process exists! Is it?

The only policy answer from a libertarian perspective is elimination of democracy, and of public property. This is unrealistic, say the Hoppeans. But even more so is the collectivist notion of the state keeping out immigrants in any way that emulates the market decisions and choices of the taxpayers. Since it is unrealistic, why even consider asking the government to do so? Between two unrealistic choices, why, on libertarian grounds no less, favor the one that necessitates state action?

Furthermore, although it is a solidly sound point that taxpayers who do not wish to see their expropriated wealth go to social services or even roads used by immigrants are in a sense further invaded if their wealth is distributed in this manner, it is equally true that taxpayers who oppose border restrictions, or who in fact especially *want* immigrants to enter their communities, are victimized by any tax-financed restrictions on immigration of which they do not approve. The question remaining is whether we should err in favor of an inclusive policy or an exclusive one, knowing that as long as there are socialized sectors of the economy, any use of the tax-funded resources will be invasive toward those who have been forced to pay. Since the restriction of immigration, carried out by states, is itself a state-enhancing government program, libertarians should err on the side of rejecting state activity and therefore opposing state border controls and immigration laws. All the talk about free immigration empowering the state seems to fly in the face of the logic. It is not as if the state will enforce immigration controls in ways that lessen its power.

Although Hoppe and others have argued that free immigration, for cultural and political reasons, will ultimately lead to bigger government and more statism than restricted immigration, it is vastly problematic to allow and empower the state to protect the borders and enforce immigration laws in the hope that the state will do so in a manner that limits its future size, expense and power. Many policy proposals, in fiscal and monetary policy especially, have constituted attempts to use immediate state action with the intention that it will reduce state action in the long run. Price controls have been rationalized as a way to preempt total bailouts. Deficit spending and education subsidies have been defended on the grounds that they will allow for lower tax rates in the future. Foreign interventions have been advocated as ways to preclude the need for greater interventions and wars at a later time. Government intrusion in the healthcare market has been championed as a method to save future healthcare costs to the public sector. State action in the present to reduce the overall socialization of society, growth of the state, and threat to liberty and private property, has a failed record when put to practice and is therefore pernicious for a libertarian to endorse. In the case of immigration, if it is indeed true that certain types of immigrants tend to serve the interests of the state and further its growth, then it is rather unlikely that the state would ever exclude such immigrants effectively. The incentive simply is not there. But giving the state the power and authority to exclude immigrants does necessarily give it the power, at least potentially, to exclude those immigrants who even the taxpayers most protective of liberty and private property might not want to exclude.

Indeed, Hoppe is correct that open borders compounded with large sectors of socialized society constitute a de facto "forced integration" upon all the native taxpayers and inhabitants who would not normally invite in the immigrants who enter the country. Forced integration is a violation of private property rights and free association, and must be rejected on libertarian grounds. But what of the fact that immigration controls of any feasible sort similarly constitute a de facto forced discrimination? Some taxpayers want immigrants in their country, and to do business with them and associate with them in the framework of the peaceful, if hampered, market economy. Any immigration quota, limiting the number of immigrants per nation of origin, would be riddled with all the same problems as affirmative action in public universities: although it is true that private universities might want to discriminate against white applicants, for example, and accept a disproportionately high number of minority candidates, based on quotas — and although such discrimination is undeniably the private property right of free association of any private organization — it is highly problematic to allow already socialized institutions to practice the same discriminatory entrance policy on the basis of what the market would allow. The state cannot emulate the market; there is the classic economic calculation problem. To allow the state to discriminate on the grounds that private individuals should be allowed is to pave the way to a perverse enervation of free association in its own name.[4]

The free market is a unique mechanism of human organization that respects property rights and freedom of association. No matter what the state does, it cannot simulate the market process. Attempting to shift its policies to better reflect the market preferences of one group will always compromise and invade upon the preferences of another.

---

[4]Private road owners would probably prohibit drunk driving (Block, 2005a). But see Rockwell (2000) who opposes government acting like a private business in this regard.

Increasing the state's power and monopoly over human travel cannot improve the overall level of liberty in society.

IV. The Majoritarian Restitution Argument, Forestalling and the Tragedy of the Commons

Some libertarian theorists, drawing on and expanding upon the Hoppean analysis, have developed additional arguments against open borders. N. Stephan Kinsella has invoked an argument that immigration controls, or at least prohibition of illegal immigrant usage of "public lands," can constitute a form of restitution. Since tax payers have been victimized by the state's forcing them to finance public spaces, they are owed something by the state as victims. Kinsella argues that "restitution need not be made only in dollars. It can be made by providing other value or benefits to the victims." Since "99% of my fellow taxpayers would simply prefer some immigration restrictions, and therefore probably would prefer some kinds of rules of the road that discriminate against outsiders — given this preference, which does not seem per se unlibertarian — it is obvious that far more restitution is made overall if such rules are enacted." (See Kinsella, 2005).

There are several problems with this analysis. First of all, we have similar reductio ad absurdums as we do in the case of the general Hoppean argument. Just because most victims of the state would prefer that the state do something with government property does not make it a priori more just than how the minority would prefer it to be used. If immigration controls are coercive against the innocent in their own right, majority support for them is no more valid than any other program of majoritarian social democracy. A majority of Americans might believe in trade restrictions. A majority might not want people carrying guns in the public sphere — which would just as effectively prevent them from legally having a gun in their homes (by preventing them from transporting them to their homes) as would Kinsella's road restitution proposal effectively prevent people from hiring illegal immigrants to do their housework. Majority rule is no way to determine the justness of public policy, even within the realm of the socialized commons.

Immigration controls, far from being restitution to the victims of tax aggression, only expand the coercive activity of the state against another class of people, including both natives and foreigners. In principle, a victim of robbery has no "right" to direct his assailant to aggress against others as a matter of "restitution," for doing so violates yet other people's natural rights. Being a victim of the state in no way entitles someone to use the state against anyone else. Since socialist policies are such an inefficient drain on the economy, it is inevitable that people's grievances will far outweigh the capacity of the state to compensate them. Moreover, the state does not have its own resources, and it can only "compensate" people by robbing from others.

Another serious problem arises with Kinsella's majoritarian restitution argument. This author correctly points out that restitution "need not be made only in dollars." Also true is the fact that *harm done* by the state against people need not only be in terms of dollars. In other words, taxpayers are not the only victims of the state who have a moral claim to restitution. Most strikingly, the US government has imprisoned hundreds of thousands of people for victimless crime laws, and has, over the years, maimed and killed millions through its wars and interventions in other countries. The victims of war are especially worth considering, for there are millions of such people who have lost their homes, families, livelihoods, and everything they have to the US government's acts of aggression — nearly none of which can be measured in a dollar amount. If anyone has a legitimate claim to restitution from the government, it is the victims of US foreign policy. But should they be granted non-monetary restitution for what they have suffered? If paying taxes into the government gives American taxpayers the right to "benefits" from the government in the form of more restrictive immigration controls, should foreigners who have been even more seriously victimized by the state be given the right to direct US government polices as a compensatory measure? Should they be allowed to dictate the form of immigration controls, trade controls, or regulation of the American economy? It can be argued that they would never have a right to use the US government to enforce policies that are per se unlibertarian, whereas immigration controls and other regulations regarding the already collectivized commons are not per se unlibertarian. But it is hard to imagine Kinsella endorsing the view that innocent Iraqis and other foreigners who lost everything to the US war machine should be compensated in the form of government restrictions on his own freedom of movement on public property, or prohibitions on transferring alcohol on public property, or laws that force women to wear certain traditional Muslim clothing on public property — but that is the logical consequence of his line of argument.

One important consideration in attempting to direct the state to compensate its victims as much as possible through non-monetary "restitution" is whether in doing so the state will victimize others. Hoppe and Kinsella are both eminently correct that no one has a natural right to travel onto property owned by anyone else. But for the state to act on public property in a way that supposedly minimizes the de facto trespass of forced integration, it can only do so justly insofar as it does not violate the libertarian principle of forestalling (Block, 2004, 2005a, 2005b; Block and Whitehead, 2005). Under the concept of forestalling, one has no natural right to do anything that prevents others from exercising their natural rights. If the government were to nationalize all of the land other than private residences, it would be a further invasion of liberty to prevent people from entering the socialized land if doing so means keeping them locked only onto their own private property and precluded from entering that of other consenting, inviting owners.

It is against natural rights for the government to "homestead" the land completely surrounding the privately owned land of another homesteader thus rendering the latter homesteader a prisoner on his own land. The government cannot legitimately maintain all the roads and public spaces in a city, for example, and prevent some people from entering them. Thus we see another problem with Kinsella's argument that the government, or even the taxpayers as a collective entity, can "own" all the roads.[5]

Furthermore, Kinsella's critique of open borders, much like Hoppe's, rests on the notion that the taxpayers own government property. They cannot own it, however, because ownership rights cannot logically conflict in the way that preferences for public resource use inevitably conflict in the tragedy of the commons. Indeed, from Hoppean (1993) and Rothbardian (1998) homesteading theory, we can deduce that much of the land government claims to own is neither private nor public property, but rather no property at all. There is another reason for this conclusion as well: There cannot be two legitimate owners of one and the same property (Hoppe, 1998) at the same time and in the same respect.[6] If the government is the legitimate owner of the property in question[7] then the citizenry cannot also have proper title it; and, of course, the reverse is true as well.

V. Homesteading

If the government nationalized 90% of the land, we would not want the state to keep people off of it who we would not welcome in our own private property. Neither greater public space nor smaller public space warrants immigration controls by the state.

What are the facts of the matter? The brute undeniable situation is that the federal government alone owns 29% of the land mass of the country (Property Rights Alliance, 2005). In some states, this figure is much higher. For example, these are the comparable figures for the states with the highest percentage of governmental (at all levels) land ownership[8]: Alaska (89%), Nevada (81%), Utah (70%), Idaho (67%), Wyoming (55%), Arizona (54%), and California (42%). In general, the so-called "public sector" accounts for far more land ownership west of the Mississippi than east of it.

Now suppose an immigrant, one whom Hoppe thinks is "uninvited," takes it upon himself to "invade" some of this unowned land. How could this author react to such an occurrence? On the one hand, he (Hoppe, 2001, 2002) is on record as opposing any such behavior as "forced integration." On the other hand, he is *also* (Hoppe, 1993) noted for championing the right of homesteading of virgin or hitherto unowned land. There would appear to be somewhat of a "tension" between these two viewpoints, not to say an actual down right contradiction between them.

One possible way, however, to reconcile these seemingly disparate perspectives is to take the position that the government land is really owned by the citizens of the

---

[5]Thanks to BK Marcus for pointing out this application of Block's forestalling principle.

[6]We, of course, abstract from partnerships in making this claim.

[7]A claim that is awkward in the extreme when made in the context of the libertarianism shared by Hoppe and the present authors.

[8]These figures are not that much below the soviet level of some 97%. See on this Gregory and Stuart (1980, 30-33); see also Wadekin (1973).

respective states, or, possibly, all of it by the occupants of the entire country. At first glance, this "works." For, if *all* the land in the U.S. is really owned, the private along with the public, then when an outsider takes over a plot of land in either category, he is not homesteading but rather trespassing. But a moment's analysis will show that this attempted reconciliation cannot really suffice. For, as anyone who has ever been in an airplane above the Rocky Mountains, or practically anywhere in Alaska knows, there are vast stretches that have *never* been so much as touched by human beings. If so, it is hard to see how a staunch homesteading theorist as well as anarchist such as Hoppe can countenance the claim that this land is really owned by a government none of whose agents has even been within miles of their supposed "property."

VI. Punishment Theory

All men of good will can empathize with Hoppe's goals. No one wants to be overrun by hordes of new criminals from abroad. Then, too, immigrants have been implicated in a myriad of other, lesser, transgressions against a civilized order. For example, trespass, public urination and defecation, sexual harassment of the "hey babe" variety, sexual solicitation and littering.[9]

To this we can only say that just as libertarians do not seek punishments against drug use or gun ownership, but only belligerent drug users and gun owners, we should apply the same standard to immigrants.

VII. Conclusion: The Coercive and Socialistic Nature of Immigration Controls

Despite the tragedy of the commons, highlighted so well by Hoppe's trenchant critique with the problem of open immigration in a party socialized society, government immigration controls are per se coercive and socialistic. A border guard necessarily has the power to invade the private property owners along the border, to step on their land searching for illegal aliens. Black markets in immigration will emerge, leading to further crackdowns on illegals, including proposed civil and economic liberties violations such as the National Identification Card and invasions of the privacy of accused employers of illegal immigrants. The enforcement of immigration controls will invariably be burdened by many of the flaws of any other socialist program, which means they will likely fail and be followed by further enhancements of state power. The infrastructure along the borders poses a potential threat to the right of Americans to *emigrate*, as well. Of course, any and all state enforcement of immigration controls will require taxation, takings, government inflationism, and/or some other forms of coercive extraction of wealth (Block, 1998; Gregory, 2005a, 2005b). Because of the socialist economic calculation problem (Boettke, 1991, 1993; Dorn, 1978; Ebeling, 1993; Foss, 1995; Gordon, 1990;

---

[9]Although on this last transgression, see Block (1976, pp. 210–216).

Hoff, 1981; Hoppe, 1989, 1991, 1996; Horwitz, 1996; Keizer, 1987, 1997; Kirzner, 1988; Klein, 1996; Lavoie, 1981, 1985; Lewin, 1998; von Mises, 1975, 1981; Osterfeld, 1992; Pasour, 1983; Reynolds, 1998; Rothbard, 1971, 1976, 1991; Salerno, 1990, 1995; Steele, 1981, 1992), there is neither any way for government immigration controls to work in keeping out the "uninvited," letting in the "invited," or even determining who would fall into each category. The state simply cannot mimic the market, and directing its coercive mechanism in such an attempt will prove ineffective in achieving desirous goals, wasteful of wealth created in the private sector, and destructive to liberty.

Inevitably, of course, immigration controls violate the property rights of those inside America as well as outside, who wish to exchange with each other, and who can indeed maintain the costs of the immigrant's stay.

While our take should not be one of guilt by association, it is more than passing curious that the Hoppe position has been embraced by none other that Democratic Senator Hillary Clinton of New York, who stated: "(I do not) think that we have protected our borders or our ports or provided our first responders with the resources they need, so we can do more and we can do better... I am, you know, adamantly against illegal immigrants." ("Hillary Eyes Immigration", 2004)

The best practical argument is that the people who will implement and enforce any anti-immigration policy will surely be statists. No sort of Hoppean propertarian principles will be implemented by their border police. The state created many problems by socializing half the economy. One of these has to do with immigration. More state socialism conducted by fascists like Hillary is not the answer. State agents who would enforce immigration controls are living on taxed wealth, and thus are acting in ways invasive toward the taxpayers who *do* want to associate with immigrants. It is hard to see how this fact can be reconciled with Hoppe's opposition to government, which he has so aptly described as the "expropriating property protector" as it relates to every other issue.[10]

References

Block, W (1979). Free market transportation: Denationalizing the roads. *Journal of Libertarian Studies: An Interdisciplinary Review*, III(2), 209–238.

Block, W (1998). A Libertarian Case for Free Immigration. *Journal of Libertarian Studies: An Interdisciplinary Review*, 13(2), 167–186. http://www.mises.org/journals/jls/13_2/13_2_4.pdf [22 August 2007].

Block, W (25 May 2004). The State Was a Mistake. [Review of the book *Democracy, The God that Failed: The economics and politics of monarchy, democracy and natural order.*], http://www.mises.org/fullstory.asp?control=1522 [22 August 2007].

Block, W (2005a). *Terri Schiavo: A libertarian analysis*. Unpublished.

Block, W (2005b). *Homesteading, ad coelum, owning views and forestalling*. Unpublished.

---

[10]See, for example, Hoppe's introduction in his *The Myth of National Defense* (2003, p. 8).

Block, W and G Callahan (2003). Is there a right to immigration? A libertarian perspective. *Human Rights Review*, 5(1), 46–71.

Block, W and R Whitehead (2005). Compromising the uncompromisable: A private property rights approach to resolving the abortion controversy. *Appalachian Law Review*, 4(2) 1–45.

Block, W, W Barnett II and G Callahan. The paradox of Coase as a defender of free markets. To appear in *NYU Journal of Law & Liberty*.

Boettke, PJ (1991). The Austrian critique and the demise of Socialism: The Soviet case. In *Austrian Economics: Perspectives on the past and prospects for the future*, Vol. 17, RM Ebeling (ed.), pp. 181–232. Hillsdale, MI: Hillsdale College Press.

Boettke, PJ (1993). *Why Perestroika failed: The politics and economics of Socialist transformation*. London: Routledge.

Brimelow, P (1995). *Alien Nation: Common sense about America's immigration disaster*. New York: Random House.

Dorn, J (1978). Markets true and false in Yugoslavia. *Journal of Libertarian Studies*, 2(3), 243–268.

Ebeling, RM (1993). Economic calculation under Socialism: Ludwig von Mises and his predecessors. In *The Meaning of Ludwig von Mises*, Jeffrey Herbener (ed.), pp. 56–101. Norwell, MA: Kluwer Academic Press.

Epstein, R (1980). *A Theory of Strict Liability: Toward a reformulation of tort law*. San Francisco: Cato Institute.

Foss, N (1995). Information and the Market Economy: A note on a common Marxist fallacy. *Review of Austrian Economics*, 8(2), 127–136.

Gordon, D (1990). *Resurrecting Marx: The analytical Marxists on freedom, exploitation, and justice*. New Brunswick, NJ: Transaction Publishers.

Gregory, A (2005a). In defense of open immigration. *Freedom Daily*. Fairfax, VA: The Future of Freedom Foundation. http://www.fff.org/freedom/fd0410e.asp [22 August 2007].

Gregory, A (2005b). The trouble with 'cracking down on immigration.' http://www.lewrockwell.com/gregory/gregory79.html [22 August 2007].

Gregory, P and R Stuart (1980). *Comparative Economic Systems*. Boston: Houghton Mifflin.

Hillary Eyes Immigration as Top 2008 Issue (21 November 2004). *NewsMax.com*. http://archive.newsmax.com/archives/ic/2004/11/21/233417.shtml [31 August 2007].

Hoff, TJ (1981). *Economic Calculation in a Socialist Society*. Indianapolis: Liberty Press.

Hoppe, H-H (1989). *A Theory of Socialism and Capitalism. Economics, Politics, and Ethics*. Boston: Kluwer Academic Publishers.

Hoppe, H-H (1991). De-socialization in a united Germany. *Review of Austrian Economics*, 5(2), 77–106.

Hoppe, H-H (1993). *The Economics and Ethics of Private Property: Studies in Political Economy and Philosophy*. Boston: Kluwer Academic Publishers.

Hoppe, H-H (1996). Socialism: A property or knowledge problem? *Review of Austrian Economics*, 9(1), 147–154.

Hoppe, H-H (2001). *Democracy, the God that Failed: The economics and politics of monarchy, democracy and natural order*. New Brunswick, NJ: Transaction Publishers.

Hoppe, H-H (2002). Natural order, the state, and the immigration problem. *Journal of Libertarian Studies*, 16(1), 75–97. http://www.mises.org/journals/jls/16_1/16_1_5.pdf [22 August 2007].

Hoppe, H-H (ed.) (2003). The myth of national defense: Essays on the theory and history of Security Production. Auburn, AL: The Ludwig von Mises Institute. http://www.mises.org/etexts/defensemyth.pdf [22 August 2007].

Hoppe, H-H, with G Hulsmann and W Block (1998). Against Fiduciary Media. *Quarterly Journal of Austrian Economics*, 1(1), 19–50. http://www.mises.org/journals/qjae/pdf/qjae1_1_2.pdf [22 August 2007].

Horwitz, S (1996). Money, money prices, and the Socialist calculation debates. *Advances in Austrian Economics*, 3, 59–77. http://www.lewrockwell.com/rockwell/drunkdriving.html [22 August 2007].

Keizer, W (1997). Schumpeter's Walrasian stand in the socialist calculation debate. In *Austrian Economics in Debate*, W Keizer, B Riben and RV Zijp (eds.), London: Routledge.

Keizer, W (1987). Two forgotten articles by Ludwig von Mises on the rationality of Socialist economic calculation. *Review of Austrian Economics*, 1, 109–122.

Kinsella, NS (1 September 2005). A simple libertarian argument against unrestricted immigration and open borders. http://www.lewrockwell.com/kinsella/kinsella18.html [22 August 2007].

Kinsella, NS (1998–1999) (Why 2 years?). Inalienability and punishment: A reply to George Smith. *Journal of Libertarian Studies*, 14(1), 79–93. http://www.mises.org/journals/jls/14_1/14_1_4.pdf [22 August 2007].

Kinsella, NS (1997). A libertarian theory of punishment and rights. *Loyola of Los Angeles Law Review*, 30, 607–645.

Kinsella, S (1996). Punishment and proportionality: The Estoppel approach. *The Journal of Libertarian Studies*, 12(1), 51–74. http://www.mises.org/journals/jls/12_1/12_1_3.pdf [22 August 2007].

Kirzner, IM (1988). The economic calculation debate: Lessons for Austrians. *Review of Austrian Economics*, 2, 1–18.

Klein, PG (1996). Economic calculation and the limits of organization. *Review of Austrian Economics*, 9(2), 3–28.

Lavoie, D (1981). A critique of the standard account of the Socialist calculation debate. *The Journal of Libertarian Studies*, V(1), 41–88.

Lavoie, D (1985). *Rivalry and Central Planning: The Socialist calculation debate reconsidered.* New York: Cambridge University Press.

Lewin, P (1998). The firm, money and economic calculation. *American Journal of Economics and Sociology*, 57(4), 499–512.

von Mises, L (1975/1933). Economic calculation in the Socialist Commonwealth. In *Collectivist Economic Planning*, FA Hayek (ed.)., pp. 87–103Clifton, NJ: Kelley.

von Mises, L (1981/1969). *Socialism.* Indianapolis: LibertyPress/LibertyClassics.

Osterfeld, D (1992). *Prosperity versus planning: how government stifles economic growth.* New York: Oxford University Press.

Pasour Jr, EC (1983). Land-use planning: Implications of the economic calculation debate. *The Journal of Libertarian Studies*, 7(1), 127–139.

Pirie, M (1986). *Privatization in Theory and Practice.* London: Adam Smith Institute.

Property Rights Alliance (2005). Federal Government Land and Building Ownership: Physical Property. http://www.propertyrightsalliance.org/index.php?content=fedlndown [31 August 2007].

Reynolds, MO (1998). The impossibility of Socialist economy. *The Quarterly Journal of Austrian Economics*, 1(2), 29–43.

Rockwell, LH (3 November, 2000). Legalize drunk driving. http://www.mises.org/sory/2343 [22 August 2007].

Rothbard, MN (1971). Lange, Mises and Praxeology: The retreat from Marxism. In *Toward Liberty,* Vol. II, pp. 307–321. Menlo Park, CA: Institute for Humane Studies. Reprinted in *The Logic of Action One: Method, Money, and the Austrian School,* pp. 384–396. Cheltenham, UK: Edward Elgar Publishing Ltd., 1997.

Rothbard, MN (1976). Ludwig von Mises and economic calculation under Socialism. LS Moss (ed.), *The Economics of Ludwig von Mises,* pp. 67–77. Kansas City: Sheed & Ward.

Rothbard, MN (1991). The end of Socialism and the calculation debate revisited. *Review of Austrian Economics*, 5(2), pp. 51–70.

Rothbard, MN (1993). *Man, Economy, and State,* Vols 1 & 2. Auburn, AL: Ludwig von Mises Institute. http://www.mises.org/rothbard/mes/chap15d.asp [22 August 2007].

Rothbard, MN (1994). Nations by consent: Decomposing the nation-state. *The Journal of Libertarian Studies,* 11(1), 1–10.

Rothbard, MN (1998/1982). *The Ethics of Liberty.* Atlantic Highlands, NJ: Humanities Press. http://www.mises.org/rothbard/ethics/ethics.asp [22 August 2007].

Salerno, JT (1990). 'Postscript: Why a Socialist economy is impossible'. In *Ludwig von Mises, Economic Calculation in the Socialist Commonwealth,* pp. 51–71. Auburn, AL: Ludwig von Mises Institute.

Salerno, J (1995). A final word: Calculation, knowledge and appraisement. *Review of Austrian Economics,* 9(1), 141–142. http://www.mises.org/journals/rae/pdf/rae9_1_8.pdf [22 August 2007].

Steele, D (1981). Posing the problem: The impossibility of economic calculation under Socialism. *The Journal of Libertarian Studies,* V(1), 7–22. http://www.mises.org/journals/jls/5_1/5_1_2.pdf [22 August 2007].

Steele, DR (1992). *From Marx to Mises: Post Capitalist society and the challenge of economic calculation.* La Salle, IL: Open Court.

Stromberg, J (17 May 1999). World War I. http://www.mises.org/fullstory.aspx?control=224&id=77 [22 August 2007].

Wadekin, K-E (1973). *The Private Sector in Soviet Agriculture.* Berkeley: University of California Press.

V. Redistributive Justice

On Reparations to Blacks for Slavery[1]

Walter Block

Introduction

This paper is an attempt to shed light on the legitimacy of some recent claims by prominent black leaders and scholars that reparations are owed to members of their race, and should be paid for by the U.S. government out of tax revenues. I shall critically consider in the light of libertarian theory both the views in favor of this position put forth by Robinson[2] and those against it of Horowitz[3] <http://www.salon.com/news/col/horo/2000/05/30/reparations/index.html>. See also Myron Magnet, *The Dream and the Nightmare*, New York: Manhattan Institute for Policy Research, especially the chapter,"Race and Reparations."

Libertarianism

Libertarianism is a political philosophy with private property rights at its core. Its axiom is that physical invasions of persons or property are unjustified,[4] and should be punished. It is based on a variant of Lockean[5] homesteading theory, according to which mixing one's labor with the land justifies ownership of it, whether or not Locke's proviso of"enough and as good still being available"is met.

This proviso is all well and good when there are vast lands unsettled, as in the U.S. frontier of historical memory. But what is to be done when virtually all usable land has been taken up? There are only several possibilities. We can resort to government ownership[6]. But why should land socialism[7] work any better than the economic variety? In any case, members of the apparatus of the state, by stipulation, did not mix their labor with the land, or do anything else which would remotely justify their ownership status over it. Why, then, should it be granted? True, the government can auction off the land to the highest bidders, or on a first come first served basis, but why would the owners who eventuate from such a process—initially unjustified—be preferable to those whose ownership is based on homesteading? This process would indeed see more money placed into the coffers of the state, but it is easy to make the case that they already have far too much wealth and control over the economy as it is.[8] The only other candidate is claim theory; ownership, here, is based on a mere affirmation. But this also fails to establish any connection between the owner and that which is owned. In addition, there is the problem of vast over determination, as anyone would be free to claim anything he wishes.

Alterations in Property Titles

Having established initial ownership in property, the next step in determining justice in property titles is to outline a theory of how they can legitimately change hands from one person to another. This may be done in any non-invasive manner possible, e.g., trade, gifts, inheritance or gambling, for these are the only options compatible with ownership in the first place.[9] That is, if I give you my ring in exchange for your car, this is logically consistent with property rights; if I merely seize your auto, it is not. Nozick calls this "legitimate title transfer."[10]

It must be emphasized that the key element of libertarian punishment theory[11] is an attempt to make the victim "whole," preeminently by compensating him. While this is never fully possible, the goal is to attain this state of affairs insofar as possible. Crimes, in this perspective, are not committed against some general "society," and the main emphasis is not on incarceration, much less reform. Rather, a crime such as assault and battery, murder, rape, etc., is seen as aimed primarily at the victim. Jail, to the extent it arises in a libertarian society, is merely a way of forcing hard labor upon the perpetrator in an attempt to get him to compensate the victim.

Reparation Theory

Justified reparations are nothing more and nothing less than the forced return of stolen property—even after a significant amount of time has passed. For example, if my grandfather stole a ring from your grandfather, and then bequeathed it to me through the intermediation of my father, then I am, presently, the illegitimate owner of that piece of jewelry. To take the position that reparations are always and forever unjustified is to give an imprimatur to theft, provided a sufficient time period has elapsed. In the just society, your father would have inherited the ring from his own parent, and then given it to you. It is thus not a violation of property rights, but a logical implication of them, to force me to give over this ill gotten gain to you.[12] "In short, we cannot simply talk of defense of 'property rights' or of 'private property' *per se*. For if we do so, we are in grave danger of defending the 'property right' of a criminal aggressor—in fact, we logically must do so." Of course, "possession is nine tenths of the law." It is not sufficient, on your part, merely to claim that the ring now on my finger is

Block   **55**

rightfully yours. You must be forthcoming with specific evidence undergirding this demand. A dated picture of your grandfather wearing it, or a bill of sale, would do just fine. Moreover, it is only I who owe you this piece of jewelry, not my neighbor or the general taxpayer,[13] and it is owed only to you, not to any person who wants it, or to those of a given race or ethnicity. Further, I am not a criminal for innocently possessing the ring before you came to claim it, but I am guilty of a criminal act once it is proven that the ring was really your grandfather's and I refuse to give it up to you.

Precisely the same analysis applies to slavery. Owning a slave is a crime under libertarian law. The Nuremberg Trials have established the validity of ex post facto law. Those people who owned slaves in the pre civil war U.S. were guilty of the crime of kidnaping, even though such practices were legal at the time. A part of the value of their plantations was based on the forced labor of blacks. Were justice fully done in 1865, these people would have been incarcerated, and that part of the value of their holdings attributable to slave labor would have been turned over to the ex slaves. Instead, these slave masters kept their freedom, and bequeathed their property to their own children. Their (great) grandchildren now possess farms which, under a regime of justice, would have never been given to them. Instead, they would have been in the hands of the (great) grandchildren of slaves. To return these specific lands to those blacks in the present day who can prove their ancestors were forced to work on these plantations is thus to uphold private property rights, not to denigrate them.

Horowitz

Horowitz seems to have had a knee-jerk reaction to the claims of Jesse Jackson, John Conyers, Randall Robinson and their confreres.[14] Since they have been wrong in just about everything they have ever said in the areas of economics, politics, discrimination, race relations, ethics, etc., he presumes that this applies in this case as well.[15] But here, perhaps through sheer good luck, they have finally hit upon a principle compatible with libertarianism and the free society. Because of his inability to discern a pro free enterprise viewpoint when it comes from so unlikely a quarter, Horowitz is then unable to tax these people with their logical inconsistency; they are Johnny come latelies to the banner of capitalism. If they really wish to press their reparations claims, which are based on the doctrine of private property rights (returning possessions to their rightful owners), then they must renounce all of their previous positions which are incompatible with this vision, e.g., their support for welfare, unions, government intervention into the economy, regulation, business nationalization, etc. Alternatively, if they insist on maintaining these spurious views, then upon pain of contradiction, they must withdraw their demand for reparations, based on stolen (labor) property.[16]

Paradoxically, even if Jackson, Conyers, Gates, Robinson, Farakhan, Lewis, Afrik, Thornton, et al. do change their tune on property rights in general, and thus are logically enabled[17] to press for reparations, they will derive no great measure of comfort from it, as a practical matter. This is because it is notoriously difficult to trace back· property titles back in history for any great length of time, particularly if there were no written records kept (as was the case with Indians and other pre literate people).[18] This is why, again only as a practical matter, there are no implications of the libertarian theory of reparations to far off events such as Mongol hordes, competing claims in Jerusalem from 2000 years ago, etc. Reparations theory comes into its own regarding more recent occurrences such as land stolen in the USSR, Cuba, East Germany, etc., where scrupulously accurate records are available. Black slavery in the U.S. occupies an intermediate position; it took place a century and a half ago, and while there were written records, many have been lost in the sands of time. It is only from the position of an all-knowing God that reparations, from as far back in history as you wish to go, written records or no, are relevant to property titles in the present time.

There is another reason black leaders cannot take too much comfort from libertarian reparations theory. Suppose there were 500 slaves on a plantation, but the grandchildren of only one of them can be found. They are entitled to split amongst themselves not all the contributions made by all the slaves, but rather only one-five-hundredth of that, the estimate of the productivity of their own ancestor alone.

Why is this? Would it not be more reasonable to award the children of this one slave the fruits of the labor of all the slaves? At first blush, this is a tenable idea. After all, at the time of the freedom of the slaves, were justice to have reigned at that time, the product of their entire output would have been given to them; none of it at all would have remained in the slave master's hands.[19] And if the ex slave owner would not have been able to keep any of this property, he would not have been able to hand it down to his own progeny. Instead, it would have been under the control of the ex slaves, and, with time, wended its way into the hands of blacks now alive.

Although not an altogether unreasonable scenario, it is simply incompatible with libertarian law. This is because we must look at this matter not from the point of view of 1865, and on the assumption that the offspring of all 500 slaves can be found, but rather from the perspective of the case we are assuming; that is, it is now the modern era, almost a century and a half after these historical events have unfolded, and we can demonstrate a connection between only one slave and persons now living. Yes, the property in question, in justice, never should have remained in the hands of the slave master; but it did. He handed it on to his innocent children, and they

Block **57**

to theirs. Now, as judges, we are faced with blacks who can trace their roots back to only one of the 500 slaves. Why should they be entitled to land to which they have no connection. The extant owners, at least, are not themselves guilty of any land theft or slave holding, and have established homestead rights to that which they occupy.

Rothbard explains:

But suppose that Jones[20] is *not* the criminal, not the man who stole the watch, but that he had inherited or had innocently purchased it form the thief. And suppose, of course, that neither the victim nor his heirs can be found.[21] In *that* case, the disappearance of the victim means that the stolen property comes properly into a state of no-ownership. But we have seen that any good in a state of no-ownership, with no legitimate owner of its title, reverts as legitimate property to the first person to come along and use it, to appropriate this now un-owned resource for human use. But this 'first' person is clearly Jones, who has been using it all along. Therefore, we conclude that even though the property was originally stolen, that *if* the victim or his heirs cannot be found, *and if* the current possessor was not the actual criminal who stole the property, then title to that property belongs properly, justly, and ethically to its current possessor.

To sum up, for any property currently claimed and used: (a) if we *know* clearly that there was no criminal origin to its current title, then obviously the current title is legitimate, just and valid; (b) if we *don't* know whether the current title had any criminal origins but can't find out either way, then the hypothetically 'unowned' property reverts instantaneously and justly to its current possessor; (c) if we *do* know that the title is originally criminal, but can't find the victim or his heirs, then (c1) if the current title-holder was not the criminal aggressor against the property, then it reverts to him justly as the first owner of a hypo-thetically unowned property. But (c2) if the current title-holder is himself the criminal or one of the criminals who stole the property, then clearly he is prop-erly to be deprived of it, and it then reverts to the first man who takes it out of its unowned state and appropriates it for his use. And finally, (d) if the current title is the result of crime, *and* the victim or his heirs can be found, then the title properly reverts immediately to the latter, without compensation to the crimi-nal or to the other holders of the unjust title.[22]

There are three reasons why black leaders[23] should jettison their social-ist leanings, and begin to support capitalism. One, it will inure to the ben-efit of their followers right now in ways unrelated to reparations. Two, they will be able to logically maintain their position on reparations as a matter of principle. And three, some few black grandchildren might actually be able to trace their claims back in time to the pre civil war era, and thereby obtain some compensation, under libertarian law.

A Critique

In one sense, I have nothing critical to say about Horowitz (2000). He opposes "the idea that taxpayers should pay reparations to black Ameri-cans for the damages of slavery and segregation" and so do I. My argument is that no one owes anything to anyone for segregation, since the law of free association guarantees (or should guarantee) that anyone can discriminate against anyone else for any reason, or no reason at all. Secondly, I maintain that although reparations are indeed owed to some blacks, from some whites, for slavery, all blacks should not be creditors in this regard, nor all (non-black) taxpayers, debtors.

In another sense, I look upon Horowitz (2000) with profound disquiet. For one thing, this essay "proves" far too much; in the view of this writer, there are no blacks at all who are owed reparations from anyone. That is, presumably, the entire concept of reparations for past crimes is somehow invalid, or at least when applied to black slavery in the U.S. For another, most of his arguments are contrary to the libertarian doctrine of reparations in general; they are not limited to all blacks being owed reparations by all (non-black) taxpayers. As such, Horowitz is arguing against a bunch of philosophically very weak straw men: so-called "civil rights" leaders such as Rep. John Conyers, D-Mich., Henry Louis Gates of Harvard, Jesse Jackson, and Randall Robinson, the author of "The Debt: What America Owes to Blacks." By laying waste to their arguments, he concludes that no reparations are owed in this case, and that is a fallacy.

I do not quarrel in every respect with Horowitz's decision to take these people to task for their many and serious lapses of logic and intellectual coherence. Their views may be dead from the neck up, but they are certainly politically powerful. However, I would not want the impression left that the only arguments for reparations[24] are the ones put forth by these representatives of the "civil rights" establishment. And since Horowitz (2000) argues against *all* reparations for black slavery in the U.S., seemingly as a matter of principle, not mere expediency, it is important that a necessary corrective of his views be entertained. Further, he overlooks one important criticism of his opponents that they richly deserve to have rubbed in their faces.

Among the charges made by Horowitz is that the call for reparations for black slavery will negatively "impact on race relations and [lead to] the self-isolation of the African-American community." To my mind, these are purely peripheral issues. In this reply I shall instead focus on whether these claims are *just*. After all, it is entirely possible that to hang an innocent man will have positive effects on race relations and reduce isolation of the black community. Even if this merely utilitarian consideration is true, it is still almost unworthy of consideration. Of far more importance is the *justice*, or lack of same, underlying these claims.

Ten Reasons

Horowitz considers and rejects ten separate claims for reparations. In what follows, I shall comment on each, following his order of presentation.

Block   **59**

1. States Horowitz:"Assuming there is actually a debt, it is not at all clear who owes it."Our author is perfectly correct in objecting to the claim made by our friends in the"black studies"departments of our major universities that all Americans owe blacks a debt for slavery. No one living now, clearly, was alive during that unhappy epoch; not everyone in the U.S. has illegitimately inherited property not properly belonging to their ancestors. But just because not everyone owes blacks for enslaving their forebears does not mean no one does. As we have seen, the present possessors of wealth handed down to them through the generations, emanating from slavery, do indeed owe a debt to those who can prove they are the direct decedents of the slaves involved.

Horowitz argues"It was not whites but black Africans who first enslaved their brothers and sisters. They were abetted by dark-skinned Arabs ... who organized the slave trade. Are reparations going to be assessed against the descendants of Africans and Arabs for their role in slavery? There were also 3,000 black slave owners in the antebellum United States. Are reparations to be paid by their descendants too?"He asks these questions as if there were no possible affirmative answers. Slave holding, or slave capturing are crimes. This is so regardless of the skin color of masters or victims. Yes, a hundred times yes, if a black person A can prove that money now held by another member of his own black race, B, was inherited improperly by B, and that the grandparents of A were the victims, then this property *should* be transferred from B to A. The races of A and B are strictly irrelevant.

2. It may well be that the socialist black advocates of reparations rely on "the idea that only whites benefited from slavery."As Horowitz avers, this claim"is factually wrong...."But this criticism is the reddest of red herrings as far as the libertarian case for reparations is concerned. I might"benefit" in some direct or indirect sense from any number of goings on. For example, if I own a detective agency, then my profits rise with juvenile delinquency rates. But this does not make me responsible for the crime wave in the first place. An orange grower in Florida benefits from a frost which kills this type of fruit in California; but no one is rash enough to blame him for the bad weather on the other side of the country.

In like manner, Horowitz scores heavily against his straw men opponents by noting that"American blacks on average enjoy per capita incomes in the range of 20 to 50 times those of blacks living in any of the African nations from which they were kidnapped,"and asks,"What about this benefit of slavery? Are the reparations proponents going to make black descendants of slaves pay themselves for benefiting from the fruits of their ancestors' servitude?"Yes, these are telling arguments against those of Robinson, et. al. But they are irrelevant to the libertarian case on behalf of returning stolen property to the modern descendants of African slaves. It is not a matter of adding up benefits and costs, and subtracting the one from the other. On the contrary, there are specific people who now own property they should not have inherited. This physical property and land, and it alone, is vulnerable to transfer. Since "benefits" play no role in the justification, the question of comparing those which helped blacks against those which hurt them does not arise.

Suppose a man rapes a woman, and it is later somehow proven that had he not molested her in this way, she would have instead been run over by a bus and killed. Should this fact mitigate the punishment imposed on him? Not a bit of it. He is a rapist, and should be punished to the full extent of the law. It is entirely irrelevant that in some sense blacks gained from their association with whites, through slavery and kidnaping. The enslavers and the kidnapers should still be punished.

3. Horowitz plaintively and very tellingly asks: "Why should the descendants of non-slaveholding whites owe a debt? What about the descendants of the 350,000 Union soldiers who died to free the slaves? They gave their lives. What possible morality would ask them to pay (through their descendants) again?" This is all well and good insofar as is concerned the claim that all (white) Americans owe a debt to blacks. But it does not at all relieve of obligation the descendants of specific white plantation[25] slave owners to give up their ill inherited gains.

4. Our author in this section makes the point that the children of post slavery immigrants to the U.S. owe blacks nothing for the subjugation of their ancestors, since their ancestors had nothing to do with it, and I am entirely in accord with him on this.

5. Not so, unfortunately, with regard to the precedents established by reparations to Jews from Germany and Japanese-Americans from the U.S. Horowitz rejects these analogies on the entirely spurious ground that "The Jews and Japanese who received reparations were individuals who actually suffered the hurt." But what does it matter whether the payments go to the people actually brutalized, or to their children, who would have, in any case, inherited from them? In taking this position, our author is implicitly arguing that my grandfather owes his grandfather the ring he stole from him, but that when they both die, all bets are off between himself and myself regarding the return of this stolen property. But Horowitz gives no reason for believing that there is some sort of natural statute of limitations for crimes which calls justice to a complete halt when the specific victims and victimizers vanish from the scene.

Horowitz continues: "Jews do not receive reparations from Germany simply because they are Jews. Those who do were corralled into concentration camps and lost immediate family members or personal property. Nor have all Japanese-Americans received payments, but only those whom the government interned in camps and who had their property confiscated." Yes, this argument will suffice against the black "leaders" who argue for compensation for all people of a certain skin color, but it falls by the wayside as far as the more powerful libertarian case is concerned.

6. According to Horowitz, "Behind the reparations arguments lies the unfounded claim that all blacks in America suffer economically from the consequences of slavery ..." His "exhibit A" to the contrary is the case of wealthy blacks such as Oprah Winfrey. But this is not only insufficient to undermine the libertarian case, it doesn't even lay a glove on the case put forth by Jesse Jackson, et al. Even the latter do not claim a transfer of wealth on the ground that blacks are poorer than whites; they do so out of (somewhat misguided) claims of *justice*. Is it not possible that rich people can be oppressed? It seems to be Horowitz's argument that this cannot occur; it deserves to be characterized, and rejected, as the "Oprah Winfrey fallacy." This is part and parcel of the *left wing* philosophy which sees the poor as impoverished because of the wealth of the rich,[26] of the view that the well off cannot, by definition, be victims.

It matters not one whit how affluent is Oprah. If she is the great-granddaughter of a slave who worked on the XYZ plantation in Alabama, and she can prove this, then she is entitled, as a matter of libertarian law, and justice, not welfare statism, to a portion of that which she would have received from her great-grandparents upon their release from bondage, had full justice occurred at that time. Her "extraordinary achievement" does not at all "refute ... the reparations argument," contrary to Horowitz. All of his discussion about the prosperous black middle class, and the underclass, and the relative success of "West Indian blacks in America (who) are also descended from slaves," is all beside the point. No one asked how rich were the Japanese Americans who had their property stolen, nor about the wealth of Jews in Germany. This was entirely irrelevant, and properly so. Why is it apropos in the case of blacks?

7. In this section Horowitz argues that black claims will be resented by other ethnic groups. This is but another red herring. We are here concerned with the *justice* of the claim for reparation, not what others will think of it.

8. Horowitz's next sally concerns the "'reparations' to blacks that have already been paid. Since the passage of the Civil Rights Acts and the advent of the Great Society in 1965, trillions of dollars in transfer payments have been made to African-Americans, in the form of welfare benefits and racial preferences (in contracts, job placements and educational admissions)—all under the rationale of redressing historical racial grievances."

This is not at all a bad argument against the case of the left wing blacks for reparations. After all, they want taxpayers' money, and they have already had quite a bit of that.

However, it will not suffice against the libertarian claim. Yes, there were "transfer payments" galore, but not a penny of this was in the form of true reparations. This is because the latter can only come *from* the illegitimate holders of what is in effect stolen property, not from entirely innocent tax-payers as a whole, and none of these"transfers"came courtesy of the present owners of plantation grounds and buildings. Further, proper reparations can only go to the grand children of those were slaves. The welfare system fails as a candidate for reparations on both these grounds. In addition, the so-called"Great Society"payments actually harmed blacks, rather than help them. As Charles Murray[27] has shown, giving blacks money under provisos which discouraged the formation of their families only undermined their economic and social conditions. The Jews and the Japanese-Americans were given reparations with no such strings attached. Why should blacks be any different, were these true reparations? Further, it was not only blacks who suffered from welfare: so did all recipients, of whatever hue.

9. Horowitz is also mistaken with regard to the relevance of"the debt blacks owe to America—to white Americans—for liberating them from slavery."Contrary to this author, the Civil War was *not* fought to end enslavement; it was undertaken in order to quell secession.[28] However, there is no doubt, as our author eloquently attests, that"there never was an anti-slavery movement until white Englishmen and Americans created one."That ought to put quite a spoke in the wheels of those who deride the culture and philosophy of"dead white males." But of what relevance is the fact that *some* whites acted in a manner which greatly benefited blacks (e.g., the Union soldiers) to the claim that *other* whites (the grand children of plantation owning slave holders) owe them a debt of reparations? Horowitz is free with his condemnation of"racism"against his interlocutors, but here it would appear he is guilty of precisely this himself. That is, whites are not all alike. Some of them owe a debt to blacks for slavery, others actually helped blacks. But the latter only erases the former if, somehow, members of the same race (the white one, in this case) are interchangeable. And I take this implicit claim to be incorrect and objectionably racist.

10. According to Horowitz,"The final and summary reason for rejecting any reparations claim is recognition of the enormous privileges black Americans enjoy as Americans, and therefore of their own stake in America's history, slavery and all."

What are these"privileges"? If they are the ability to participate in a (semi) free enterprise system, then they are rights, not privileges. If this refers to the disproportionate number of blacks on welfare, or who have "benefited"from programs such as affirmative action, then these are not privileges either; they are rights violations. In either case, however, how do these "benefits"relieve the whites who are now in possession of the farm grounds and buildings from their responsibilities to give up their ill gotten gains?

Our author warns"the African-American community [against] isolat[ing] itself even further from America [for this] would be to embark on a course whose consequences are troublesome even to contemplate. Yet the black

Block **63**

community has had a long-running flirtation with separatists and nationalists in its ranks, who must be called what they are: racists who want African-Americans to have no part of America's multiethnic social contract."

But what is so wrong with separatism? If blacks wish to isolate themselves from Horowitz and his ilk, i.e., whites, it is their right. Does this author believe in forced integration with unwilling blacks? Are we to have school busing again, only this time with an impetus from a different direction? Will he oppose black secession, if that is what (some) blacks want? And if so, on what ground? Does he not believe in freedom of association, which would allow blacks (or anyone else) to withdraw from unwanted social (or other) contact?

And which "social contract," pray tell, is he referring to? Did anyone ever sign this contract? If not, why is it incumbent upon anyone to respect it? According to libertarian law, the only thing that all people must respect are the persons and property of everyone else. Anything more is made up by Horowitz as he goes along.

Yes, many of the leftists who urge reparations to all blacks from all non-black U.S. taxpayers are admirers of "Fidel Castro, one of the world's longest-surviving and most sadistic dictators." But this does not by one iota denigrate the rights of the black grandchildren of slaves to compensation from the white grandchildren of slave owners, for the property and value created through this brutal human chattel system.

States Horowitz: "For all their country's faults, African-Americans have an enormous stake in America and above all in the heritage that men like Jefferson helped to shape. This heritage—enshrined in America's founding and the institutions and ideas to which it gave rise—is what is really under attack in the reparations movement." But this is a non sequitur of the highest order. Just because some advocates of reparations couple their demands with anti Americanism does not at all mean that their goal is not justified; it does not mean that all arguments for reparations must be coupled with a denigration of this country. Certainly, that applies to the libertarian case.

Unfair?

The thought will perhaps occur to some that I have been unfair to Horowitz. After all, here he is criticizing the argument for reparations from all (non black) citizens to all blacks, coupled with a healthy dose of socialism and anti Americanism, and I am attacking him for overlooking the possibility that reparations from some whites to some blacks is justified. It might appear that I am guilty of launching a bit of an intellectual ambush upon this author.

I plead innocent of these charges. Horowitz is not merely rejecting one version of reparations, he is attacking the entire concept. I join with him for the most part, of course, in his dismissal of the case put forth by Randall Robinson, Jesse Jackson, John Conyers, Henry Louis Gates, Dorothy Lewis, Hannibal Afrik, Albert Thornton, and numerous professors of black"studies." But it is Horowitz's view that, having dealt with these straw men, he has demolished the entire argument for reparations. He has not, and this was the burden of my reply to him. Horowitz simply cannot be allowed to maintain, as he does, that his devastation of a weak and faulty case for reparations undermines all argument for this conclusion. And his unwarranted attack upon the rich, his mistaken embrace of what I have characterized as the Oprah Winfrey fallacy, must be addressed. This is no ambush, but rather a measured reply to an essay which in many ways is correct, but also contains numerous and grave flaws.

Objections

Let us consider a series of objections to the foregoing, some of which might be launched by Horowitz, others not.

1. "Most land in the south was stolen by carpetbaggers. Why wouldn't they owe compensation? Suppose the carpetbagger stole the land from the guilty slave owner, and then sold it to its present owner. Surely, the present owner would escape liability in this case."

If the carpetbagger (or his heirs, to whom he gave his ill gotten gains) can be located, then the grandchildren of the slave have no case against the present occupier, but instead must obtain their compensation from the grandchildren of the carpetbagger. However, suppose, as is more likely, that the carpetbagger and his brood have vanished without a trace. Then we have only the grandchild of the slave, and the present (innocent) owner. The question is, which of them is the legitimate title holder? The libertarian answer is clear: the property must go to its rightful owner, the children of the slave.

According to Rothbard[29] :"suppose that a title to property *is* clearly identifiable as criminal, does this necessarily mean that the current possessor must give it up? No, not necessarily. For that depends upon two considerations: (a) whether the *victim* (the property owner originally aggressed against) or his heirs are clearly identifiable and can now be found; *or* (b) whether or not the current possessor is *himself* the criminal who stole the property. Suppose, for example, that Jones possesses a watch, and that we *can* clearly show that Jones's title is originally criminal, either because (1) his ancestor stole it, or (2) because he or his ancestor purchased it from a thief (whether wittingly or unwittingly is immaterial here). Now, *if* we can identify and find the victim or his heir, then it is clear that Jones's title to the watch is totally invalid, and that it must promptly revert to its true and legitimate owner. Thus, if Jones inherited or purchased the watch from a man who stole it from Smith, and if Smith or the heir to his estate can be

Block **65**

found, then the title to the watch properly reverts back to Smith or his descendants, *without* compensation to the existing possessor of the criminally derived 'title.'"

In the context of our example, Jones is the present owner of the land, the carpetbagger is the thief, and Smith is the grandchild of the slave. To allow Jones to keep his land in the face of proof from Smith that he is the rightful owner, is to not uphold legitimacy in private property rights; it is to denigrate it. If A is the rightful owner, B steals property from A, sells it to C and then disappears, there is only one correct answer to the question of who should keep it, according to libertarianism: A. C is out of luck, unless he can somehow locate B.

2. "If a person is an unjust owner, can he legitimately bequeath his property? No. If not, then he cannot properly sell it either. Thus, the libertarian search for the successor of the original (and rightful) owner will also have to include those who purchased the land. Every land transaction after 1860 will have to be declared null and void. And why stop with 1860, since that date is arbitrarily based on the beginning of the war? In theory, there is nothing to preclude the requirement from going back in history until the beginning of time."

This sounds like a telling criticism of the libertarian theory of property rights, but it is not. Again we call upon Rothbard (1998, p. 57) to elucidate: "if we do not *know* if Jones's title to any given property is criminally derived, then we may assume that this property was, at least momentarily, in a state of non ownership... and therefore that the proper title of ownership reverted instantaneously to Jones as its 'first' (i.e., current) possessor and user. In short, where we are not sure about a title but it cannot be clearly identified as criminally derived, then the title properly and legitimately reverts to its current possessor."

The point is, the libertarian theory is actually a very conservative one, not at all calling for research in property titles back to the dawn of history. Unless proven otherwise, every extant property title is to be considered legitimate. The burden of proof, that is, rests squarely on the shoulders of those who wish to *overturn* duly registered property. If there is any historical research to be undertaken, this must be done by those who wish to refute already accepted claims, not those who wish to defend them.

In sum, this objection puts things in the exact opposite position from where they are, as far as practicality is concerned. It misconstrues the burden of proof.

3. "The white Southerners who permitted blacks to live on their land should not be held more responsible for reparations than Northerners who refused to even allow blacks to pass through their borders."

There are several problems here. First, it is not the case that before the Civil War slavery was confined to the south; yes, it predominated there, but this section of the country had no monopoly over the"curious institution." Second, compare all those who engaged in slavery vs. all those who did not, but would not allow blacks entry into their territory. Surely the former would be regarded by all, not just libertarians, as by far the more serious crime. That is, kidnapping is a major violation of human rights, while prohibiting immigration is at best only a minor one.[30] If so, then if heavy reparations are due to those who were enslaved, only light ones are due to those who were forcibly prevented from engaging in immigration.

4."The slave owners fed and sheltered their property. These expenses should be offset against any debt owed by their progeny to the grandchildren of slaves. With any reasonable discount rate, since so many years have already passed, there will be little or nothing due to the children of slaves in reparations."

Suppose I kidnap you and keep you prisoner for a year. Whereupon I am caught by the forces of law and order, and argue, in my defense, mitigation in that I fed and sheltered you for this duration of time. This should be dismissed out of hand. For had I not fed and sheltered you in this context, I would have been guilty of the more serious crime of murder. That I am "only" charged with kidnapping results from the fact that at least I did keep you alive during your period of capture. In other words, feeding and housing you does indeed mitigate the charge of *murder*. But, as I am not being charged with that crime, only with kidnapping, this offset has already been made. To put this in the parlance of accounting, this objection incorporates the fallacy of double counting.

5."White landowners paid freed blacks a form of private welfare after their emancipation. Many former masters allowed former slaves to stay on their land, and fed them too, as a matter of charity. These monies should be subtracted from any reparations owed."

The difficulty with this objection is that there were two separate acts, one of slaveholding, the other of charity, and it is not clear that the two are connected in any way. Let us return to the case where my grandfather stole a ring from your grandfather. In addition to this act of theft, this ancestor of mine also gave some charity to your grandfather. Does the second act mitigate the first? Not at all. Were my grandfather hauled into court on charges of stealing a ring from your grandfather, it would avail him nothing in defense, nor should it, that in a completely separate incident, he acted charitably toward your grandfather.

So, charity will not do. However, there is a kernel of truth in this objection, in that while mere charity will not suffice, reparations will. That is, had my grandfather given yours money not as an act of charity, but out of contrition and reparations, then and only then would this count in ameliorating the debt I now owe you.

Let us now return to the slave reparations case. Suppose that Oprah Winfrey can prove her great grandfather was a slave on the XYZ plantation, and that the descendants of this owner owe her $X. Then from this, using the same rate of discount, should be subtracted whatever reparations—not charity—the master of XYZ gave to her ancestor. However, it should be noted that the burden of proof is now reversed. Oprah has, let us say, satisfied her burden of proof. But now the present holders of the XYZ lands, whoever they are, must prove that the original owner supported his ex slaves not as a matter of charity, but out of a motive of repatriating money owed to them. If so, then the' one can indeed be an offset against the other.

6. "The whites in the U.S. who bought the slaves purchased them not from freedom but from slavery, the state in which they were found in Africa. Shouldn't the original enslavers, then, bear a great deal of the burden? Perhaps there should be a search through the tribal history of Africa to discover the identities of all the descendants of enslavers dating back centuries?"

This is meant as a reductio ad absurdum of the libertarian position. The proper answer is to take the bull by the horns and accept the premise: yes, if the black slave catchers in Africa who first captured the slaves who were later to be sent to the U.S. can be identified, then the money they left to their grandchildren would become vulnerable to a lawsuit brought by the grandchildren of these U.S. slaves. This is meant to sound preposterous, but instead it has all the earmarks of a just solution. Libertarian law cares not at all for the color of a person's skin. If he is a slave master, of whatever race, he is guilty of the crime of slave holding or slave capturing.

7. "Many blacks, after the war, were drafted into the army and sent west to slaughter Indians. Public slavery in effect replaced private slavery. Why focus on crimes further back in time rather than crimes more recent in time?"

There are several difficulties with this objection. For one thing, while a military draft and slavery are both crimes against humanity, there are surely relevant differences. For another, the answer to the question of which crimes to pursue, earlier or later ones, is "both." Just because the U.S. government drafted newly freed blacks to kill Indians (for which compensation is due, from those responsible) does not for a minute allow a free ride to those guilty or earlier crimes. Libertarian punishment theory is a timeless concept. It matters not when or where or why a crime was committed. If it can be proven, there is a case for pursuing the victimizers, and for turning over the fruits of the crime to their rightful owners.

8. "The demand for slavery reparations is a diversion to put all Americans—black, white, brown—off their guard as tax slavery emerges on a global scale.

> "It sounds farfetched, but hear my tale. Consider first how nonsensical is the reparations argument. Whites, none of whom are or were slave owners, would be making transfer payments to blacks, none of whom are or were slaves.

"Not even a'sins of the fathers'rationale justifies race reparations. A majority of white Americans are descendants of people who immigrated to these shores after slavery had come to an end. Similarly, many blacks are descendants of people who arrived in the United States in the post-slavery era. Indeed, the millions of"preferred minorities"who have arrived in the last three decades are legally privileged compared to native born whites.

"To find descendants of slave owners and descendants of slaves would require more extensive racial genealogies than Nazi Germany and South Africa were able to assemble for their race-based policies. And how would we classify people with ancestors in both camps? Would they pay reparations to themselves? Would descendents of black slave owners pay reparations?"[31]

There is a difficulty with Roberts' characterization of taxes as akin to slavery, or worse, an instance of it. A more accurate description of taxation is not slavery but theft. Both the tax gatherer and the holdup man demand money from you against your will. In contrast, the kidnapper and the enslaver, but neither the tax-man nor the robber, take over your physical body—if you give in to their demands. An objection might be that the tax collector comes to you at the behest of a majority vote, and offers services in return for the money he mulcts from you. Neither suffices. First, assume that not one but *two* highwaymen accost you, demanding your money. When you demur on grounds of democracy (they are philosophical thieves and are willing to discourse with you) they hold an election as to whether you should keep your money or give it to them, and they win by a majority two to one vote. The point is, you have not *agreed* to be part of this"election," no more than you have consented to be part of the"robber gang called the U.S. government."[32] Second, posit that these stickup men offer the"service"of giving you a paper clip in return for your money. Again, the point is not whether those who take your money by force give you something in return, but whether or not you have assented to the deal.

There are problems, too, with Roberts' dismissal of the arguments for reparations to today's blacks for the enslavement of their ancestors. First, I am not arguing this case on the basis of"sins of the fathers."Rather, reparations are justified because"sins of the sons,"namely, not giving up the property they never should have received from their sinning fathers in the first place, to the sons of those from whom it was stolen. This is a subtle difference, but an important one nonetheless.

Further, if indeed it is true that reparations would"require more extensive racial genealogies than Nazi Germany and South Africa were able to assemble for their race-based policies,"so what. Perhaps we can do better than those two countries, genealogical-wise. And if we cannot, this is *still* not an argument against the *justice* of reparations, as maintained by Roberts, but only of the inability of the sons of slaves to prove their case. Based on the principle that"possession is nine tenths of the law,"these disputed properties would stay in the hands of the white grandchildren of slave owners.

Block **69**

And last but not least, the phenomenon of mixed race children places no insuperable barrier to our analysis. For it is unlikely in the extreme that any slave master would have given his *black* children any of his land or possessions. If so, then there will·be no *black* descendents of slave owners who owe other children of slaves any reparations. But let us suppose that this is indeed the case. That is, Mr. X, the mixed race child of a white male slave owner and a black. female slave,[33] is given some land by his father, which he duly hands down to his own descendants. These particular blacks would have not case for reparations in the present day since, by stipulation, they have *already* been given them. Roberts' problem is that he is analyzing the argument for reparations as if it is made on a racial basis. While this is indeed true for some proponents of this program,[34] this certainly does not apply to libertarianism.

Notes

1. The author holds the Wirth Endowed Chair of Economics at Loyola University New Orleans. He wishes to thank Jeff Tucker of the Mises Institute for helpful comments and criticism, and Hannah Block for editing.
2. Robinson, Randall, *The Debt: What America Owes to Blacks*. See also Richard America *Paying the Social Debt: What White America Owes Black America*.
3. Horowitz, David, "The latest civil rights disaster: Ten reasons why reparations for slavery are a bad idea for black people—and racist too."
4. On libertarianism, see Anderson, Terry and Hill, P.J., "An American Experiment in Anarcho-Capitalism: the *not* so Wild, Wild West," *Journal of Libertarian Studies* Vol. 3, No. 1, 1979, pp. 9-29; Barnett, Randy E., *The Structure of Liberty: Justice and the Rule of Law*, Oxford: Clarendon Press, 1998; Benson, Bruce L., 1989, Enforcement of Private Property Rights in Primitive Societies: Law Without Government," *The Journal of Libertarian Studies*, Vol. IX, No. 1, Winter, pp. 1-26; Benson, Bruce L., "The Spontaneous Evolution of Commercial Law," *Southern Economic Journal*, 55: 644-661, 1989; Benson, Bruce L., *The Enterprise of Law: Justice Without the State*, San Francisco: Pacific Research Institute for Public Policy, 1990; Cuzán, Alfred G., "Do We Ever Really Get Out of Anarchy?," *Journal of Libertarian Studies*, Vol. 3, No. 2 (Summer, 1979); De Jasay, Anthony, *The State*, Oxford: Basil Blackwell, 1985; Friedman, David, *The Machinery of Freedom: Guide to a Radical Capitalism*, La Salle, IL: Open Court, 2nd ed., 1989; Friedman, David, "Private Creation and Enforcement of Law: A Historical Case," *Journal of Legal Studies*, 8: 399-415, 1979; Hoppe, Hans-Hermann, *A Theory of Socialism and Capitalism: Economics, Politics and Ethics*, Boston: Kluwer, 1989; Hoppe, Hans-Hermann, *The Economics and Ethics of Private Property: Studies in Political Economy and Philosophy*, Boston: Kluwer, 1993; Hoppe, Hans-Hermann, "The Private Production of Defense," *Journal of Libertarian Studies*, Vol. 14, No. 1, Winter 1998-1999, pp. 27-52; Hummel, Jeffrey Rogers, National Goods Versus Public Goods: Defense, Disarmament, and Free Riders, 4 *Rev. Austrian Econ.* 88 (1990); Morriss, Andrew P., "Miners, Vigilantes and Cattlemen: Overcoming Free Rider Problems in the Private Provision of Law," *Land and Water Law Review*, Vol. XXXIII, No, 2, 1998, pp. 581-696; Peden, Joseph R., 1977, "Property rights in Celtic Irish law," *The Journal of Libertarian Studies*, Vol. 1, No. 2, Spring, pp. 81-96; Rothbard, Murray N., *For a New Liberty*, Macmillan, New York, 1978; Rothbard, Murray N., *The Ethics of Liberty*, Humanities Press, Atlantic Highlands, N.J., 1982; Rothbard, Murray N., "Society Without a State." J. R. Pennock and J. W. Chapman (eds.), *Anarchism: Nomos XIX*. New York: New York University Press, 1978, pp. 191-207; Rothbard, Murray N., *Man, Economy and State*, Auburn AL: Mises Institute, 1993;

Skoble, Aeon J."The Anarchism Controversy," in *Liberty for the 21stCentury: Essays in Contemporary Libertarian Thought*, eds. Tibor Machan and Douglas Rasmussen, Lanham MD: Rowman and Littlefield, 1995, pp. 77-96; Sechrest, Larry J.,"Rand, Anarchy, and Taxes," *The Journal of Ayn Rand Studies*, Vol. I, No. 1, Fall 1999, pp. 87-105; Spooner, Lysander, *No Treason*, Larkspur, Colorado, (1870) 1966; Stringham, Edward, "Justice Without Government," *Journal of Libertarian Studies*, Vol. 14, No. 1, Winter 1998-1999, pp. 53-77; Tinsley, Patrick,"With Liberty and Justice for All: A Case for Private Police," *Journal of Libertarian Studies*, Vol. 14, No. 1, Winter 1998-1999, pp. 95-100; Tannehill, Morris and Linda, *The Market for Liberty*, New York: Laissez Faire Books, 1984; Woolridge, William C., *Uncle Sam the Monopoly Man*, New Rochelle, N.Y.: Arlington House, 1970.

5.   John Locke, *An Essay Concerning the True Original, Extant and End of Civil Government, in* SOCIAL CONTRACT 17-18 (E. Barker ed., 1948)

6.   The U.S. Bureau of Land Management's control over much of the land west of the Mississippi has been more of a disaster than anything else. See on this Anderson, Terry L., and Hill, Peter J.,"Property Rights as a Common Pool Resource," *Bureaucracy vs. Environment: The Environmental Costs of Bureaucratic Governance*, John Baden and Richard L. Stroup, eds., Ann Arbor: University of Michigan Press, 1981; Anderson, Terry L., and Leal, Donald R., Free Market Environmentalism, San Francisco: Pacific Research Institute, 1991; Block, Walter,"Ownership will save the environment," *New Environment*, First Quarter, 1991, 41-43; Block, Walter,"Protection of property rights key to maintaining resources," *Environment Policy and Law*, Vol. 1, No. 3, June 1990, p. 28; Block, Walter, ed., *Economics and the Environment: A Reconciliation*, Vancouver: The Fraser Institute, 1990; Block, Walter,"Earning Happiness Through Homesteading Un-owned Land: a comment on 'Buying Misery with Federal Land' by Richard Stroup," *Journal of Social Political and Economic Studies*, Vol. 15, No. 2, Summer 1990, pp. 237-253; Hill, Peter J., and Meiners, Roger E., eds., *Who Owns the Environment?*, New York: Rowman and Littlefield, 1998; Hoppe, Hans-Hermann, *The Economics and Ethics of Private Property: Studies in Political Economy and Philosophy*, Boston: Kluwer, 1993; Horwitz, Morton J., *The Transformation of American Law: 1780-1860*, Cambridge: Harvard University Press, 1977; Margaret N. Maxey and Robert L. Kuhn, eds., *Regulatory Reform: New Vision or Old Curse*, New York: Praeger, 1985; Rathje, William L.,"Rubbish!," *Atlantic Monthly*, Vol. 264, No. 6, December 1989, pp. 99-109; Ray, Dixie Lee, 1990, *Trashing the Planet*, Washington D.C.: Regnery Gateway; Rothbard, Murray N.,"Law, Property Rights, and Air Pollution," *Economics and the Environment: A Reconciliation*, Walter Block, ed., Vancouver: The Fraser Institute, 1990; Stroup, Richard L., and John C. Goodman, et. al., (1991) *Progressive Environmentalism: A Pro-Human, Pro-Science, Pro-Free Enterprise Agenda for Change*, Dallas, TX: National Center for Policy Analysis, Task Force Report; Stroup, Richard L., and Baden, John A.,"Endowment Areas: A Clearing in the Policy Wilderness," *Cato Journal*, 2 Winter 1982, pp. 691-708

7.   The latest managerial failure of the bureaucrats, the U.S. Park Service in this case, has been a forest fire set by the authorities themselves as a preventative; the only difficulty is that it raged out of control, creating millions of dollars of damages. See on this "Los Alamos Under Siege," *Newsweek*, 5/22/00, p. 35. Were any private enterprise guilty of so massive a blunder, it surely would have become enmeshed in bankruptcy proceedings. But by the very nature of things this fail-safe mechanism is available to markets, not governments.

8.   Gwartney, James, Robert Lawson and Walter Block, *Economic Freedom of the World, 1975-1995*, Vancouver, B.C. Canada: the Fraser Institute, 1996

9.   Nozick, Robert, *Anarchy, State, and Utopia*, New York: Basic Books Inc., 1974; N. Stephan Kinsella,"A Theory of Contracts: Binding Promises, Title Transfer, and Inalienability" (paper presented at Auburn, Alabama, April 1999, Ludwig von Mises Institute's Austrian Scholars Conference 5; published version forthcoming).

10. Nozick, Robert, *Anarchy, State, and Utopia*, New York: Basic Books Inc., 1974, p. 150.

11. On libertarian punishment theory, see Barnett, Randy, and Hagel, John, eds., *Assessing the Criminal*, Cambridge MA: Ballinger, 1977; Block, Walter,"Toward a Libertarian Theory of Guilt and Punishment for the Crime of Statism," Huelsmann, Guido, ed., *The Rise*

Block **71**

> *and Fall of the State,* forthcoming; Block, Walter,"National Defense and the Theory of Externalities, Public Goods and Clubs," Hoppe, Hans-Hermann, ed., *Explorations in the Theory and History of Security Production,* forthcoming King, J. Charles, A Rationale for Punishment, 4 J. Libertarian Stud. 151, 154 (1980); Kinsella, Stephan N.,"A Libertarian Theory of Punishment and Rights," (volume) 30 Loy. L.A. L. Rev. 607-45 (1997); Kinsella, Stephan N.,"New Rationalist Directions in Libertarian Rights Theory,"12:2 J. Libertarian Studies 313-26 (Fall 1996); Kinsella, Stephan N.,"Punishment and Proportionality: The Estoppel Approach," 12:1 J. Libertarian Studies 51 (Spring 1996); Kinsella, Stephan N., "Estoppel: A New Justification for Individual Rights," Reason Papers No. 17 (Fall 1992), p. 61; Kinsella, N. Stephan,"Inalienability and Punishment: A Reply to George Smith,"*Journal of Libertarian Studies,*Vol. 14, No. 1, Winter, 1998-1999, pp. 79-93; Rothbard, Murray N., *The Ethics of Liberty,* Humanities Press, Atlantic Highlands, N.J., 1982.

12. States Rothbard, Murray N., *The Ethics of Liberty,* New York: New York University Press (1998 [1982], pp. 51-52):"Suppose we are walking down the street and we see a man, A, seizing B by the wrist and grabbing B's wristwatch. There is no question that A is here violating both the person and the property of B. Can we then simply infer from this scene that A is a criminal aggressor, and B his innocent victim?

    "Certainly not—for we don't know simply from our observation whether A is indeed a thief, or whether A is merely repossessing his own watch from B who had previously stolen it from him. In short, while the watch had undoubtedly been B's property until the moment of A's attack, we don't know whether or not A had been the legitimate owner at some earlier time, and had been robbed by B. Therefore, we do not yet know which on of the two men is the *legitimate* or *just* property owner. We can only find the answer through investigating the concrete data of the particular case, i.e., through 'historical' inquiry.

    "Thus, we *cannot* simply say that the great axiomatic moral rule of the libertarian society is the protection of property rights, *period.* For the criminal has not natural right whatever to the retention of property that he has stolen; the aggressor has no right to claim any property that he has acquired by aggression. Therefore, we must ... clarify the basic rule of the libertarian society to say that no one has the right to aggress against the *legitimate* or *just* property of another.

13. The freed slaves were presumably promised "40 acres and a mule." If from the slave master, well and good; this is at least an approximation of what they were owed. If this was to have been derived, however, courtesy of the taxpayer, then it would be unjust, since the average citizen was not responsible for slavery. For the view that voting does not make one responsible for the acts of politicians, see Spooner, Lysander, *No Treason,* Larkspur, Colorado, (1870) 1966.

14. Others who have made similar reparations claims in behalf of blacks include Dorothy Lewis and Hannibal Afrik of the National Coalition of Blacks for Reparations in America, and Albert Thornton, chairman of the political science department at Howard University, a historically black college. See on this *The National Post,* 6/7/00, p. A3.

15. I confess this is a very tempting conclusion to draw. Why else do we have induction if not for things just of this sort?

16. Their probable response to this sally would be to attempt to have their cake and eat it, presumably by claiming that "linear" logic is an invention of dead white males, and doesn't apply to them. It is impossible to take such people seriously. In any case, my argument is with the far more coherent Horowitz, not them. Here, I am only pointing out that Horowitz does not carry forth as fully as he might his attack on these black "scholars," due to his inability to see that for once they are on the side of the angels.

17. In Kinsella's terminology, they are no longer "estopped" from making these claims. See on this Kinsella, Stephan N.,"A Libertarian Theory of Punishment and Rights," (volume) 30 Loy. L.A. L. Rev. 607-45 (1997); Kinsella, Stephan N.,"New Rationalist Directions in Libertarian Rights Theory,"12:2 J. Libertarian Studies 313-26 (Fall 1996); Kinsella, Stephan N.,"Punishment and Proportionality: The Estoppel Approach,"12:1 J. Libertarian Studies 51 (Spring 1996); Kinsella, Stephan N.,"Estoppel: A New Justi-

fication for Individual Rights," Reason Papers No. 17 (Fall 1992), p. 61; Kinsella, N. Stephan, "Inalienability and Punishment: A Reply to George Smith," *Journal of Libertarian Studies*, Vol. 14, No. 1, Winter, 1998-1999, pp. 79-93

18. A court in Canada has ruled that written records are not required for proof of ownership. The recollections of tribal elders will suffice in their stead. Here is the court's finding in R. v. Van Der Peet, [1996] 2 S. C. R. 507, a case concerning Indian fishing rights, from the summary: "A court should approach the rules of evidence and interpret the evidence that exists, conscious of the special nature of aboriginal claims and of the evidentiary difficulties in proving a right which originates in times when there were no written records of the practices, customs and traditions and customs engaged in. The courts must not undervalue the evidence presented by aboriginal claimants simply because that evidence does not conform precisely with the evidentiary standards applied in other contexts." This finding played a role in the decision concerning a land reparations case, Delgamuukw v British Columbia, [1997] 3 S.C.R. 1010, which cited paragraph #68 of R. v. Van Der Peet, [1996] 2 S. C. R. 507, which has been just summarized. To say the least, this determination is not at all compatible with libertarian requirements of proof. For one thing, the testimony may be a lie. For another, it is not disinterested. For a third, it may be honestly believed, but mistaken. This seems to be the conclusion in many cases of recovered memory of girlhood incest charges on the part of adult women. But there is a practical implication as well. If this unwarranted decision were to become a precedent, then, truly, the reductio ad absurdum charge against the libertarian position that it would open the floodgates of land reparations cases back before the beginning of recorded history could then be sustained. But this is only a utilitarian consideration, unworthy, probably, of our attention.

19. He, himself, would have been very seriously punished for the crime of slave holding.

20. In our context, Jones is the white grandchild of the slave owner, who is now in possession of the property under dispute.

21. Remember, in our example, the 499 victims and their heirs cannot be found.

22. Rothbard, Murray N., *The Ethics of Liberty*, Humanities Press, Atlantic Highlands, N.J., 1982, pp. 58-59

23. We must never lose sight of the fact that there are *some* black scholars who are avid supporters of free enterprise. Unfortunately, while this honor roll is very distinguished, it is also very short. Among economists it includes Williams, Walter, E., *The State Against Blacks*, New York, McGraw-Hill, 1982; Williams, Walter E., *South Africa's War Against Capitalism*, New York: Praeger, 1989; Williams, Walter E., and Walter Block, "Male-Female Earnings Differentials: A Critical Reappraisal," *The Journal of Labor Research*, Vol. II, No. 2, Fall 1981, pp. 385-388; Sowell, Thomas, *The Vision of the Annointed*, New York: Basic Books, 1995; Sowell, Thomas, Race and Economics, New York: Longman, 1975; Sowell, Thomas, *Patterns of Black Excellence*, Washington D.C.: Ethics and Public Policy Center, 1976; Sowell, Thomas, *Pink and Brown People*, San Francisco: The Hoover Institution Press, 1981; Sowell, Thomas, Ethnic America, New York: Basic Books, 1981; Sowell, Thomas, "*Weber* and *Bakke* and the presuppositions of 'Affirmative Action,'" *Discrimination, Affirmative Action and Equal Opportunity*, Walter Block and Michael Walker, eds., Vancouver: The Fraser Institute, 1982; Sowell, Thomas, *The Economics and Politics of Race: An International Perspective*, New York, Morrow, 1983; Sowell, Thomas, *Civil Rights: Rhetoric or Reality*, New York: William Morrow, e1984; Sowell, Thomas, *A Conflict of Visions: Ideological Origins of Political Struggles*, New York: William Morrow, 1987; Sowell, Thomas, "Preferential Policies," in *Thinking About America: The United States in the 1990s*, Annelise Anderson and Dennis L. Bark, eds., San Francisco: The Hoover Institution Press, 1988; Sowell, Thomas, *Inside American Education: The Decline, the Deception, The Dogmas*, New York: The Free Press, 1993; Sowell, Thomas, *Race and Culture: A World View*, New York: Basic Books, 1994.

24. For the libertarian case in behalf of reparations, see Block, Walter and Yeatts, Guillermo "Land Reform," *University of Kentucky Journal of Natural Resources and Environmental Law*, forthcoming; Block, Walter, "Review Essay of Bethell, Tom, *The Noblest Triumph: Property and Prosperity Through the Ages*, New York: St. Martin's Press, 1998," in *The*

Block **73**

Quarterly Journal of Austrian Economics, Vol. 2, No. 3, Fall 1999, pp. 65-84; Walter Block, David Friedman, Milton Friedman, Philip Wogaman, Kenneth Boulding, Walter Berns, Edmund Opitz, Paul Heyne and Geoffrey Brennan, *Discussion,* in MORALITY OF THE MARKET: RELIGIOUS AND ECONOMIC PERSPECTIVES, Walter Block, Geoffrey Brennan & Kenneth Elzinga, eds., Fraser Institute: Vancouver, 1985, pp. 495- 510; Rothbard, Murray N., *The Ethics of Liberty,* Humanities Press, Atlantic Highlands, N.J., 1982. Levin, Michael, *Why Race Matters: Race Differences and What They Mean,* New York: Praeger, 1997, pp. 229-289 opposes reparations to blacks, but not for reasons relevant to our present concerns.

25. There is a temptation to assume that all defendants in black reparations cases would be southerners. This must be resisted. Strictly speaking, of course, slavery took place not only in the South, but in the North as well.
26. Horowitz may have had "second thoughts" about some of his previous political economic philosophy, but, judging from his analytic framework in the present case, there are still vestiges of Marxism infecting his perspective.
27. Murray, Charles, *Losing Ground: American Social Policy from 1950 to 1980,* New York: Basic Books, 1984
28. Hummel, Jeffrey Rogers, *Emancipating Slaves, Enslaving Free Men: A History of the American Civil War,* Chicago: Open Court, 1996; DiLorenzo, Thomas, *The Real Lincoln,* Roseville, CA: Forum/Random House, 2002.
29. Rothbard, Murray N., *The Ethics of Liberty,* New York: New York University Press (1998 [1982], pp. 57-58); for more on the libertarian theory of private property rights, see Hoppe, Hans-Hermann, *A Theory of Socialism and Capitalism: Economics, Politics and Ethics,* Boston: Kluwer, 1989; Hoppe, Hans-Hermann, *The Economics and Ethics of Private Property: Studies in Political Economy and Philosophy,* Boston: Kluwer, 1993
30. For a libertarian debate on whether immigration laws are per se criminal, see *Journal of Libertarian Studies: An Interdisciplinary Review,* Vol. 13, No. 2, summer 1998. If true, then all those who can prove that their grandparents would have immigrated to a given country have a reparations case under libertarian law not against all citizens of that nation, but only against the grandchildren of those few people specifically responsible for this law.
31. Roberts, Paul Craig, "Taxing Away Freedoms," TownHall.Com Columnists, 7/21/01. See http://www.townhall.com/columnists/paulcraigroberts/pcr20010719.shtml, accessed on 7/21/01
32. See Spooner, Lysander, *No Treason,* Larkspur, Colorado, (1870) 1966)
33. The opposite case is a possible albeit less historically likely occurrence.
34. E.g., see Robinson, op. cit.

THE ECONOMICS AND ETHICS OF LAND REFORM: A CRITIQUE OF THE PONTIFICAL COUNCIL FOR JUSTICE AND PEACE'S "TOWARD A BETTER DISTRIBUTION OF LAND: THE CHALLENGE OF AGRARIAN REFORM"

WALTER BLOCK* AND GUILLERMO YEATTS**

I. INTRODUCTION

Land reform can be defined as the forced transfer of the ownership of land from one person to another. This must be distinguished from the *voluntary* transfer of land from one person to another, such as ordinary buying and selling. In the latter case, there can be no true third party. All those involved in land purchases -- agents, insurers, lawyers -- are mere agents of one party or the other to the agreement. But in the former there *must* be a third actor, the one who forces one party to give up his land to a second party. This third party is typically the government. Were this third party not involved in the process, there could be no land reform.

In most land reform debates only two sides are represented. These two, together, dominate all such discussions. On the one hand there are those, usually called conservatives or right wing activists, who oppose land reform per se.[1] They maintain that such forced transfers are in total opposition to private property rights. From this perspective land reform must be stopped, as private property rights are the very bedrock of civilization. Once these rights are breached all law, to say nothing of the economic well being of the populace, is in grave danger.

On the other hand there are those, usually called socialists or left wing activists, who maintain that land reform is justified *if and only if* the donors of the land are rich or powerful and the recipients are poor

---

*Chairman, Department of Economics and Finance, University of Central Arkansas.

**Chairman, Phoebus Energy, Ltd. B.S., M.A. New York University, OPM Harvard University Graduate School of Business Administration.

[1] *See* LUDWIG VON MISES, SOCIALISM 45 (J. Kahane trans., Yale University Press 1951); *see also* DAVID FRIEDMAN, THE FRASER INSTITUTE, MORALITY OF THE MARKET: RELIGIOUS AND ECONOMIC PERSPECTIVES 497-98, 506-07 (1985).

or powerless. Their argument is one of equity. Socialists maintain that the privileged of the world are few; they own vast tracts of land which are often not cultivated. In contrast, many people are poor and on the verge of starvation. They have the labor power necessary to feed themselves, but not the land upon which to do this.

One of the purposes of this paper is to explore a third alternative in this debate, the libertarian or classical liberal position. This analysis is important for three reasons. First, the other two perspectives are at loggerheads. There is, seemingly, no possible reconciliation between them. It is possible that the introduction of a third philosophy of property rights and land reform may serve as an ameliorating device, allowing each of the other two parties to compromise with one another. Here, we speak not of compromise in the sense of adding up the differences and dividing by two, but rather of a *principled* compromise where each is able to retain the best part of its vision. Second, we shall maintain that while both the socialist and conservative view have aspects of justice on their side, neither has a monopoly in this regard. As presently constituted, both appear fatally flawed. Each philosophy, however, shares certain libertarian elements. If these can be brought out into the clear light of day, we can more closely approximate the truth of the matter. In this essay, we shall attempt to offer the view that only the classical liberal vision is in accord with justice.

Finally, even if we cannot fully succeed in defending this rather ambitious claim, this exercise will still prove beneficial by providing a third alternative for society to consider. It may well be that this dialogue, now intellectually dead in its tracks, may take on some new light with the advent of a third viewpoint.

II. Libertarian Land Reform

A. The Libertarian Philosophy of Land Reform

What, then, is the libertarian philosophy of land reform? Such a theory maintains both that forced transfers of land from one person to another *are* justified, and that far from being incompatible with a strict regime of private property rights, land transfers are sometimes *required* by this doctrine. While these transfers are often from rich to poor, this is not required. It is even possible that justified land reform *may* enrich the wealthy and impoverish the poor. Conservatives may therefore find comfort in the fact that the libertarian position strongly *upholds* private property rights, while socialists may exult in the fact that land reform is not prohibited by the libertarian doctrine.

How can we reconcile these seemingly irreconcilable claims? It is simple. Valuable items are sometimes stolen. When these items are forcibly returned to their rightful owners, private property rights are protected. The following example supports this position:

> Suppose we are walking down the street and we see a man, A, seizing B by the wrist and grabbing B's wristwatch. There is no question that A is here violating both the person and property of B. Can we then simply infer from this scene that A is a criminal aggressor, and B his innocent victim?
>
> Certainly not -- for we *don't know* simply from our observation whether A is indeed a thief, or whether A is merely repossessing his own watch from B who had previously stolen it from him. In short, while the watch had undoubtedly been B's property until the moment of A's attack, we don't know whether or not A had been the legitimate owner at some earlier time, and had been robbed by B. Therefore, we do not yet know which one of the two men is the *legitimate* or *just* property owner. We can only find the answer through investigating the concrete data of the particular case, i.e., through 'historical' inquiry.
>
> Thus, we *cannot* simply say that the great axiomatic moral rule of the libertarian society is the protection of property rights, *period.* For the criminal has no natural right whatever to the retention of property that he has stolen; the aggressor has no right to claim any property that he has acquired by

aggression. Therefore, we must modify or rather clarify the basic rule of libertarian society to say that no one has the right to aggress against the *legitimate* or *just* property of another.

In short, we cannot simply talk of defense of 'property rights' or of 'private property' *per se.* For if we do so, we are in grave danger of defending the 'property right' of a criminal aggressor -- in fact, we logically must do so.[2]

The libertarian, then, must favor this "wristwatch" reform, as A is the proper owner of the stolen property, while B is merely its criminal possessor. Consider, however, the following situation: B, the thief, passed the watch down to B', his son, whereupon B", the grandson, inherited the watch and currently has possession. Presume also that had A not had the time piece stolen from him, it would have ended up the property of A", through a similar inheritance process. At this point A" goes to the police and demands the return of the watch from B". The latter objects that it is his *private property,* and to engage in the "wristwatch reform" described above constitutes socialism or communism. With our libertarian insights, however, it is easy to see the error of this response. Protecting B" under these conditions would not uphold private property rights, but rather would denigrate them. B" simply has no leg to stand upon. While he is not a thief, he is the possessor of what must be considered stolen property. To allow him to continue holding the watch would be to keep it from its rightful owner, A".

Of course, in this scenario the burden of proof rests squarely with A", the person attempting to alter and abolish present property titles. As the old legal adage goes, "Possession is nine tenths of the law," as it is the best evidence, in our uncertain world, of legitimate title. But this is only a presumption. With appropriate evidence, it can be defeated. If A" can prove he is the just owner, then the watch should be returned to him. If somehow the forces of law and order could travel

---

[2]MURRAY ROTHBARD, THE ETHICS OF LIBERTY 51-52 (1982).

back in time to capture B, the robber, they would be entitled to do more than merely relieve him of his ill gotten gains.[3] While B" is entirely innocent of the original crime, it would be a *denial* of private property rights to allow him to keep this stolen watch.

The problem with the conservative position is that it sometimes amounts to a defense of thieves. Since the conservative philosophy is linked with capitalism in the public mind, the free enterprise system is widely dismissed as merely upholding the "survival of the fittest," or of the claims of the rich against the poor. But nothing could be further from the truth. Free enterprise does not at all condone land,or any other kind, of theft; rather, this doctrine advocates the return of stolen property.

If the libertarian position deviates from the conservative philosophy, it also differs with the socialist view of redistribution from rich to poor. To put this another way, libertarians agree with the conservative advocacy of private property rights, but not with the claim that it constitutes a legitimate defense of all extant property titles. Rather, the libertarians agree with the socialists that forced transfer of property titles are sometimes justified when it is necessary to effectuate justice.

B. Property Redistribution

What is morally wrong with the socialist vision of coercively taking possessions from rich people and giving them to the poor? First, it is *theft*, and civilized societies have always looked askance upon such actions. To the extent that massive stealing becomes the order of

---

[3]*See generally* RANDY BARNETT & JOHN HAGEL, ASSESSING THE CRIMINAL 309-21 (1977); Charles J. King, *A Rationale for Punishment*, 4 J. LIBERTARIAN STUDIES 151, 154 (1980); Stephan N. Kinsella, *A Libertarian Theory of Punishment and Rights*, 30 LOYOLA L. REV. 607, 613 (1997); Stephan N. Kinsella, *New Rationalist Directions in Libertarian Rights Theory*, 12 J. LIBERTARIAN STUDIES 313, 321 (1996); Stephan N. Kinsella, *Punishment and Proportionality: The Estoppel Approach*, 12 J. LIBERTARIAN STUDIES 51, 57 (1996); Stephan N. Kinsella, *Estoppel: A New Justification for Individual Rights*, Reason Papers No. 17 (Fall 1992); Stephan N. Kinsella, *Inalienability and Punishment: A Reply to George Smith*, 14 J. LIBERTARIAN STUDIES 79-93 (1998); MURRAY ROTHBARD, THE ETHICS OF LIBERTY 51-55 (1982).

the day, a social breakdown is the inevitable result. The Ten Commandments prohibit not only robbery, but even coveting the property of others.[4]

Additionally, egalitarians are unable to follow the Kantian imperative to make their principles the basis upon which all men act;[5] if we may redistribute land and presumably money from rich to poor, then why may we not steal from the rich and give to the poor those characteristics which have allowed them to become wealthy in the first place? That is, assuming we had the ability to do so, we could take away the musical ability of Mozart, the athletic accomplishments of Michael Jordan, the entrepreneurial and innovational attainments of Bill Gates, the intelligence of Steven Hawking and the sense of humor of Jay Leno. Were we to engage in such egalitarian redistribution, and continue the process until there were no difference between those rich and poor in talent and characteristics, we would have truly reduced the human race to the undifferentiated blob of a *Brave New World*.[6]

Third, there is simply no practical reason to engage in land reform from rich to poor if the goal is to help the poor attain a better standard of living. If there is one thing that most people have learned in the last few decades (or, in any case, *should* have learned) it is that the last best chance for lifting the poor out of poverty is not to give them the property of others, but rather grant them economic freedom.[7] In comparison to other factors, the redistribution of land does little to increase overall wealth.

Support for this position can be found in economist David Friedman's critique on the redistribution of Native American land in the United States, wherein he wrote,

---

[4] *See Exodus* 20:1-17.

[5] *See* Immanuel Kant, Theory and Practice II; *Introduction to the Theory of Right, in* KANT'S POLITICAL WRITINGS (Hans Reiss ed. & H.B. Nisbet trans., Cambridge Univ. Press 1970).

[6] ALDOUS HUXLEY, BRAVE NEW WORLD (1946).

[7] *See* JAMES GWARTNEY ET AL., ECONOMIC FREEDOM OF THE WORLD 1975-1995 151-70 (1996); *see also* HERNANDO DE SOTO, THE OTHER PATH: THE INVISIBLE REVOLUTION IN THE THIRD WORLD 214-29 (1989); ANTHONY DE JASAY, THE STATE 208-32 (1985); HANS-HERMANN HOPPE, A THEORY OF SOCIALISM AND CAPITALISM: ECONOMICS, POLITICS AND ETHICS 56 (1989).

In the U.S. at the moment, if you gave the country back to the Indians, in some fair way where you didn't give them the buildings that are built on it, but just the land; and divided it fairly evenly among the Indians, it would not noticeably affect the distribution of income in the U.S. It wouldn't much affect how well off I am.[8]

III. HOMESTEADING

How, in the libertarian view, is an original claim to land justified? We begin our analysis with an excerpt of John Locke's homesteading theory, wherein he states:

[E]very man has a *property* in his own *person*. Thus nobody has any right to but himself. The *labour* of his body and the *work* of his hands, we may say, are properly his. Whatsoever, then, he removes out of the state that nature hath provided and left it in, he hath mixed his labour with it, and joined it to something that is his own, and thereby makes it his property. It being by him removed from the common state nature placed it in, it hath by this labour something annexed to it that excludes the common right of other men. For this labour being the unquestionable property of the labourer, no man but he can have a right to what that is once joined.[9] [Emphasis added]

The above serves only as a starting point. Locke applies his theory only in cases where there is a superfluity of land. In other

---

[8] David Friedman, *Discussion, in* MORALITY OF THE MARKET: RELIGIOUS AND ECONOMIC PERSPECTIVES 505 (Walter Block et al. eds., 1985)

[9] JOHN LOCKE, *An Essay Concerning the True Original, Extant and End of Civil Government, in* TWO TREATISES OF GOVERNMENT (P. Laslett, ed. 1960).

words, where there is enough virgin land so that no man can homestead it all, there is plenty left for the use of others. Libertarians, in contrast, believe the Lockean theory should be applied in *all* cases, whether or not there are sufficient land resources available for other people.[10]

What is the alternative to Locke in general or, more specifically, for the cases where excess land is not available? Several possibilities exist, the first of which is the claim theory. Under this approach, a man gets to own land or other natural resources merely by assertion. "I claim the sun, the moon, the stars, and the oceans," a man might say, and, if he is the first to do so, he thereby becomes the legitimate owner. The problems with this are legion. Who will know who has claimed what? There could be numerous people claiming the same thing at the same time; this theory offers no way to choose between them. According to folk wisdom, "talk is cheap." Anyone can say anything he wants, and as such, mere speech should not entitle anyone to own anything.

A second approach utilizes the apparatus of the state. With regard to unowned land, the government can give this property to its favorites, sell it to the highest bidder in an auction, or parcel it out on a first-come-first-serve basis.[11]   But why should politicians and bureaucrats own unused land? What did they do to deserve proprietary status? Even if the government sells the land instead of giving it away, the government already has far too many resources; why should they possess any more?

Third, there is the doctrine of *equal shares for all individuals*. To illustrate, if there are six billion of us, then we all own one six billionth of all virgin land. This is rather arbitrary, as we do not all own equal shares of anything else, up to and including human capital. The following example by Professor Rothbard illustrates the problem inherent in this theory: "It is difficult to see why a newborn Pakistani baby should have a moral claim to a quotal share of ownership of a

---

[10] *See* Rothbard, *supra* note 3, at 53.
[11] *See, e.g,* U.S. Homesteading Act, 12 U.S. Statutes at Large, §75 et seq. (1862).

piece of Iowa land that someone has just transformed into a wheatfield -- and vice versa of course for an Iowan baby and a Pakistani farm."[12]

A fourth possibility is the theory of communal ownership with proportionate distribution of shares. In this view, we do not each own one six-billionth of all virgin land on an *individual* basis; rather, we own it all, communally. In other words, before anything can be done with specific pieces of land, all of us, or at least a majority, must agree. This sounds like either a recipe for interminable committee meetings or, more likely, the concentration of power in a small set of the population.

A fatal flaw in all of these theories can be illustrated not with the ownership of land, but of something even more important, the human person. Civilized people agree to individual ownership of ourselves. This is consistent with homesteading. Initially, when we are born, we are unable to "homestead ourselves," or rather, mix our labor with our own persons. Our parents perform tasks for us and thus, in a sense, "own" us throughout infancy.[13] Then, as we reach adulthood, we gradually take control over our bodies, or "homestead" them in effect, and thus come to own them fully. Perhaps an application of the above mentioned theories of property ownership to the ownership of people will prove insightful.

Under the *claim theory*, anyone could claim control over anyone else, provided only that he was the first to do so. This is a recipe for slavery on a grand scale, and must be rejected on that basis alone. Government ownership of people is similarly unacceptable. This merely shifts the slave master from the first claimant to the state. Or, to put this in another manner, it would hearken back to medieval days in Europe, or to Oriental despotism, where the king was considered the literal owner, to do with as he wished, with all his subjects. Likewise, the theory of *equal shares for all individuals* is also unsatisfactory. Obviously, if we were to do things in this way, Michael Jordan and Bill Gates would have been given far more than an equal share. They would, at least partially, have to do the bidding of the rest

---

[12] MURRAY N. ROTHBARD, FOR A NEW LIBERTY 35 (1973).

[13] More exactly, own the right to keep raising, or homesteading us. For the libertarian perspective on children, property and ownership. *See id.* at 97-112.

of us. Who could argue in favor of ordering around some people merely because they were born with or developed extra talents and abilities? Finally, communal ownership also fails when it comes to control over human beings. If interpreted literally, no individual could so much as scratch his nose without the permission of at least a majority of six billion people. Under such a system, the human race would surely perish.

IV. THE PONTIFICAL COUNCIL'S DOCUMENT[14]

A. Concentration and Misappropriation of Land

The Pontifical Council ("hereinafter Council") often refers to the phrase, "concentration and misappropriation of land" in a negative manner.[15] But why must high concentration necessarily be equated with misappropriation? In the United States, there is high concentration in the computer, automobile and movie industries, which are dominated by a handful of very large and prosperous entities. The same phenomenon occurs with regard to the ownership of agricultural land in the United States, where the farming sector is a significant contributor to overall economic wealth. There are sizable farms and ranches, such as in Texas and Montana, which are measured not so much in acres, nor even in hectacres, but rather in square miles. Even so, apart from some *Nader*-ite attacks on "agribusiness," the large size of these farms has not led to, nor does it constitute, "misappropriation."

---

[14]Roger Card et al., *Toward a Better Distribution of Land: The Challenge of Agrarian Reform*, NAT'L CATH. REP., Vol 34, No. 22 (April 3, 1998). [hereinafter *Presentation*]. Our commentary is based on the organization of the Presentation. We shall not discuss each and every section, as there is much repetition between them, and some are irrelevant to our concerns. Indeed, the organization of this document, or rather lack of it, makes a reply to it rather difficult. There are several recurrent themes of interest, and we will comment on all of them, but some are spread all throughout the document, stated in slightly different formats each time.

[15]*See id.* para. 1, 2, 4. Indeed, there are very few paragraphs in the entire document which do not criticize "land concentration."

1999-2000]   LAND REFORM   47

B. Developing Economies

The Council continually uses the term "developing economies" to refer to the poor nations of the third world. The use of this phrase is highly inaccurate. Characterizing an economically deteriorating nation as "developing" cannot possibly materially help any nation or its citizens. An acknowledgment of the truth is the first step in correcting the real problems these nations are facing. Denial is of little use in either economics or psychology. The point is that many of these countries are not developing at all; some are barely holding their own, while others are actually retrogressing from an economic development point of view. The Council's choice of words camouflages this reality.

1. Factors of Production

In the Council's economic analysis, "land ... given the predominantly agricultural nature of the economy ... constitutes the fundamental production factor, *together with labor*, and the chief source of national wealth."[16] This is a curious conclusion. The document under consideration is concerned solely with *land* reform. This would imply, given the goal of economic development, that *land* is the key factor in production. Why, then, is the phrase "together with labor" added to the statement? Could it be that the writers of this document realize that *land* reform is not all that important, even to a predominantly agricultural nation? This realization comes from the fact that *labor*, on the margin, contributes far more to GDP than land.[17] Even in the undeveloped countries people own their own labor. Hence, this admission by the Council seriously undercuts their thesis.

This, however, is only the beginning of the problem. There are other factors of production beyond land and labor, most notably capital. But capital, based on savings and engendered by security of property

---

[16]*Id.* para. 1 (emphasis added).

[17]In 1992, labor's distribution of the United States National Income was 74.1%. In 1980 this figure was 75.9%, indicating only a slight variation. *See* Morgan Reynolds, ECONOMICS OF LABOR 55.

rights, tends to avoid nations which engage in the socialist, egalitarian "reform" of land holdings, thus further leading to impoverishment. While no one can move land from one place to another, this immobility certainly does not apply to capital. The point is that the Council's thrust is in the direction of killing the goose which will, at least potentially, create the golden eggs.

2. Preferential Option for the Poor

The clergymen begin this section on a sound footing, with their embrace of the "preferential option for the poor."[18] It has been well said that a society may fairly be judged by how its poor are treated. If there is any obvious conclusion which must be drawn from an international economic comparison of countries, it is that the poor in rich nations are treated far more decently than those in the underdeveloped part of the world. If anything, the poor in places such as Switzerland, the United States and Canada boast of more material possessions than even the middle class in many parts of the globe. Were Roger Cardinal Etchegaray, the author of the Council's land reformation document, serious about improving the lot of the poor in the third world, he would advocate economic development of the sort enjoyed in Europe, North America, Japan and other such countries; that is, a great reliance on private property and free markets.

How did these wealthy nations achieve their present enviable economic status? Clearly, it was by embracing economic freedom, private property rights, and the rule of law.[19] Therefore, anything the clergy advocate in this direction will be helpful to the poor. But

---

[18] *Presentation, supra* note 14, para. 2.

[19] *See generally* PETER T. BAUER, REALITY AND RHETORIC: STUDIES IN THE ECONOMICS OF DEVELOPMENT 38-52 (1984); *see also* PETER T. BAUER, EQUALITY, THE THIRD WORLD, AND ECONOMIC DELUSION 86-102 (1981); HERNANDO DE SOTO, THE OTHER PATH: THE INVISIBLE REVOLUTION IN THE THIRD WORLD 108-20 (1989); JAMES GWARTNEY ET AL., THE FRASER INSTITUTE, ECONOMIC FREEDOM OF THE WORLD, 1975-1995 36-42 (1996); F.A. HAYEK, LAW, LEGISLATION AND LIBERTY (1973); HANS-HERMANN HOPPE, A THEORY OF SOCIALISM AND CAPITALISM: ECONOMICS, POLITICS AND ETHICS 53-60 (1989); ADAM SMITH, AN INQUIRY INTO THE NATURE AND CAUSES OF THE WEALTH OF NATIONS 52-63 (Edwin Cannon, ed. 1965) (1776).

anything in the opposite direction will be harmful, and thus contrary to the preferential option for the poor, the presumed motivating axiom of the entire document. How well then does the Cardinal's statement do when measured against these criteria?   While there are some exceptions, the overall assessment, unfortunately, is negative. The analysis and evidence behind our assessment is explored below.

3. Latifundia

The Pontifical Council defines "latifundia" as "large land holdings, often belonging to absentee òwners, where the land is worked on by hired labour, using out-dated farming techniques."[20] This would be unexceptionable but for the fact that the Council sees the latifundia as a basic cause of the problems of the third world.[21] What is the basis of this opinion? After all, as we have seen, "large land holdings" are characteristic of many countries with a vibrant agricultural sector, such as the United States and Canada.

Nor are "absentee owners" a barrier to economic development. Much real estate in advanced industrial nations is owned by people who do not live in their own high-rises. The corporation itself, emblem of economic success if ever there was one, is a paradigm case of "absentee ownership." Here, millions of people buy shares in companies which own land, mines, farms, resources all over a country, and, in the case of multinationals, all around the world. Surely stockholders are all "absentees." Are the clergy attacking the very idea of the corporation as economically unviable? One would have thought that with the toppling of the Soviet Union this sentiment would no longer be expressed quite so blatantly. It appears, however, that such thinking is wrong. Despite the Council's position, it is clear that an objective examination of economic reality shows that the criticism of the latifundia is misplaced.

---

[20] *Presentation, supra* note 14, para. 2, n.2.
[21] *See id.*

4. Competence

The Council states that its land reformation document "is not a document of political intent, for that lies outside the Church's field of competence."[22] This issue would not have been brought up but for its entanglement with economics, since to do so would smack of *argumentum ad hominem*. It is not for us to state that only those with doctorates in economics are competent to analyze current economic issues, such as those involving the development of third world countries. However, since the Pontifical Council brought up this issue, perhaps it is appropriate to discuss it. In the Council's view, they are *not* competent to discuss political intent, but, presumably, they *are* competent to discuss economic development. But why the difference? Are they not *theologians*, therefore unqualified to also act as political scientists and economists? If so, what is the argument for competence in the latter but not the former?

5. Mortgage of the Past

The Council states that "[p]rivate appropriation of the land ... introduced serious distortions into the land market."[23] This sounds suspiciously like an advocacy of Soviet or Cuban style collectivized farming, systems that have never been accused of being based upon "private appropriation." The Council proceeds to note five market distortions.[24] Before examining these in detail, it is necessary to set into context the bishops' charges. Clearly located in the socialist camp, they are stating that markets, capitalism, free enterprise, etc., are to blame for the plight of the poor, and that government action, such as taking land from the rich and giving it to the poor, is the solution.

The first market distortion is the government's forcing of indigenous populations into artificial groups and locating them on reservations. The reservations are often located in "infertile areas, far

---

[22] *Id.* para. 2.

[23] *Id.* para. 4.

[24] *See id.*, para. 4, n.5.

from [economic] markets or poor in infrastructures."[25] The problem is that the *government,* and *only* the government, has the legal power to force the geographic relocation of the populace. Additional examples of forced relocation include the creation and maintenance of a German ghetto for the Jews during World War II, zoning laws in general, and the "Jim Crow" legislation in the American South following the Civil War. The common denominator is that it is the *state* which imposes such mandates. How, then, can the blame rest with the *market*? Imprisoning specific portions of the population in infertile or otherwise inhospitable areas represents a "government failure," not a "market failure."

The second market distortion is "the imposition of discriminatory taxes on the produce of small indigenous farmers."[26] It can hardly be denied that this too is due to government action, not to market action. Ironically, this charge is most often raised by clergymen. The Church itself advocates discriminatory taxes in the form of land redistribution. Although the Church's redistribution plans generally favor the poor rather than the rich, who are traditionally the government's beneficiaries, *both* practices are discriminatory.

The third market distortion occurs when a pricing system is adopted that works in favor of large producers and to the disadvantage of small farmers.[27] Price controls create this distortion and are clearly created by the government, not private citizens. Only the government has the legal power to impose price controls. A fourth distortion can be found in "the imposition of import barriers in order to protect the produce of large landowners from international competition."[28] Again, only the state has the capacity to set up barriers to international trade. Such barriers are indeed serious impediments to economic growth and are in need of control or elimination.

The final market distortion involves "the provision of public services and subsidies from which only large landholdings could, in

---

[25]*Id.* at 5(a.).
[26]*Id.* at n.5(b).
[27]*Id.*at n.5(e).
[28]*Id.* at n.5(d).

actual practice, benefit."[29] At first glance, this seems consistent with *laissez faire* capitalism. However, it is the government and not the market that is providing the services and subsidies. Taken on its face, it seems the argument should be to end business subsidies because they represent a market distortion. Then, the so-called "public services" would be provided privately. This comports with the libertarian philosophy.[30] However, the present authors believe that the Pontifical Council's document reflects the likelihood that the clergy may agree with this result more in principle than in practice.

C. Industrialization at the Expense of Agriculture

The Council's document is, for the most part, a socialist creed written in opposition to private property rights. Nevertheless, there are portions which support market theory and criticize government regulations. For example, the document attacks protectionism, exchange rate manipulation and price controls.[31] Again, this comports with the classic liberal perspective. However, it is worth noting that the document appears to represent the product of a diverse committee composed of theologians with both pro- and anti-government sentiments.[32] It is apparent that there was little communication between the divergent groups.[33]

---

[29] *Id.* at n.5(e).

[30] *See* Rothbard, *supra* note 12, at 194-241.

[31] *See Presentation, supra* note 14, para. 6.

[32] This has given rise to what can only be called the "Daddy likes me best" school of commentary. Since the PC has both pro and anti market elements, commentators from both sides have each stressed the parts of the document which favors their own position, and then claimed that the Church, or the Cardinal in this case, or the Pope in the case of several of the Papal Encyclicals, is "really on their own side." In sharp contrast, the present commentary has no ax to grind in this regard. If we say so ourselves, it is a straightforward political economic analysis of "Toward a Better Distribution of Land" which notes that it contains both support for, and criticism of, capitalism.

[33] Sometimes libertarians are accused of a similar sort of inconsistency, for we favor both economic (laissez faire capitalism) and political liberty (civil liberties, free speech). In contrast, the overwhelming majority of commentators favor either economic but not civil liberty (conservatives) or civil, but not economic liberty (socialists). To champion both is thus to be held internally self-contradictory. However, in the libertarian philosophy, freedom is a

The Pontifical Council concludes that unwise government interventions in free enterprise have led to a "fall in farm income [which] has affected small producers so badly that many have been forced to give up farming."[34] The Council assumes that there are negative consequences when people are "forced to give up farming." However, this is not necessarily so. In the United States, the market has similarly "forced" people out of farming. The result has been an increase in productivity within both the agricultural sector and the industrial sector. This has led to enhanced income and wealth.

Strictly speaking, of course, the market can never "force" anyone to do anything. The market simply consists of no more and no less than the concentration of all voluntary economic interactions.[35] Perhaps a more accurate way of describing the process is that people are led by Adam Smith's "invisible hand"[36] to do that which is in the best interests of the economy. It is a result of the market that people have moved from the farm to the city where productivity is higher. The market may exert economic pressure on farmers to move to the city, but any given farmer in the United States is free to remain where he is. The result has been simply that the farmer has had to accept a lower income because consumers place a higher value on goods made in the cities.

The Council, on the other hand, views the move toward industrialization as problematic *per se*. The Church clearly favors reduced poverty. However, at the same time, the Church seems to favor the maintenance of a farm-based economy. There appears to be a logical inconsistency here. The Church wants to improve the conditions of the poor but not at the expense of farming.

---

seamless web, which must be pursued in any and all directions. The contradiction, then, takes place on both of these other place on the political economic spectrum, not on our own patch of it.

[34]*Presentation, supra* note 14, para. 6.

[35]*Cf.* ROBERT NOZICK, ANARCHY, STATE, & UTOPIA 163 (1974) (stating the market consists of "capitalist acts between consenting adults").

[36]*See* SMITH, *supra* note 19.

D. Failure of Agrarian Reform

The Council discusses what it views as the reasons for the failure of past efforts in land reform.[37] In addition to the concentration problem, there is a failure of "preventing the expulsion of large masses of peasant farmers from the land that is still free, but which may be marginal...."[38] However, this argument can be shown to be incorrect. Given that free enterprise is the key to economic development of third world countries and that private property rights are integral to free enterprise, then anything which promotes private property rights should in turn reduce poverty. The right to expel farm workers, if they are no longer needed, is a component of private property. Therefore, be it ever so counterintuitive, the right of expulsion is actually a means of *enhancing* overall wealth.

Additionally, the Pontifical Council's opposition to "migration to urban centers"[39] may be counterproductive. For example, the migration of black farm workers from rural states such as Mississippi and Alabama to urban areas such as Chicago, Detroit and Philadelphia during the 1940s and 1950s resulted in economic growth. Surely the Council can recognize that it would have been economically disadvantageous for people to have stayed on the farms.

Economists place great value on the phenomenon of "voting with the feet" as an indicator of economic well-being. If there is a migration lasting several years in duration then the newcomers can tell those still on the farm of the benefits of city life. If those who had remained subsequently migrate to an urban center, then that migration provides strong evidence that life is better in the new venue. Historically this pattern has occurred under various circumstances. For example, there was traffic by Jews *out* of Nazi Germany, not into it. Similarly, the Industrial Revolution was beneficial for poor peasants who migrated to cities in spite of the factory hardships portrayed by

---

[37]*See Presentation, supra* note 14, para. 7.
[38]*Id.*
[39]*Id.*

socialists such as Dickens.[40] This pattern can also be found in pre-Mandela Apartheid South Africa. It can be deduced that South Africa was *better* for blacks than the alternatives available elsewhere because migration was *into* South Africa. The migration patterns criticized in the Presentation demonstrate this pattern. When peasants continue to flock from farm to city, it is evidence that urbanization is an *improvement* in their lives. It is worth noting that churches are also clustered in cities and are less common in rural areas. The clergy too are "voting with their feet" by remaining largely urban.

E. Contradictions

The Pontifical Council seems to contradict itself when it calls for more government contributions to infrastructure as a necessary component of land reform and economic development. This is contrary to their notion that large landholders, not poor peasants, tend to benefit from these social services and subsidies.[41] Infrastructure generally includes items such as roads, harbors, lighthouses, electricity, other power sources, libraries and public universities. Socialism is hardly the best way to provide such items. There is little doubt that private capital would be able and willing to invest in the infrastructure. However, the Council does not explain why it believes socialism will succeed and the market will fail in this regard.

The Pontifical Council also calls for "fixing prices" in order to enhance economic development.[42] This is in direct contradiction to its opposition to "control of food prices."[43] It is clear that both cannot be correct. Either price controls are a force for economic development or they are not. Clearly, only the latter is correct. Prices have a role to play in economic growth. Prices coordinate the decentralized behavior of millions of consumers, sellers, entrepreneurs, workers, resource

---

[40] *See* CHARLES DICKENS, OLIVER TWIST (Kathleen Tillotson ed., Oxford Press 1966) (1842).

[41] *See Presentation, supra* note 14, para. 7.

[42] *Id.* para. 8.

[43] *Id.* para. 6.

56   J. NAT. RESOURCES & ENVTL. L.   [VOL. 15:1

owners and middlemen.[44] In this manner, uncontrolled prices based on private property coordinate the market. The only other alternative is central planning, which has always proven unsuccessful. The economic collapse of the Soviet Union is evidence of the negative effects of cental planning.

The Council intends that "land reform" extend beyond taking the property of some and giving it to others. In addition to land redistribution, the Council calls for subsidies, subsidized credit, grants, government funded improvements in infrastructure, "social services," and price fixing or price controls.[45] Clearly such programs require a great deal of money. The burden is likely to be borne by the taxpayers. Not only have taxpayers not been at all implicated in the (possible) Latifundista land theft that contributed to land inequality, they are also less likely to benefit from the proposed improvements. Furthermore, the programs proposed are incompatible with free enterprise. Finally, the Council recognizes that past attempts at land reform have been accompanied by massive corruption; yet, the Council favors this course of action without providing any means to prevent corruption from again occurring.

F. The Management of Agricultural Exports

1. Debt Repudiation

As in the case of land reform, the libertarian can agree with the socialist as to the propriety of debt repudiation, at least in principle, while disagreeing as to the specific justification. In contrast, the conservative, with an imprecise understanding of private property, rejects this course of action outright. The left wing individual defends both land reform and debt repudiation on egalitarian grounds believing that as long as the poor gain, the case is made.

---

[44]*See* MISES, *supra* note 1, at 53; *see also* F.A. HAYEK, THE CONSTITUTION OF LIBERTY 112 (1960).

[45]*See Presentation*, *supra* note 14, para. 8.

Consider for example an illegitimate government, such as in Cuba. Suppose that a new libertarian regime somehow takes over this troubled island nation. Does this new government have the responsibility to pay off the creditors of Fidel Castro and his fellow criminals? People who have purchased Cuban bonds fall into two classes: those who did so under duress, and those who made willing contributions, either out of ideological support for Cuban style communism, and/or because the interest rates were attractive to them. The determination of whether to pay the latter is quite simple. Those willing lenders were cooperators, or perhaps even conspirators, with a government that is composed of criminals. Therefore, no repayment is necessary; indeed, it would be highly improper. The payment obligation is not as easy to determine in the situation where innocent people were forced to buy Castro's bonds. Yes, they should be able to sue Fidel and his minions in order to collect their debt. Their lien against these people should be a high one. In that sense, there is no debt repudiation. On the other hand, it would be totally impermissible for the new regime to tax innocent Cubans to repay these debts. To do so creates a new set of innocent victims to compensate the first set of victims.

Of course, a practical question to consider along with repudiation is the future impact on borrowing. Repudiation makes it is less likely that new borrowing will occur. Further, a substantial difference exists between socialists repudiating debt because the lenders are rich and the borrowers poor, and the libertarians doing so because the lenders have cooperated with thieves. In the former case, the bond market may close off entirely or charge such high rates of interest as to impede future borrowing. With the latter case there is not enough historical evidence to render an opinion about this essentially empirical issue.

2. Agrarian Dualism

It is almost an article of faith amongst clerical commentators on business subjects that selling goods for the export market is uneconomic. According to the Council, "[i]f the market prompts small farmers to grow export crops, this often takes place at the expense of production intended mainly for their own consumption, thus putting farming families at considerable risk."[46]   This view amounts to economic illiteracy. The only way the market can "prompt" anyone to do anything, farmers or industrialists, large or small, is by holding out the prospect of greater profits. Should American wheat farmers stop their production for export, and instead produce something consumed in the U.S., such as apple pie? Should German exporters of Volkswagens cease and desist, and instead manufacture wiener-schnitzel, something beloved of local consumers? Should a small family firm in the tulip bulb industry of Holland cancel production for the world market because "unfavorable climatic or market conditions can lead to a vicious circle of hunger, so that such families contract debts that then force them to give up ownership of their land"?[47] To ask these questions is to answer them.

The business world is unpredictable. Indeed, it is the essence of the entrepreneurial function to bear risk. There are no guaranteed safe havens - any businessman, no matter how powerful, must risk financial loss. This applies to exporters and importers, as well as those who produce for the local market. Self-sufficient family farms, beloved of the clergy, like small family groceries, can also prove to be uneconomic.

It is one thing to use the cloak of theology to give good business consulting advice; even this is problematic, as it exceeds the competence of the authors. It is quite another to give erroneous opinions, cloaked under the authority of the Church, which might well be believed by unsophisticated people in the third world. This is no way to promote the "preferential option for the poor."[48]

---

[46] *Id.* para.10.
[47] *Id.*
[48] *Id.* para. 2.

G. Expropriation of the Land of Indigenous Populations

According to the Council, "the rights of the indigenous inhabitants have been ignored when the expansion of large scale agricultural concerns . . . (and industry) . . . have been decided . . . ."[49] The clergy are absolutely correct in claiming that Indian lands have been stolen throughout history. They are on morally firm ground in implying that these wrongs should be righted, despite claims to the contrary from conservatives who favor the status quo in terms of land ownership, believing that any forced changes would be a violation of private property rights.

However, the clergy here tread on thin ice in several regards. First of all, "large scale" agriculture, hydroelectric, mineral, oil and timber interests are not the only ones to have ridden rough shod over native property rights. Small concerns have also done so. Second, these thefts, or takings, are by no means as extensive as thought of in some quarters, at least on Lockean grounds. The fact of the matter is that we are talking about people with essentially stone-age technology who were for the most part incapable of homesteading property. Much, but not all, tribal behavior consisted of hunting and gathering, not in mixing their labor with the land as in farming. To be sure, then, such people would be the legitimate owners of all the berries they had gathered and animals they had killed, etc., as well as of their northern and southern camping grounds, and the right of access from one to the other. But a mere handful of natives could not have homesteaded anything approaching the entire territory of the continental United States before the arrival of the Europeans. The point is, Lockean-Libertarian homesteading theory requires that the Indians mix their labor with the land; that they transform it in some way. This, for the most part, they failed to do, and therefore cannot be considered its legitimate owners. While they would be considered proper owners of the relatively small areas where they pitched their tents and farmed, they would not be the owners of the vast areas in which they traveled and hunted.

---

[49] *Id.* para. 11.

Finally, "the burden of proof" rests with those who wish to alter property titles. Otherwise, we are reduced to claim theory for all property. An individual can claim he is the rightful owner of another's land or wristwatch, but unless he can offer proof of his contention there is no warrant for transferring these items from the one to the other. Yet, the Council goes so far as to admit that "the origins of" Indian ownership claims to land "are lost in memory."[50] If so, how can any rational court grant these claims? They state, further that "[i]ndigenous populations can also run the absurd but very real risk of being seen as 'invaders' of their own land."[51] But where is it engraved in stone that this is "their own land"? If they never in justice owned the land in the first place, and then others arrived who did homestead it, then it is not at all incorrect to characterize the natives as "invaders," not of their own land, but of the land of these later arrivals.

The libertarian theory is not merely an excuse for the status quo, as is that of the right wing. It is not merely the dressing up in more sophisticated clothing of the conservative head in the sand attitude toward land reform. Yet it may well *appear* this way when we consider very ancient wrongs, or those done to people with stone-age technology who thus have no written records to support their claims. However, with regard to more recent events where there is physical evidence of robbery (e.g., the Nazis and the Jews, the Americans and the Japanese in the early 1940s, and even black pre-civil war slavery in the United States) the libertarian and conservative conclusions as well as underlying analyses are very different.

H. Violence and Complicity

The clerics quite properly oppose "terror." But one man's terror may be another man's self defense. Property rights are the only way to tell if A's fist is imposing on B's face, or if B's face is attacking A's fist. Consider again Professor Rothbard's analysis of "A, seizing B by the

---

[50] *Id.*
[51] *Id.*

wrist and grabbing B's wristwatch."[52] Who is aggressing against whom in this action? It all depends upon who is the rightful owner of the timepiece. If the property belongs to B, then A is indeed the terrorist.

The same thing appears to have occurred between the latifundistas and the landless peasants. The clergy see the former as using violence on the latter, and too quickly assume they are guilty of terror. But if the land properly belongs to the large latifundistas, then all those who squat on their property, or trespass upon it, or attempt to take it over, are the real criminals. Property rights determine terror; without this understanding, we cannot possibly distinguish an invasion from self defense.

Suppose that landless peasants "sat in" on church property, attempting to turn these edifices devoted to worship to housing for themselves. Would not the reaction of the priest be to call in the police for eviction, thus resorting to what these trespassers would see as "terror"? And what of the defense of "protests of workers who are forced to work at an inhuman pace for wages that often do not cover their travel and living expenses."[53] This Council charge is economically incomprehensible. The wage depicted by the clergy would have to be below that of subsistence level. Even slaves were paid a wage greater than "living expenses" in terms of food and shelter; otherwise they would have perished. Wages are determined in markets based on productivity. This must of necessity be higher for free men than slaves, if only because there are no costs of guarding the former from escape which must be deducted from the former's productivity. Thus, *ceterus paribus*, free labor must always be paid more than slaves. And if slaves are paid enough to keep them alive, free persons must be paid even more. Thus, this charge is an untenable one.

I. Legal Recognition of Ownership Rights

The Council complains of the fact that it is difficult "for small farmers to obtain legal recognition of ownership rights over land that

---

[52] *See supra* text pp. 3 – 8.

[53] *Presentation, supra* note 14, para.12.

they have been farming for a long time and of which they are the de facto owners." The problem with this is that there can be no such thing as de facto ownership, where a different person has actual title to the land, and is thus the de jure owner.

How could such a situation arise? One possibility would be for a tenant to farm land for many years for an absentee owner, and then, suddenly, to declare that he, the farmer, was the real owner, and that the landlord to whom he had been paying rent for decades was not. Another would be the case of an owner of land who allowed a passerby to take a shortcut through his property, so much so that over the years a path was worn by his footsteps; whereupon the traveler, instead of being grateful to the owner, asserts his access rights.

Were this sort of thing countenanced by the courts, it would place into disrepute private property, and with it, capitalism, the system which has brought about modern civilized standards of living. Alternatively, no one would ever allow passerby access to one's land; and tenancy would become a thing of the past. This would run counter to the preferential option for the poor, as they are people of limited means, who cannot afford to buy homes, cars, farms, etc., and are thus the main beneficiaries of tenancy.

J. Environmental Concerns

The authors of this report worry that without seizing the land of the latifundistas, there will be an "over-exploitation of natural resources without concern for environmental sustainability or without considering the intergenerational continuity of family property."[54] Environmental degradation, running out of resources, extinction of species, and other legitimate environmental concerns simply have nothing to do with who owns the land, provided only that it is held in private, not public, hands. The reason for this is rather straight forward.

---

[54] *Id.* para. 14. In another section, the clerics express the view that "inequalities in the distribution of land ownership set in motion a process of environmental degradation that is hard to reverse." *Id.* para. 21. For a critique of this sentiment, see *Pollution Trading Permits as a Form of Market Socialism, and the Search for a Real Market Solution to Environmental Pollution*, 1 FORDHAM ENVTL. L.J. 46, 51-57 (1994).

When an individual or family farmer or large firm in the agriculture business industry owns land, or a cow, or a tractor, they tend to take care of them. They do not overuse these resources, or dissipate them. If they fail to act in this manner, they suffer the full attendant losses.

In very sharp contrast, when items such as these are owned in common through coercion, when no one really owns them but all may partake in their use, they are in great danger. Under such circumstances, the user is not confronted with the full costs of his actions. For example, a farmer would be foolish to allow his sheep to graze too long on his own land, since if he does, the grass will be eaten down to its roots, and will not regrow easily. He has every incentive to preserve this land for another day, for if he does, he is assured that his sheep will be the ones who benefit.. On the other hand, if the pasture is compulsorily held in common by all shepherds, then none of them has an economic incentive to act in this environmentally and economically sound way. Here, if he moves his sheep up the mountain to another less accessible pasture, he has no guarantee at all that when he returns to the more convenient one, the grass will be held for his animals. On the contrary, it is almost certain that someone else will come along and allow their sheep to overgraze.

K. The Credit Market

The Council takes the position that "[i]n rural areas, there is often no legal credit market, so that small farmers have to turn to money lenders if they need loans, thus exposing themselves to risks that can lead to the partial or even total loss of their land -- for property speculation is usually the real focus of such moneylenders' operations."[55] There are numerous economic fallacies espoused here. First, a confusion exists in the passage between the geographical and the legal spheres. Lack of a legal credit market results from jurisprudence and not location. When an enactment forbids credit, or sets interest rate maximums, it can apply to *either* rural or urban areas.

---

[55] *Presentation, supra* note 14, para. 15.

No presumption exists, as the clergymen contend, that this law applies with greater feverishness to the rural milieu rather then the urban.

Secondly, money lenders *are* part of the credit market. While the clergy thinks all money lenders loan at very high rates, this is not always the case. Basically, what determines the rate of interest in a given society is time preference: the impatience of the populace to consume now, as opposed to deferring gratification for the future. If the former, then a high interest rate ensues; if the latter, a low one. Imagine a society composed primarily of seven year old school children, each of whom would gladly give up five candy bars tomorrow to have one today. Very few of them would be willing to lend at almost any interest rate, and most would be willing to borrow at virtually any cost. Interest rates will thus be very high, regardless of other occurrences. At the opposite end of the spectrum are people who "save for a rainy day." They are willing to lend money even at low rates of interest, and rarely borrow, even when it is cheap to do so. A low interest rate will prevail in such a society. In other words money lenders, and the interest rates they charge, are merely the messengers for the underlying economic realities.[56]

Finally, the real cause of artificially high interest rates stems from laws prohibiting usury. Imposing ceilings upon interest payments creates a situation where only individuals with good collateral obtain loans, which tend to be large enterprises. Small businesses and other risky borrowers with little collateral and poor repayment records have to deal in the black market, at much higher interest rates because of the laws against usury; the lender, as a law breaker, cannot rely upon the courts to help collect bad debts. Only exorbitant interest rates compensate a lender enough to risk loans to such enterprises. This means that the "risks" are borne not by the small borrowers whose interests the Counsel think they are defending, but by the usurious money lenders. The small and/or risky borrower does not face risk, but

---

[56]For more on this subject, see MURRAY ROTHBARD, MAN, ECONOMY AND STATE 74-120 (1993); EUGEN BOHM-BAWERK, CAPITAL AND INTEREST 35 (George Hunke & Hans Sennholz trans., Libertarian Press 1959) (1884); Walter Block, *The Negative Rate of Interest: Toward a Taxonomic Critique*, 2 J. LIBERTARIAN STUD. 121, 124 (1978).

1999-2000]   LAND REFORM   65

rather faces either astronomical interest to attract the loan or the inability to obtain financing at all.

Property speculation, or any other kind, does not represent the evil that the Council believes. Instead, the market process of speculation places goods and services into the hands of those who value them greatest, as demonstrated by willingness to pay. The speculator helps stabilize price differentials so that the same price for the same commodity, apart from transportation and other such costs, prevails everywhere. For example, if oranges are selling for one dollar in Miami and two dollars in Boston, the speculator will take advantage of this difference by purchasing cheap southern oranges and transporting them north. The first part of this operation raises the prices of oranges in Miami, and the second part lowers them in Boston.

While money lenders may sometimes speculate in land, and land speculators may sometimes lend money, the two professions are not at all economically connected, as the clergymen espouse. The only common bond between lenders and speculators mirrors the bond between middlemen and landlords: All four are cordially hated by socialists as emblematic of capitalism.

L. Idle Land

The Counsel states, with regard to the case for breaking up the Latifundias, and redistributing them to the landless poor, that "[s]uch large landholdings are often poorly cultivated, or simply left uncultivated for speculation . . . . "[57] If these holdings represent stolen property then a strong moral case exists for redistributing them not to the poor, but to the children of the victims. However, to offer "poor cultivation" as a reason for land reform implicitly acknowledges the legitimacy of these property titles; in effect, the clergy argue that, even if the latifundistas had clear and legitimate title to their lands, they should still lose their land on grounds of poor cultivation.

If true, more land would exist to redistribute than to merely the latifundias in the third world. Anything left idle would be fair game.

---

[57]*Presentation*, *supra* note 14, para. 32.

For example, Mr. A goes on a year long holiday for a trip around the world. He leaves his home vacant and loses ownership in it. Mr. B doesn't use his fifteen speed bicycle, tennis racket or fishing rod for five years; long before that time, his property rights in these items lapse. Miss C doesn't wear her blue dress for a few years, with the same result. Farmer D utilizes his land only eight months of the year, allowing it to lie fallow for the remaining four months. A teenager leaves his childish toys in the attic; when he tries to recover them, he is told he no longer owns them. An old man sticks his money in his mattress for years, "for a rainy day," and thereby loses it.

The problem of this philosophy, apart from the blatant unfairness of seizing idle property, is that it disregards Keynes' "precautionary motive."[58] People value their possessions whether they "use" them or not. Alternatively, and perhaps preferably, people do not always use them in normally accepted standards. The point is, Mr. A's home, Mr. B's sports equipment, Miss C's dress, Farmer D's land, the teenager's toys and the old man's money are not really "idle." They only appear so to outsiders. One man's lack of cultivation or idleness is another man's valuable contemplation or financial support against the unknown future. If the goods did not provide services, if they were truly idle, they would be sold. Thus, people occasionally *sell* their homes, dresses, toys, etc. Are there no Church owned bibles, organs, churches, surplices, wine, automobiles, or wafers that go unused for any appreciable amount of time? If so, these would be forfeited under the philosophy adumbrated by the Council.

M. The Lack of Infrastructures and Social Services

In the view of the Pontifical Council, "small farmers are forced to depend on local markets to sell their produce... They are also dominated by traders whose monopolistic position means that farmers are forced to accept the price offered if they want to sell their

---

[58]JOHN MAYNARD KEYNES, GENERAL THEORY OF MONEY, INTEREST AND MONEY 215-225 (1936).

produce."[59] The clergymen illuminate an important issue, but their lack of economic sophistication distorts its perception. The clergy raise the issue of marketing boards, which are government entities that prohibit farmers from selling to anyone but themselves. Also, the boards typically purchase at prices far below levels farmers could otherwise attain, whether on local or world markets.[60] These monopsonists[61] exploit farmers, particularly small ones who cannot bribe their way out of these regulations or otherwise avoid them.

The flaw with the cardinal's analysis is that the boards are *government* entities, not a part of the market. To label them "traders" confuses markets with central planning and economic freedom with economic coercion. Nor could it be otherwise. If these underdeveloped countries were dedicated to capitalism, then any local private person would be a monopsonist who served as the sole purchasing agent for small farmers and bought at below world market prices, would earn vast profits. However, this would attract other competitors into the industry, both local and multinational, which would compete with the original monopsonistic one to buy farm produce. With greater demand, prices paid to farmers, even *small* ones, would increase until the profits earned in the industry reached market standards, abstracting from risk.

Why, then, do the clergy, being aware of monopsonistic exploitation of small farmers, not call for an end to governmental marketing boards? Presumably, the answer is that they have cast their lot, and their moral authority, with socialist central planners, not free enterprise, and marketing boards play an integral role in this philosophy. To frontally attack statist marketing boards would undermine their moral and intellectual underpinnings.

---

[59] *See Presentation, supra* note 14, para. 17.

[60] *See generally supra* note 19 and accompanying text.

[61] A monopsonist is the opposite of a monopolist. The monopolist is the only legal seller (e.g., the post office, a medallioned taxi cab) in a given market, while a monopsonist is the only legal buyer (a marketing board to which all farmers are forced to sell). Under this system, it is illegal to sell agricultural products to anyone else other than the official marketing board.

68   J. Nat. Resources & Envtl. L.   [Vol. 15:1

N. Economic Consequences

The Council complains of "the pegging of farm wages at low levels".[62] "Pegging" refers to governmental setting of wage rates. Pegging at levels below equilibrium would be equivalent to setting wage maximums. This can cause labor shortages. The clergy, perhaps unfamiliar with technical economic usage, do not refer to this situation. Instead, somewhat confusingly, they mention ordinary functioning of the labor markets and not governmental interference. To wit "this pegging is a result of the simultaneous rise in supply and fall in demand for farm labour."[63]

If true, it causes both happiness and regret. Farm workers will receive less compensation, which means they are not needed in agriculture, but, rather, they are in relatively greater demand elsewhere, presumably in industry. If so, this would roughly follow the United States pattern, where virtually the entire populace engaged in farm labor in the 18th century and virtually no one farms at the close of the 20th century.[64]

IV. Conclusion

There is massive suffering on the part of the poor in the underdeveloped countries of the world. However, this is due to central planning, socialism, and a public ethic which views with great suspicion profits, free markets, individual initiative, and entrepreneurship -- it is not at all the fault of free enterprise and private property rights.

This story exists the world over. Well meaning critics, ignorant of the niceties of economic science, see abject poverty and misinterpret its causes. They have still to learn Adam Smith's lesson that the wealth of nations stems from economic freedom, not its absence. The theological critics fail to see the deleterious effects of government

---

[62] *Presentation, supra* note 14, para. 18.

[63] *Id.*

[64] *See* The Statistical Abstract of the United States (1998).

theological critics fail to see the deleterious effects of government control and, instead, espouse the same policies that created the plight of the poor in the first place.

It cannot be denied that there has been land theft, occurring in both the economically underdeveloped part of the globe and the developed nations. Land reform, based on returning stolen property from the children of the thieves to the children of the victims, is justified. But the burden of proof must reside on those who would overturn extant property titles. This requirement vitiates against the undoing of robbery buried in the sands of time, against the very poor who do not have access to records which can prove their claims, and against native peoples lacking written records. Thus, while the libertarian message on land reform is a very radical one in theory, it is rather conservative in application. It is radical in theory because it is open to the possibility, which is rather unexpected given its almost total unanimous rejection on the part of those who ostensibly favor private property rights.[65] On the other hand, as a practical matter, the libertarian perspective on land reform cannot be expected to support changes in extant land titles between Jews, Arabs, Palestinians, etc., since their disputes go back for thousands of years; nor for Indians vis-a-vis white settlers in the United States for lack of written records. This view may, however, buttress claims from Japanese Americans stemming from World War II expropriation.

If the church wishes to promote the interests of the poor, if it desires to take to heart "the preferential option for the poor," instead of continually urging socialistic welfare schemes that only impoverish them, it must make itself more knowledgeable about economics. Then, as easy as falling off a log, the clergy will realize that the last best hope for the poor are the institutions of free enterprise, precisely the ones responsible for the relatively vast wealth enjoyed in the western industrialized nations.

---

[65] *See* TOM BETHELL, THE NOBLEST TRIUMPH: PROPERTY AND PROSPERITY THROUGH THE AGES 35 (1998). For a response, see Walter Block, *Review Essay of Tom Bethell*, 3 Q. J. AUSTRIAN ECON. 65 (1999).

VI. Fringe Benefits

Comments on Thomason and Burton, Bruce and Atkins, Anderson and Meyer, and Green and Riddell

Walter Block, *College of the Holy Cross*

I. Introduction

This commentary will focus in large part on the economic rationales offered by the four sets of authors (Thomason and Burton; Bruce and Atkins; Anderson and Meyer; Green and Riddell; all in this issue) on behalf of public-sector unemployment insurance and workers' compensation. My thesis is that their explanations for the existence of these two programs are for the most part unsatisfactory, in that they give insufficient attention to the alternative hypothesis—rent seeking on the part of special interest groups—and that their accounts are problematic on logical and empirical grounds.

In an important sense, however, such criticism is rather harsh and unfair, at least when applied to these eight specific economists. This is because—under the provisions of the Fraser Institute–Donner Foundation project on income maintenance for which their articles have been written—they have been asked to render an account of the economic rationale for these programs, and this they have done in an entirely satisfactory and accurate manner. That is to say, theirs is indeed an unerring rendition of how these programs are actually explained and justified within the profession.

By taking on the responsibility for reporting on this state of affairs within the profession, they are not thereby committed to *accepting* these traditional justifications for workers' compensation and unemployment insurance. They may or they may not hold these views, but they are certainly

I would like to thank Gary M. Anderson and Reuben Gronau for numerous helpful suggestions. Any remaining errors and infelicities are of course my own responsibility.

[*Journal of Labor Economics*, 1993, vol. 11, no. 1, pt. 2]

not obliged to do so merely for having written the articles now under discussion.[1]

This article, then, insofar as economic rationales are concerned, is more of a criticism of the profession as a whole than of these four sets of authors. My comments utilize their renditions of the arguments, however, as cases in point.

II. Thomason and Burton

1. If we were to undertake a justification of the government's role in the workers' compensation system, how might we argue? We might begin as do Thomason and Burton (p. S6) with the claim—major premise—that as long as there is a "perfectly competitive world" (in this case, all firms are fully experience-rated) there is no case at all for government intervention. However—minor premise—it might be argued (curiously, they do not carry the argument through this step or make the case in any other way; on the contrary, after beginning the syllogism in this very traditional neoclassical manner, they allow the entire matter to drop) that we in fact do not live in a perfectly competitive world, in which these claims can be made unambiguously. Therefore—conclusion—it follows that government intervention is justified since the market has been shown to be "imperfect."

There are a whole host of critics who have rejected this argument in other contexts (see, e.g., Demsetz 1968). The main point is that it does not logically follow from market imperfections that interventionism will work any better. On the contrary, there is the phenomenon of "government failure" studied by the public choice school (Buchanan and Tullock 1971; Buchanan 1975) and others. If markets fail, then so do governments. If there is to be an improvement in institutional arrangements, it must be predicated on state inefficiency being less than market shortfalls; it cannot be based on the mere demonstration of market imperfections. Everything known to man on this, the wrong side of the Garden of Eden, is imperfect: this applies to private enterprise as well as public. That the one is imperfect does not prove the preferability of the other.

The difficulty, in this case, is that it is not clear that anything less than all firms being fully experience-rated is sufficient to make the case for entrusting workers' compensation programs to the tender mercies of the public sector. As Thomason and Burton state, the riskier the occupation, other things being equal, the higher the risk premium that tends to be added to the wage. In the alternative scenario, the whole institution of

---

[1] This may be putting matters too strongly, in that there was no reason for not criticizing the traditional economic account, and, indeed, several of them on occasion have indulged in just that activity—e.g., Anderson and Meyer (in this issue), and Bruce and Atkins (in this issue).

compensation would be privatized; it would be left to the wage system, perhaps buttressed by a private insurance industry, to determine the occupational safety rate and compensation for accidents. This is the usual presumption among economists, and nothing uncovered by Thomason and Burton would appear to tip the balance in the other direction.[2]

Further, Thomason and Burton believe that the source of the market imperfection is a sort of externality: less than fully experience-rated firms "do not bear the complete cost of *ex post* compensation." Rather, they are able to shift some of these costs onto "all firms within the rating class to which the firm belongs" (p. S7). The difficulty with this is that in a full free enterprise setting insurance companies would compete with one another to distinguish between (e.g., discriminate among) the high- and low-risk firms that have so far been included within the same rating class. If they can make such fine distinctions, they can profitably base their premiums on perceived risk.

At present, unfortunately, it is "politically incorrect" (particularly in Canada) to so "invidiously" discriminate between people, particularly if the high-risk firms have more than a proportionate share of women, blacks, "visible minorities," native peoples, or "physically challenged" (e.g., handicapped) persons.[3] Thus, part of the reason that the market cannot function as well as it might emanates, not from the private sector, but from the public.

2. Thomason and Burton describe the attempt of the insurance firms to transact business with the insurees as one of the former applying "pressure" to the latter. It is possible that this is just a matter of stylistic expression. However, based on the numerous occasions this terminology is employed, it may indicate a substantive normative point, namely, that these offers are improper, exploitative, immoral, or otherwise unjustified.

But it is an axiom of economic theory that all trades (at the very least

---

[2] Milton Friedman (1990, p. 15) claims that many people accept the following argument, which he has dubbed the "welfare state syllogism." Here is the invariably accepted major premise:

A: Socialism has been an utter failure. (It's been a failure in Russia, in China, in New Zealand, in Australia, in Britain, and even in Canada and the United States.)

The widely agreed upon minor premise:

B: Free market capitalism has been a great success. (It's the only system which has been able to combine a relatively high degree of prosperity for the masses with a large degree of political and human freedom.)

And now for the conclusion, to which most people concur:

C: Therefore, we need more socialism.

This is then applied to medical care, welfare, etc., and is certainly apropos to our present concerns as well.

[3] For a critique of this antidiscrimination philosophy, see Epstein (1992) and also Block (1992).

those between consenting adults) are mutually beneficial, if only in the ex ante sense (Rothbard 1977). That is, no one engages in trade unless he expects to gain thereby. In the present case, the insurance company makes an offer to the injured party to trade an uncertain future income stream for a lump sum payable immediately. If the accident victim has a preference for risk avoidance, we would expect him to accept even an offer that is worth less than the present discounted value of what he might otherwise have received. But no less is true in the case of the sale of a lottery ticket; the price is higher than the expected value of the gain. The same is true of many cases of gambling, buying insurance in the market, and other similar activities. These offers are not "mathematically fair," at least not intentionally so, otherwise there would be no scope for administrative costs, profit, and so forth. Would we accept "pressure" as an accurate way of referring to all such offers? Hardly.

3. It is important to keep in mind the normative-positive distinction between equity and equality. In the view of Thomason and Burton, people who suffer from permanent partial disabilities (PPDs) are pressured by insurance companies to accept compromise and release agreements (C&R's). As we have seen, the lump sum payments negotiated thereby are often of less worth than the present discounted value of the income stream they might otherwise have received had they gone on to adjudication. But the adjudication process costs time, money, and effort and is riskier than the acceptance of a compromise lump sum agreement. Presumably, the lump sum was worth more to the claimant ex ante than the package deal of the prospect of higher period payments, coupled with the attendant risks and initial costs. Otherwise, he would scarcely have agreed to it.

Let us contemplate a case, however, where this is for some reason not true. Where, for example, the claimant somehow unwittingly agreed to a less preferable deal than might otherwise have been available to him. Thomason and Burton characterize such a situation as "less equitable" (p. S12). With that, they slip almost imperceptibly from the relatively firm ground of positive analysis to the quicksands of the normative realm. Inequity implies injustice, and nowhere do Thomason and Burton offer any independent support for their claim. They state, "Obviously, unequal compensation is inequitable compensation only if the injuries are equally severe" (p. S12). But this is not as obvious as it might appear. Even if injuries are absolutely identical, it is not necessarily inequitable for compensation to be unequal. Perhaps premium payments were different. Perhaps the policies accentuated risk in a different manner. Perhaps the additional amount represents a voluntary gift from the insurance company to one of the injured workers.[4] According to Robert Nozick's (1974) en-

---

[4] Would it be inequitable if two workers with identical marginal productivities were hired and the first was paid, as he must be in equilibrium, a wage equal to

titlement theory, all agreements are equitable, as long as no force or fraud is involved and as long as the property or wealth being traded is legitimately owned.

We can utilize the Stiglerian (1961) search model in this regard. Suppose that a poor man searches for a bargain for only 5 days, while a rich man searches for 10 days. The latter is able to hold out longer because he is wealthier. Now, it is one thing to conclude that the additional days of search might yield bargains that are more than offset by their expenditure and that therefore there is an advantage to being affluent. (This need not happen; it is entirely possible that the alternative cost of the additional time spent in searching might be less than the value of the additional bargains unearthed.) But it is quite another matter to label this situation as inequitable. Yes, greater wealth enables people to attain more command over all kinds of goods and services complementary to search, but to label this inequitable is tantamount to claiming that all divergences in affluence are themselves inequitable.[5]

If we want to consider a real case of inequity, we might contemplate the following. The New York State Workers' Compensation Board "will not permit agreements where medical treatment has been required. . . . Nor will the board approve compromise settlements when the worker is totally rather than partially disabled. All lump sum settlements must be approved by a member of the board" (Thomason and Burton, p. S14).

This is inequitable because it amounts to a taking (Epstein 1985) with no compensation on the part of the government. As we have seen, trade is mutually beneficial, at least in the ex ante sense. However, this applies not only to the more pedestrian cases of trade but to all commercial agreements as well, specifically, lump sum C&R's. On the assumption that claimants are not mentally but rather physically disabled,[6] it is clear that these prohibitions are injurious to them. If a settlement must be approved of by a board member, this implies that at least some mutually beneficial

---

this level, while the second was paid 50% more? Hardly. Why is it necessarily inequitable to offer the second what is in effect a gift equal to 50% of his wage? The temptation to call this inequitable arises from a comparison of the situations of the two workers. This can be ameliorated by focusing instead on the boss and the favored employee. Surely, if the former offers the latter additional monies, on a completely gratuitous and voluntary basis, this cannot be judged inequitable.

[5] Alternatively, it is possible that the rich man will engage in less search activity than the poor one. The rich man has higher opportunity costs and thus faces higher costs of search for the same item. As a result, he may search for fewer days. If so, a rich consumer is likely to pay a higher price for the same item. It is the rare person who would think this inequitable.

[6] Strictly speaking, this would be inequitable even in the case of mental disability, unless we accepted the claim that the state is a better guardian than the claimant's friends or family.

transactions will be prohibited. Certainly this applies in the cases of total disability. The two parties have agreed to a settlement, one which, in their view, represents their best interests, and the board sets it aside. This is economically indistinguishable from a nontariff barrier to trade. In both cases, consumer sovereignty is set aside.

4. Thomason and Burton appear to take the view that there is something problematical about a "higher industrial accident rate experienced during the 1960s." They must see this as a difficulty in need of rectification, otherwise they could not characterize the commission's recommendation to make alterations in the workers' compensation program as being undertaken in order to "correct these problems" or that to make changes in the law would be to "improve" it (p. S2, n. 2).

But the view that a lower accident rate is necessarily an improvement cannot be sustained without recourse to unproven normative considerations. This assumption may accord well with the view that the man in the street is presumed to hold, but it is contrary to traditional economic theory. From that perspective, it is assumed that there is an optimal rate of industrial accidents, coupled with the strong expectation that this rate is larger than zero.

More specifically, the optimal rate is that at which the expected value of the cost of safeguarding against one more accident is just offset by the expected benefit of doing so, to wit, at the point at which the marginal revenue and marginal cost curves of accidents intersect. It is only if society were already past that point that the Thomason and Burton assumption would hold. But there is no evidence put forth by the authors to suggest that such a situation actually exists. Perhaps, on the contrary, workers' compensation benefits are already pegged at too generous a rate, and the amelioration of the misallocation of resources would be in the opposite direction from the one assumed to be the case by Thomason and Burton.

It is quite possible—perhaps because of excessive amounts of safety legislation, or as a result of an elasticity of accident prevention expenditures with respect to income that is higher for rent seekers than for the general populace—that we are now at a point of underinvestment in productive processes that are accident intensive. If so, the legal alterations of 1975 would not "improve" the situation but, rather, would worsen it.

5. Thomason and Burton state, "If insurers do use delays and other tactics to pressure claimants into smaller lump sum settlements, one would expect low income claimants to be more likely to settle with a lump sum than higher income claimants since low income claimants are less able to withstand insurance company delaying tactics" (p. S11).

This is indeed the implication of a positive income elasticity, the usual assumption in such cases. As well, in support of Thomason and Burton's point, there is another economic effect that works in the same direction; when both are considered together, their suggestion that there would be

Comments S311

a negative correlation between income and acceptance of the lump sum deal is even further strengthened. The other economic effect is that time preference rates are also negatively correlated with income (Banfield 1974, 1977); that is, the poor usually have higher time preference rates than the rich. So, on the one hand, as Thomason and Burton would have it, the richer one is, the easier it is to hold out against delaying tactics and to be able to refuse the lump sum settlement; on the other hand, the richer one is, the more likely one will prefer to wait in any case because, as a result of lighter discounting of future income streams, the more likely it is that the present discounted value of money receivable in the future will be greater than the lump sum payment.

III. Bruce and Atkins

Bruce and Atkins take great pains to establish, and correctly so, that several arguments in behalf of public sector worker's compensation cannot be considered definitive. For example, under their initial set of (perfectly competitive) assumptions, they conclude that "the socially optimal levels of employment, safety and risk spreading could be achieved without government intervention" (p. S44). As well, they refuse to accept the free-rider problem, and the existence of economies of scale in the purchase of insurance, as establishing a case for nationalization of this service.

However, when it comes to information asymmetries, Bruce and Atkins are no longer able to maintain this market-oriented public policy. Instead, they hold, "The presence of informational asymmetries may provide an argument for workers' compensation" (p. S46).

Why, in their view, should the fact that employers may know more about the riskiness of a job than employees call for governmental abrogation of commercial contracts between consenting adults? This is because "there will be underinvestment in safety if workers underestimate p_i (the probability of an injury taking place). The reason for this is that workers can only be expected to bargain for wage differentials which compensate them for the *expected* risks of employment. Thus, if workers underestimate the true risks of employment, the employer will have no incentive to provide either additional information or additional safety *unless* it has been made strictly liable for accident costs" (pp. S45–S46).

Let us ask whether it is really true that there will be underinvestment in safety in the absence of employer liability. We must start out by conceding that "the employer will have no incentive to provide either additional information or additional safety." But it by no means follows that no one else in a market situation will have such an incentive.

Who else, then? One possibility is for an outside agency to take it upon itself to rate the dangers of various occupations. Moody's and Standard and Poor's rate stocks and bonds; Consumer Reports describes the risks of hundreds of products and services; Freedom House (Gastil, various years; McColm et al. 1989) and the Fraser Institute (Walker 1988*a*; Block 1991*b*) assess the risks to freedom from residing in different countries; while the Good Housekeeping Seal of Approval is not awarded to dangerous consumer goods. Closer to home, Underwriters' Laboratories assesses the safety of various consumer goods, and private commercial chemical, mechanical, and electrical laboratories certify the quality of industrial machinery. As well, the Insurance Institute publishes data regarding the relative riskiness of the various occupations.

In an era when we are accustomed to government sources accounting for some half of the gross national product, it may be difficult for most people to imagine market alternatives to government certification. This is a problem particularly besetting the newly capitalist countries on the other side of what used to be the Iron Curtain. But in the West, it ought not to be too onerous a challenge to envision a sort of Consumer Reports for industrial accidents.[7]

A second alternative resides in labor unions.[8] Surely they embody sufficient physical and human capital necessary to ascertain the dangers of the jobs of their members. As well, they may be presumed to have a strong motivation to press for wages that fully capitalize these perils.

Then there are always other, competing, employers. If employer A is underpaying his work force, because he knows the true risks of laboring for him but his employees do not, then this is a golden opportunity for employer B, C, . . . to hire A's workers out from under him. Economists typically assume that wages tend to equal marginal revenue product in equilibrium. But this is only a first approximation. In point of fact, wages tend to equal the next best option available to the worker in question (Becker 1964). Why, after all, should the profit-maximizing employer agree to pay the marginal revenue product (MRP) of the worker? Out of the goodness of his heart? Hardly. No, the reason is based on the fear that, if he fails to do so, someone else will successfully bid away the underpaid worker. But this reasoning applies whether the wages paid by employer A are less than MRP out of inadvertence, because of sheer cussedness, or, in the present case, as a result of an informational imbalance.

It is of course true, as Bruce and Atkins assert, "that workers can only be expected to bargain for wage differentials which compensate them for the *expected* risks of employment" (p. S46). The point is, however, that competition from other employers will tend to ensure that employee ig-

---

[7] I say "envision" what might be, rather than "appreciate" what already is, because in the last few decades the government has been so active in this field. Had the state not preempted the market by giving these services away for "free" (i.e., courtesy of the taxpayer), it is likely that private enterprise would now be playing a far larger role than at present.

[8] For an account of the institutional arrangements that would render labor unions compatible with a market economy, see Block (1991*a*).

norance does not vitiate the market process. Laborers in general need not be aware of their own marginal productivities in order to be compensated at those levels; the same analysis applies in the present case.

Bruce and Atkins's rationale for compulsory government workers' compensation thus requires not only a knowledge asymmetry between employer and employee. Also needed is the assumption that this divergence applies to other employers as well, that is, that such information is totally firm specific. But this must be a rare occurrence indeed, applying only to monopolized industries. It has not been shown to be of sufficient incidence to be able to justify a public policy that applies to the entire economy.

Bruce and Atkins's position is also somewhat of a precarious one to defend because of the difficulty of generalizing. Informational asymmetries are ubiquitous in our society. If they can serve as the rationale for government intervention in this instance, they can do so in all others as well. For example, manufacturers and repair mechanics can be safely assumed to know more about the products they specialize in than do the consumers who purchase them; this applies to automobiles, refrigerators, toasters, lawn mowers, and to a whole host of other such products. Yet the case for generalization from workers' compensation to a call for government intervention in all these cases is hardly justified. There is no reason for such a public policy because commercial interactions in this regard are earmarked by a great deal of efficiency. Typically, in the market, these consumer durables come in a package deal with guarantees of various sorts; private insurance, in effect, thus addresses the problem at hand. Nor is there any reason to suppose that the incentives that lead to these arrangements would not also apply to the labor market.

Further, suppose the case where the informational asymmetry were the other way around. That is, the laborers have a clearer understanding of the hazards of the job than do the bosses.[9] According to the logic of the Bruce-Atkins argument, there would still be a resource malinvestment. Only now it would be in the direction of too much protection, not too little. Would it make sense to inaugurate an "employers' compensation" program, to counteract the "exploitation" of employer by employee? Paraphrasing Bruce and Atkins, we could argue that, if "*employers* underestimate the true risks of employment, the *employee* will have no incentive to provide either additional information or *less* safety unless he has been made strictly liable for accident costs (in this case, the costs of unnecessary safety installations)." Bruce and Atkins provide no evidence as to how they would analyze such a case, but, unless they take the bull by the horns

---

[9] For example, a restaurant owner hires an electrician to fix his wiring, or a bank engages an expert to put its operation on a computer basis. With regard to such particular specialties, the employer can be expected to be far less sophisticated than the employee.

in this dilemma and call for compensation of some sort from employee to employer, we cannot credit their original application of the principle.

The Stiglerian information search model may be utilized here as well.[10] First of all, if employer information about safety is greater than that possessed by employees, it by no means follows that the former level is optimal and the latter substandard; it could as easily be the other way around. Namely, the search for safety information of the owners of the business could be excessive and that of their hirelings exactly right. Second, even if we assume for argument's sake that knowledge of safety conditions on the part of workers is insufficient, this does not necessarily indicate economic inefficiency. As Stigler has shown, this may well be a rational decision on their part.

Bruce and Atkins also briefly mention adverse selection as an argument in favor of government intervention. Yes, "if both insurers and employers are unable to distinguish among employees on the basis of their respective probabilities of injury" (p. S46), private safety insurance will not be offered by the market. But this is unlikely, and no evidence is adduced to the contrary. There is also the added complication that other statist interferences render the insurance task more difficult. For example, in some jurisdictions it is illegal to discriminate on the basis of physical disability, a phenomenon surely correlated with workplace risk. As well, there are now all sorts of legal constraints placed on the use of economical proxy variables, such as age, gender, and race.[11] But suppose that insurance fails to materialize for reasons unrelated to governmental barriers to entry. It is still not "welfare-improving to require that all employers offer a standard form of insurance—workers compensation" as claimed by Bruce and Atkins (p. S46), unless we indulge in heroic interpersonal comparisons of utility or ignore the welfare of those on whom these programs are enforced.

IV. Anderson and Meyer

There is little doubt that the unemployment insurance (UI) that was incorporated into the Social Security Act of 1935 "arose in response to the Great Depression and was motivated by a desire to stabilize the economy" (Anderson and Meyer, p. S71). As a historical account, this is unexceptionable. However, the task of the second section of the article under consideration is not historical exegesis but rather the development of an economic efficiency rationale for UI. As such, the benevolent motivations of the enactors of this legislation are as entirely beside the point as they are for any other aspect of the New Deal or any other governmental decision for that matter. What count for public policy analysis are economic effects, not intentions. Looking at matters from the perspective of the 1930s, the

[10] For a critique, however, see Kirzner (1973).
[11] See n. 3 above.

task of public policy-making then was to address the cause of the Great Depression. But the economic malaise of the 1930s was due to poor monetary policy (Friedman and Schwartz 1963; Rothbard 1974), not to lack of an unemployment insurance program.

Since UI was not even used to address the monetary imbalances of the economy, it is unlikely that it could play any large ameliorative effect. True, the monetarist thesis was not well known or appreciated at that epoch in our history, but Anderson and Meyer are writing in the present, not the past. It is unlikely, then, that the Depression can serve as an efficiency interpretation for UI.

So they try again. In the view of Anderson and Meyer, "the main rationale for UI is that it provides insurance for workers who may lose their jobs" (p. S71). But this hardly accounts for the fact that the task has been assigned to the public, not the private, sector. The rationale for fire insurance is that it provides coverage for those who may lose their homes to a conflagration; but this by no means implies a nationalization of the industry that provides the protection.

"This rationale," continue Anderson and Meyer, "is appropriate for workers whose unemployment is unexpected, but not for individuals with frequent and predictable spells of unemployment, say in seasonal jobs" (p. S71). But why should this be so? In ordinary marketplace provision of goods and services, entrepreneurs do not usually ignore any significant number of potential customers. If a person is engaged in seasonal employment, and is willing to pay premiums high enough to purchase UI, it is likely that some private firm would be willing to oblige. It would certainly not be Pareto efficient to forbid such transactions. Indeed, the usual assumption of risk avoidance offers a presumption in this direction. It is true that the worker could probably do better for himself financially by merely saving money from his time of employment to be used during his time of unemployment, but perhaps he does not trust himself to engage in such "rational" behavior. In such a case, private provision of UI may function as a way of voluntarily engaging in "forced" savings, a result for which the worker is willing to pay.

The point is, we can expect to have as much unemployment as the government is willing to pay for. It is a basic economic axiom that, ceteris paribus, the higher the reward offered, the greater the market response. That is, although individual supply curves may conceivably bend backward, the usual presumption is that the response of the entire market to greater remuneration is additional effort. Why this general law should not be seen to apply to UI must ever remain a mystery. Missing from the analysis is an appreciation for the fact that the more generous the unemployment compensation, the higher tends to be the resulting rate of unemployment.

Anderson and Meyer then cite adverse selection as a possible explanation for governmental provision of UI, at least on the firm level. But they correctly undermine that argument by noting that private firms do offer medical and workers' compensation insurance.

Anderson and Meyer characterize the diversifiability argument as "perhaps the most compelling reason for publicly provided UI," noting that "claims for UI tend to be concentrated in recessions" (p. S71). At this point in the first version of their article Anderson and Meyer quickly and quite properly undermine any such notion on the grounds that "the counterargument here is that insurance companies should be able to borrow in a recession if their reserves are insufficient." As well, they noted in the version of their paper presented at the conference that the argument applies to earthquake insurance, and yet private companies typically offer this service, thus vitiating the diversifiability argument. Unfortunately, in their final version, the one that appears in the present volume, they have deleted this very insightful footnote.

Getting back on track, Anderson and Meyer do not accept UI on the ground that it increases societal welfare, in the absence of any "reason why the value of the job match is different for society than for the individual" (p. S72). They are agnostic on the stabilization argument, noting that, if the real business cycle (technological shocks) theorists are correct, UI will "reduce welfare by decreasing efficiency" (p. S72). Anderson and Meyer maintain that it is only if the Keynesians are correct that UI may be productive. However, one can argue that the existence of inflationary recession has all but destroyed the Keynesian system, at least from the neck up. As well, even on the assumption that this model is accurate, the private provision of UI can be expected to be every bit as countercyclical as the governmental counterpart.

The final rationale for UI considered by Anderson and Meyer is as a means of redistributing income. This rent-seeking explanation may well be the best one for why the program actually exists, but it by no means follows that UI is the most efficient means of attaining this end. Some experts (Friedman and Friedman 1980, chap. 4) maintain that the negative income tax may be a far more effective way of achieving this morally problematic (Rothbard 1973, 1982; Nozick 1974; Hoppe 1987, 1989; Friedman 1989, p. 189) goal. However, the Friedmans (1980, p. 122) state: "The negative income tax would be a satisfactory reform of our present welfare system only if it *replaces* the host of other specific programs that we now have. It would do more harm than good if it simply became another rag in the ragbag of welfare programs."

The problem with UI, at least from this point of view, is that this is just one of a "host of other specific programs that we now have." Income redistribution may be accomplished in this way and thus can serve as a justification for the program (at least for those who maintain that income redistribution is a worthwhile goal); but if the negative income tax is more efficient, only on condition that it *replaces* the other programs, then UI may actually be counterproductive to these ends.

To conclude our discussion of the public policy aspects of the Anderson and Meyer article, we now turn to a consideration of their proposals for UI reform. Anderson and Meyer put forth several options. One is to provide job training and offering cash bonuses for returning to work quickly. But evidence shows that this reform merely encourages temporary layoffs (the unemployed tend to return to their old firms.) One of the states experimenting with this policy even precluded returning workers from benefits on this ground. But it is easy to imagine the market response: firm A hires back firm B's layoffs, and B reciprocates in behalf of A. Another difficulty with this proposal is that it is merely a different way of providing UI payments and thus does not address itself to the crux of the matter: subsidizing unemployed people promotes incentives toward unemployment.

A second set of strategies is to more closely tailor a firm's UI tax payments to its layoff history. But there is a problem with these initiatives: they are not neutral with respect to the market. On the contrary, industries differ with regard to their layoff history in systematic ways, and any proposal that discriminates against those with heavy layoffs, and thus in favor of those with light layoffs, misallocates resources from the former to the latter.[12] Which are the industries that tend to engage more heavily in layoff activity? Generally, firms specializing in consumer goods tend to be more stable over the business cycle than companies involved in the production of capital goods (Hayek 1931; von Mises 1963); a similar analysis applies to seasonal versus nonseasonal careers and to general versus specific training (Becker 1964). Since this plan would not be neutral with respect to these sets of industries, it is to that extent resource misallocative.

Another set of approaches discussed by Anderson and Meyer involves the improvement of the safety net provisions of UI. These, then, are not addressed to the insurance aspects of UI but instead to this policy insofar as it is a part of the welfare system. The difficulty here is that, as a matter of pure welfare economics, as we have seen, the negative income tax (Friedman and Friedman 1980) may be a preferable means toward the end of erecting a social safety net. If so, then UI should be abolished and the negative income tax should be designed to take up whatever slack implied by that act.[13]

---

[12] If it can be shown that "society" somehow has an interest in unemployment over and above the totality of the private interests in this phenomenon, then an argument for making UI tax payments reflect layoff history can be fashioned on grounds of externalities. Unless and until this can be shown, however, the transfer of wealth from one set of industries to the other must be considered misallocative. Anderson and Meyer appear to be agnostic on this point, an interpretation that arises from their comments on p. S72, and thus quite properly this argument is not made use of.

[13] For a critique of the welfare system—e.g., raising the social safety net—see Murray (1984, 1988) and also Anderson and Block (in this issue).

It is unfortunate that Anderson and Meyer did not address themselves to another type of UI reform: privatization. This is an initiative (sometimes called "private enterprise") that has been tried (not with UI, unfortunately, but rather with regard to numerous other programs) very successfully, most notably in the West in England, Canada, Australia, and increasingly in the former Soviet Union and countries in Eastern Europe and elsewhere (Ohashi et al. 1980; Poole 1980; Walker 1988*b*). In this proposal, UI would be left to the market. Private insurance companies would provide explicit coverage if there arose a demand for it, and, if not, people would rely on the implicit UI now accomplished through the system of wage differentials. As well, governmentally imposed obstacles to such a market (e.g., anti-discrimination laws and price ceilings for premiums) would be rescinded.

V. Green and Riddell

Several years ago a popular magazine ran a full-page, exquisitely colored advertisement for an insurance corporation.[14] It showed a gigantic mansion in the background, a large manicured field in the mid ground, and a luxurious yacht at anchor in the foreground. The caption of the ad stated: "Both the house and boat are insured with Chubb. Now, if only they'd insure my golf scores . . ."

This statement could have served as the motto of the Green-Riddell article as well. For both the ad copy for Chubb and the theoretical considerations offered by Green and Riddell maintain that there is something wrong with the marketplace if it does not supply insurance protecting against just about anything that anyone could conceivably claim ought to be covered. In the former case, it is golf scores; in the latter, it is unemployment. But in both instances the charge is that of "market failure."

The insurance company's advertising is of course meant as a wry joke, intended only to direct the reader's attention to the commercial offering. Green and Riddell in contrast, appear to be entirely serious. What are their arguments?

Starting out with the indubitable premise that "individuals are generally believed to be risk averse," they note that, while owners of capital can "reduce their income risk by holding a diversified portfolio," "workers are generally unable to diversify their human capital wealth" (p. S98).

There are, however, several ways in which employees can largely escape income variations that result from unemployment, if their desire for a low-risk life is strong enough. For example, they can eschew employment in such risky endeavors as Broadway plays, wild-cat oil drilling, mineral exploration, and so forth, and instead focus on the civil service; the post office; utilities such as telephones and electricity; on large, long-existing

---

[14] *Smithsonian* 20, no. 6 (September 1989): 101.

and well-established firms in the "blue chip" areas; or on conglomerates, where if things go bad in one division, a transfer to another may be economically advantageous.

Alternatively, they can work for the very owners of capital, pointed out by Green and Riddell, who "reduce their income risk by holding a diversified portfolio" (p. S98). As well, wages are likely to be higher for employees taking part in endeavors where the chances of losing their jobs are higher (capital goods, seasonal industries, specific training), and the differential between the compensation levels can be interpreted as the implicit market insurance premium paid for relatively secure employment.

Green and Riddell are fully aware of this market mechanism. They even remark on other "forms of self-insurance such as saving, career diversification, and family labor supply" (p. S101). Nevertheless, they conclude that "these private market responses are unlikely to obviate the need for comprehensive UI" (p. S101).[15] Unfortunately, they give no justification for this contention,[16] leaving the reader to wonder what criterion they are employing in order to reach their conclusion.

Green and Riddell (p. S98) also claim that "comprehensive private insurance markets that would enable most workers to purchase insurance against the risk of loss of labor market income . . . have generally failed to emerge . . . despite the demand that *evidently* exists for such insurance" (emphasis added).

But this is rather problematic. First, while it is perhaps conceivable that a demand can exist for a product not now being supplied on the market, one might maintain that the presumption is in the opposite direction. One can never close the door to the Green-Riddell thesis, but it would appear that the burden of proof rests on them.

Second, we have just seen that the market does indeed provide (implicit) unemployment insurance through the intermediation of wage differentials between risky and nonrisky employment. It is therefore somewhat hazardous to maintain that such institutions have "failed to emerge."

Why are we to conclude that any demand for this service *evidently* exists, given the contention of Green and Riddell that little or none has been successfully supplied? In their view, this is because of the problems of moral hazard and adverse selection. But these obstacles occur with regard to several other instances where explicit insurance policies are commonly written. For example, persons insured against fire or motor vehicle accidents can affect the probability of the event against which they are protected from taking place, and both the owners of less fireproof dwellings and

[15] That is to say, public-sector UI.

[16] Apart from their critique that the " 'systemic risk' associated with the business cycle is difficult to reduce through portfolio diversification" (p. S101), an argument to which we reply below.

dangerous drivers, other things being equal, are likely to most interested in purchasing such protection.[17]

To return to our previous example, there is at present no demand whatsoever for golf score insurance. But this "failure," too, encounters the hurdle of moral hazard and adverse selection. For example, were they to be insured, golfers would be in a position to purposefully score badly, and such insurance, were it to be offered, would be of particular interest to duffers. But this is hardly why golf insurance is not explicitly offered on the market.[18] The market has "failed" in this way because the demand for this service, at a price sufficient to defray the attendant costs, evidently does not exist. Moral hazard and adverse selection may thus be obstacles to the occurrence of formal insurance, but they are neither necessary nor sufficient conditions for its absence. If the demand for the requisite insurance is sufficiently strong, and if people are willing to pay high enough premiums, these obstacles can evidently be overcome, as in the cases of fire and accidents.

Why is it that the market offers only implicit unemployment insurance (through wage differentials) and not the explicit variety (where contracts are formally drawn up)? Although this is only speculative, the following considerations appear to be relevant. First, when government subsidizes a service, and, even more so, makes it compulsory, it is very difficult for private enterprise to earn a profit selling the same product. When people are forced by law to take part in the state's UI program, this alone would tend to preclude operation of the market.

Second, the basic philosophy of insurance is everywhere under attack, particularly in the United States and Canada. A recent, popularly supported California ballot initiative would have placed a ceiling on premiums. In the Canadian provinces of British Columbia and Ontario, auto insurance has long been a monopoly of government (Grubel 1985). These institutional arrangements have not been upset by the recent trend toward privatization; in the opinion of many commentators, tax-financed insurance in this area has become somewhat of a sacred cow. Even if government were not providing "stiff competition" against private UI plans, this would be a particularly inopportune time to try to launch any such new initiative.

Third, as we have seen, there are now laws on the books in the area of discrimination that mitigate against the private provision of insurance in general; these would have especially strong ramifications in the area of UI. For example, in setting premium policy auto insurance companies

[17] It is for these reasons that Anderson and Meyer (in this issue, p. S71) explicitly and properly reject the market failure argument on the basis of moral hazard and adverse selection.

[18] There are however informal market mechanisms that can help "insure" against bad golf scores: private lessons, self-monitoring through video techniques, investment of time and green fees in putting practice, driving ranges, etc.

have come under attack for "discriminating against" young males merely because they have worse driving records than females and older males; life insurance companies have been similarly criticized for demanding higher payments of women, merely because they tend to live longer than men. It is only to be expected that severe restrictions would be placed on any insurance company attempting to deal with the problems of moral hazard and adverse selection by trying to distinguish, that is, discriminate, among potential clients on this basis. It is likely that endeavors to ascertain past employment histories would run afoul of "privacy" provisions. Nor would any attempt to attain and apply psychological profiles to prospective customers likely be tolerated in these "politically correct" times.

The only way to determine if private formal UI is viable is to repeal governmental UI and also other such laws that may have the effect of restricting private provision. Then, if the demand for the market variety of UI is strong enough, the creation and patronage of private sources of supply will occur.[19] If not, not. In neither case, however, would we be entitled to argue for government intervention on the ground of market failure. For if successful market operation occurs, even Green and Riddell will have to retract their claim.[20] If it does not, it still will not have been shown by them that the informal market insurance provisions (through wage differentials) are insufficient.

The failure of the Career Guard scheme is also not evidence of the market's inability to provide this service (p. S99). This is no more true than that the failure of the Edsel or the Studebaker shows that private commercial arrangements cannot provide automobiles. Green and Riddell quite correctly point to the efforts of insurance firms to deal with moral hazard in life insurance by refusing to pay in the case of suicide or murder by beneficiary and by declining the business of high-risk persons such as sky divers. But if allowed to by law, private insurance firms would likely avail themselves of similar options in the case of UI. For example, they could arrange matters so as not to be responsible for payouts to unemployed clients unless the overall unemployment rate in an industry or in the economy reached a specific level. Green and Riddell seem to accept this point when they state, "consequently, private companies are unlikely to offer insurance against the risk of unemployment except under very narrowly specified circumstances" (p. S99).

Next, consider their comments on nondiversifiability (p. S99): "An additional reason why private insurance companies would be reluctant to

[19] It will occur, i.e., if people are willing to pay premiums high enough to offset those aspects of moral hazard and adverse selection not overcome by the insurance industry's attempt to discriminate on this basis and to encourage coinsurance through deductibles, less than 100% coverage, etc.

[20] Although they could still argue that in the market there would be less than the optimal amount of resources invested in UI.

provide UI is that the risk of cyclical unemployment is largely nondiversifiable except over time. That is, the risk of unemployment due to the business cycle is positively correlated across members of the labor force so that pooling many insurees does not substantially reduce the aggregate risk."

But how could the alternative to private provision, namely, government-organized UI, overcome this problem? If we interpret Green and Riddell literally on this issue, society is in no better circumstance with regard to diversifiability under socialized UI. If the problem stems from something intrinsic to the cycle, unless it can be shown that the public sector has a comparative advantage in dealing with this challenge—an issue Green and Riddell fail to address—it does not follow that governmental provision of UI will be an improvement over market alternatives.

Their argument, moreover, proves too much. For there are other negative occurrences besides unemployment that tend to affect large numbers of people all at the same time, where private provision of service has long been in operation. Examples include hurricanes, firestorms, and earthquakes. In any case, a multinational firm (such as Lloyds of London) that provided private UI might be able to overcome this problem, at least to the extent that the business cycle is not uniform over the entire globe.

As well there is the problem that government itself, in the view of many commentators, is either the cause of the business cycle, and/or its attempts to ameliorate this condition tend to exacerbate this problem instead of solving it (Friedman and Schwartz 1963; Rothbard 1974). If this is so, then Green and Riddell argue in a curious fashion. The government causes/exacerbates the business cycle; the business cycle decreases the diversifiability of insurance, rendering private provision problematical; therefore, this industry should be placed in the hands of the initial causal agent of the harm, for example, the government. One might as well argue that the fox should be put in charge of the chicken coop because he has already gnawed half way through the protective wire fence surrounding it.

Next, consider their treatment of the efficiency rationale for UI. Green and Riddell assert that "the probable failure of private insurance markets to exist on a comprehensive basis provides the fundamental efficiency rationale for a publicly provided UI scheme" (p. S99). According to Pareto optimality conditions, however, a governmental program may be implemented on efficiency grounds if no one is made worse off and the welfare of at least one person is improved. Yet in this case it is easy to point to not one but several groups of individuals who will be made worse off by tax-financed UI: all of those who would not otherwise have been willing to pay taxes for this purpose; risk preferrers, whose welfare ex ante is worsened by insurance; and those entrepreneurs who would have earned profits in a privatized UI industry. Green and Riddell call for "compulsory participation" for those "who believe themselves to be low risk from opting out of insurance coverage" (p. S100). By their own admission, then, there are people who will be made worse off by such a policy.

There is of course a weak sense of this argument in which it is not necessary that there be no people harmed by a policy in order for it to be efficient in the Kaldor-Hicks sense. All that is necessary, in this case, is that the gains of the winners more than offset the harms to the losers, so that the former, at least in principle, are able to compensate the latter for accepting the new program. Green and Riddell, unfortunately, provide no evidence on this issue. In any case, because of the subjectivity of costs (Buchanan 1969; Buchanan and Thirlby 1981), and the theoretical dangers of utilizing interpersonal comparisons of utility (Rothbard 1977), this argument rests on somewhat shaky grounds.

Green and Riddell set a difficult task for themselves: deriving normative conclusions (government should set up a UI program; a tax-financed UI program is justified; there exists a rationale for compulsory UI) from premises that do not contain a normative dimension (UI programs generated in the market will be less comprehensive than alternatives, mainly because of moral hazard and adverse selection). That they have not entirely succeeded in generating that conclusion from these premises does not undercut the fact that they have used all of the relevant tools of modern economic analysis in a competent and creative manner. There is another way to characterize the problem faced by Green and Riddell. As Harold Demsetz's "The Grass is Greener" (1988) fallacy has shown, even if the market were "imperfect," it does not logically follow that supplanting government for private institutions will improve matters. Green and Riddell have indeed indicated that private insurance markets are imperfect (e.g., through adverse selection and moral hazard). But they have failed to show that government bureaucrats would enhance economic welfare.

References

Anderson, Gary M., and Block, Walter. "Comment on Hum and Simpson." In this issue.

Anderson, Patricia M., and Meyer, Bruce D. "Unemployment Insurance in the United States: Layoff Incentives and Cross Subsidies." In this issue.

Banfield, Edward C. *The Unheavenly City Revisited*. Boston: Little, Brown, 1974.

———. "Present Orientedness and Crime." In *Assessing the Criminal*, edited by Randy Barnett and John Hagel. Cambridge, Mass.: Ballinger, 1977.

Becker, Gary. *Human Capital*. New York: National Bureau of Economic Research, 1964.

Block, Walter. "Labor Relations, Unions, and Collective Bargaining: A Political Economic Analysis." *Journal of Social, Political, and Economic Studies* 16 (Winter 1991): 477–507. (*a*)

———, ed. *Economic Freedom: Toward a Theory of Measurement*. Vancouver: Fraser Institute, 1991. (*b*)

———. "Discrimination: An Interdisciplinary Analysis." *Journal of Business Ethics* 11 (1992): 241–54.

Bruce, Christopher, and Atkins, Frank. "Efficiency Effects of Premium-setting Regimes under Workers' Compensation: Canada and the United States." In this issue.

Buchanan, James M. *Cost and Choice: An Inquiry into Economic Theory*. Chicago: Markham, 1969.

———. *The Limits of Liberty*. Chicago: University of Chicago Press, 1975.

Buchanan, James M., and Thirlby, G. F. *L.S.E. Essays on Cost*. New York: New York University Press, 1981.

Buchanan, James M., and Tullock, Gordon. *The Calculus of Consent: Logical Foundations of Constitutional Democracy*. Ann Arbor: University of Michigan, 1971.

Demsetz, Harold. "Why Regulate Utilities?" *Journal of Law and Economics* 11 (1968): 55.

Epstein, Richard A. *Takings: Private Property and the Power of Eminent Domain*. Cambridge, Mass.: Harvard University Press, 1985.

———. *Forbidden Grounds*. Cambridge, Mass.: Harvard University Press, 1992.

Friedman, David. *The Machinery of Freedom*. Peru, Ill.: Open Court, 1989.

Friedman, Milton. *Fraser Forum*. January 1990, p. 15.

Friedman, Milton, and Friedman, Rose. *Free to Choose*. New York: Harcourt Brace Jovanovich, 1980.

Friedman, Milton, and Schwartz, Anna J. *A Monetary History of the U.S., 1867–1960*. New York: National Bureau of Economic Research, 1963.

Gastil, Raymond D. *Freedom in the World: Political Rights and Civil Liberties*. New York: Freedom House, 1978–89.

Green, David A., and Riddell, W. Craig. "The Economic Effects of Unemployment Insurance in Canada: An Empirical Analysis of UI Disentitlement." In this issue.

Grubel, Herbert G. *On the Insurance Corporation of British Columbia: Public Monopolies and the Public Interest*. Vancouver: Fraser Institute, 1985.

Hayek, Friedrich A. *Prices and Production*. London: Routledge, 1931.

Hoppe, Hans-Herman. "On Praxeology and the Praxeological Foundation of Epistemology and Ethics." Mimeographed. Las Vegas: University of Nevada, Las Vegas, Department of Economics, 1987.

———. *A Theory of Socialism and Capitalism*. Boston: Kluwer, 1989.

Kirzner, Israel M. *Competition and Entrepreneurship*. Chicago: University of Chicago Press, 1973.

McColm, R. Bruce; Finn, James; Payne, Douglass W.; Ryan, Joseph E.; Sussman, Leonard R.; and Zaryky, George. *Freedom in the World: Political Rights and Civil Liberties*. New York: Freedom House, 1989–90.

Murray, Charles. *Losing Ground: American Social Policy from 1950 to 1980*. New York: Basic, 1984.

———. *In Pursuit: Of Happiness and Good Government*. New York: Simon & Schuster, 1988.

Nozick, Robert. *Anarchy State and Utopia.* New York: Basic, 1974.
Ohashi, T. M.; Roth, T. P.; Spindler, M. L.; McMillan, M. L.; and Norrie, K. H. *Privation: Theory and Practice.* Vancouver: Fraser Institute, 1980.
Poole, Robert. *Cutting Back City Hall.* New York: Universe, 1980.
Rothbard, Murray N. *For a New Liberty.* New York: Macmillan, 1973.
————. *America's Great Depression.* Kansas City, Mo.: Sheed & Ward, 1974.
————. "Toward a Reconstruction of Utility and Welfare Economics." Occasional Paper no. 3. San Francisco: Center for Libertarian Studies, 1977.
————. *The Ethics of Liberty.* Atlantic Highlands, N.J.: Humanities Press, 1982.
Stigler, George. "The Economics of Information." *Journal of Political Economy* 69 (June 1961): 213–25.
Thomason, Terry, and Burton, John F., Jr. "Economic Effects of Workers' Compensation in the United States: Private Insurance and the Administration of Compensation Claims." In this issue.
von Mises, Ludwig. *Human Action.* Chicago: Regnery, 1963.
Walker, Michael A., ed. *Freedom, Democracy and Economic Welfare.* Vancouver: Fraser Institute, 1988. (*a*)
————, ed. *Privatization: Tactics and Techniques.* Vancouver: Fraser Institute, 1988. (*b*)

Comment on Hum and Simpson

Gary M. Anderson, *California State University, Northridge*

Walter Block, *College of the Holy Cross*

I. Introduction

The idea of a government-provided income floor—a Guaranteed Annual Income (GAI), sometimes termed a "negative income tax" (NIT)—has generated considerable support from social scientists across the ideological spectrum. Conservative and liberal supporters alike maintain that replacing the current patchwork quilt of welfare programs with a simpler NIT would both minimize the bureaucratic costs of welfare and maximize the freedom of choice for recipients. The series of large-scale NIT field tests, or "experiments," held throughout the United States and Canada in the early to mid-1970s has consequently received considerable attention. In fact, a veritable mountain of social science research into the effects of NIT on the behavior of recipients has resulted.

Hum and Simpson (in this issue) present a clear, concise, very thorough, and painstaking survey of the experiments conducted in the United States and Canada on the direct labor supply effects of the negative income tax or Guaranteed Annual Income programs. They report on the NIT experiments held in New Jersey; Seattle-Denver; Gary, Indiana; several rural areas of the United States; and in the cities of Winnipeg and Dauphin of the Canadian province of Manitoba (termed the Manitoba Basic Annual Income Experiment, or "Mincome"). They describe the political and historical antecedents of these experiments, their design format, and the underlying models and summarize the empirical results. Altogether, they do a fine job of reviewing the basic findings of the (considerable) literature reporting labor supply effects found in the experiments.

However, their effort is more ambitious than this alone. While refraining from overt polemic, they attempt to portray the NIT experiments as bolstering the case for a GAI. The experiments, they aver, demonstrated that the adverse consequences associated with NIT payments were only minor.

[*Journal of Labor Economics*, 1993, vol. 11, no. 1, pt. 2]

Therefore, critics of a nationwide NIT have been shown to have been misguided.

Hum and Simpson state their major conclusion concisely: "Few adverse effects have been found to date. Those adverse effects found, such as work response, are smaller than would have been expected without experimentation" (in this issue, p. S287). In their view, the principal concern of NIT critics—that cutting the link between income and work effort would severely reduce the willingness of recipients to hold regular jobs—has been demonstrated by the NIT experimental results to have been unfounded. While those experiments did indeed show some labor disincentive effects to be associated with the NIT, these responses were less than some had claimed (p. S288).

Our purpose here is to review their argument about the purported dearth of "adverse effects" found to be associated with a NIT. In reality, evidence of a variety of "adverse effects" following from the availability of a GAI abound.

II. Rewarding Unemployment with a Salary

In fact, the NIT experiments have shown a wide variation in labor response results. The labor market responsiveness of husbands appears to have been less severe than that of wives or single female heads of families. Hum and Simpson (in this issue, table 2) report that two studies that surveyed the reduction in annual hours worked in all U.S. NIT experiments taken together found reductions of 5% (Robins 1985) and 7% (Burtless 1986) in the case of husbands. Those same authors found reductions of 21% and 17%, respectively, for wives and 13% and 17% reductions for single female heads of households. In different studies of separate experiments, Hum and Simpson (in this issue) report that labor supply response for husbands varied from a low reduction of 1% in New Jersey (Burtless 1986) to a high of 8% in Seattle-Denver (Keeley 1981); for wives, the responses varied from a 3% reduction in Gary, Indiana, to a 33% reduction in New Jersey (Keeley 1981).[1] Results for single female heads of households varied from a low of −7% in the Canadian Mincome experiment to a high of −30% reported by Burtless (1986) in Gary.

Translating the reductions in annual hours worked into dollar figures puts the question of relative magnitude into a more manageable perspective. The proportional reductions in earnings were "quite close" to the reductions in hours found in these studies (Burtless 1986, p. 28). In the Seattle-Denver experiment, for example, eligible two-parent families receive transfer pay-

[1] In his analysis of the Gary experimental data, Burtless found a labor supply increase of 5% for wives associated with the NIT. But all other labor supply results, for both sexes in married couples and for single female heads of households, were consistently found to be negative. See Hum and Simpson (1991, table 2).

                                                    Anderson/Block

ments $2,700 higher than what members of the control group received. But in the Seattle-Denver experiment, the combined earnings reductions of both spouses was $1,800. Since the average tax rate, or welfare payment reduction following outside earnings, in that study was about 50%, transfer payments to recipients were about $900 above what they would have been with no work effort response. Thus, the experiment spent nearly $2,700 on transfers but succeeded in raising incomes in two-parent families by only $900! Burtless concludes with a trenchant observation: "Even if the earnings reductions are taken to be modest, it is reasonable to ask whether most taxpayers would be willing to spend $3 in order to raise the incomes of poor, two-parent families by only $1" (1986, p. 28).

This reduction in labor effort partly took the form of prolonged periods of unemployment. One result from the Seattle-Denver experiments not reported by Hum and Simpson was that unemployment drastically increased. Robins et al. found that such periods lengthened by 9 weeks (27%) for husbands, 50 weeks (42%) for wives, and 56 weeks (60%) for female heads of households, all by comparison with the control group (see Robins, Tuma, and Yeager 1980, p. 566).

The effects of NIT on the labor market behavior of some nonheads of household ("nonheads") were even more marked. In their survey, Hum and Simpson (in this issue) report the "nonstructural labor supply response estimates" of the effect on annual hours worked for husbands, wives, and single female heads of household. However, male "nonheads" were also included in the recipient pool, and female "nonheads" included two distinct groups: those who married at some time during the course of the experiments and those who remained single. West (1980) reports that NIT induced reduction in work effort among these other categories of "nonheads"; both hours of work per week and proportion of time worked were affected. This reduction was found to be 43% for male "nonheads" and only slightly lower (a reduction of 33%) for males who subsequently became husbands; females who remained "nonheads" experienced a 42% reduction in hours worked. "Nonheads" are typically adolescents or young adults, the age-group most likely to be attending school. West found no evidence of increased school attendance in this group resulting from the NIT payments, so this increased leisure time at the taxpayers' expense was not being invested in the development of the recipients' employment-enhancing human capital stock. He further explains that the possibility exists that the negative response would increase with an experiment of longer duration and thinks that the response to a permanent program might be even larger still (West 1980, pp. 587–88).

In other words, the experiments revealed that low-income individuals respond rationally to incentives—if they receive a subsidy to their leisure choice, at the margin they will choose more leisure. The average taxpayer would probably question why a series of multimillion dollar "experiments," combined with untold thousands of dollars of research time (most also paid for by taxpayers), was necessary before this conclusion could be reached. Perhaps Murray's (1984) suggestion is close to the mark: the NIT experiments were mainly a political consolation prize awarded to the supporters of a GAI after their political defeat in the early 1970s. Otherwise, it is difficult to see just what the point of the whole exercise was supposed to be.

III. Big, Small, or In-between?

So, the NIT experiments "revealed" that such a welfare scheme tended to generate negative labor supply effects with an absolute value of greater than zero. But how big is "big"? Obviously, before we can describe any magnitude as "large" or "small," we must first define the relevant criterion. In other words, "large" or "small" relative to what standard? One possible criterion is the set of precisely defined prior expectations of some specific group of observers.

Hum and Simpson (in this issue) make much of the supposedly "less than expected" magnitude of the measured effects. Unfortunately, the experiments were simply not designed to test any specific hypotheses regarding relative magnitude of labor supply effects, or any other kind of effects for that matter. In reality, there seem to have been no specific predictions at all concerning the magnitude of such effects. Critics and supporters of NIT alike merely generalized and carefully avoided quantifying their expectations.

Certainly, many early critics of NIT schemes expected even more severe negative labor supply effects. However, the observed effects at first appeared to have exceeded even the rosiest scenarios outlined by NIT enthusiasts before and during the experiments.

In a classic illustration of the danger associated with assigning the foxes to guard the chicken coop, the first reported "experimental results" of the U.S. NIT experiments were published by the U.S. Office of Economic Opportunity (OEO), the same organization that devised the experiment scheme and openly lobbied for a nationwide NIT. The OEO claimed, at first, to have found virtually no labor disincentive effects and even a positive labor market response in some cases! That is, the OEO was initially reporting that paying people a wage unrelated to work magically caused them to choose to work more, not less. Coyle and Wildavsky (1986, p. 179) note the OEO's glowing conclusion: "There is no evidence that work effort declined among those receiving income support payments. . . . On the contrary, there is an indication that the work effort of participants receiving payments increased relative to the work effort of those not receiving payments." Political supporters of NIT such as Senator Fred Harris and Ways and Means Committee Chairman Wilbur Mills were quick to pick up on this welcome news (Coyle and Wildavsky 1986). These bogus initial findings "confirmed" the expectations of NIT proponents; contrary to Hum and Simpson's claim, the expectation that the experiments would reveal *no* significant disincentive effects was quite popular at the time. In short, the labor supply effects ultimately found were indeed very large compared to the stated expectations of prominent NIT supporters.

Hum and Simpson (in this issue) vaguely refer to a supposed "emerging consensus" among economists to the effect that the NIT experiment results turned out smaller than expected—again, without defining *whose* expectations they refer to or what those expectations were. However, they are somewhat uncomfortable with this claim of "smaller than expected" and repeatedly reject the possibility of generalizing about the size of the labor disincentive effect. They shift between dismissing the measured labor supply effects as negligible and rejecting the very possibility of evaluating the relative magnitude of such effects. At several points they dismiss the issue of the severity of the disincentive effects as empty semantics. They find it "difficult to conclude whether the experimental effects . . . are large or not" (p. S281); elsewhere, they suggest that the question of whether or not these costs are high or low "depends upon the assessor and . . . political assessment of public support for GAI" (p. S288).[2]

Burtless (1986) does a better job of sensibly summing up the controversy over the relative magnitude of the NIT experiment labor supply results by noting that the observed work reductions were "probably smaller than most opponents . . . had feared, but larger than advocates had hoped" (p. 45). These results may, indeed, have proven smaller than Hum and Simpson expected, but it is nonsense to generalize, as they do, that the results were smaller than would have been expected without experimentation.

This leads to an important point. Members of the welfare (and welfare research) establishment tend to lose sight of a simple, but logically uncontestable, proposition—most of the poor are in poverty because they do not work. As of March 1989, in U.S. families below the poverty level, a majority of householders—3,509,000 of 6,845,000—did not work a single week.[3] Only a small portion of this total was physically or mentally disabled to the extent that labor market participation was impossible. The NIT experiments demonstrate that a national GAI would only exacerbate an

[2] Interestingly, the press release issued by the Economic Council of Canada describing Hum and Simpson's findings in their 1991 study on the NIT experimental results is headlined "Findings Allay Concerns about Income Program." In that monograph, there seems no trace of Hum and Simpson's "agnosticism" about the question of the severity of labor market disincentive effects; instead, they consistently describe those effects as "small" or "very small" and fears of the NIT leading to reduced work incentives "largely misplaced." See Hum and Simpson (1991, pp. xvi, 91).

[3] U.S. Census Bureau (1990, p. 461, table 749).

already serious problem: able-bodied adults who rationally prefer leisure to labor and receive a government subsidy for this preference.

IV. Marital Dissolution and the NIT

Hum and Simpson (in this issue) entitle their article "Economic Response to a Guaranteed Annual Income: Experience from Canada and the United States" but commendably acknowledge a broad definition of "economic" by including a brief discussion of various behavioral responses sometimes described as "sociological," despite having economically relevant consequences. Unfortunately, this discussion is highly misleading with respect to one important nonlabor supply problem—the effect of a GAI on marital status among recipients.

While Hum and Simpson do indeed mention the problem of the impact of a GAI on marital disruption, their discussion emphasizes the purported ambiguity of research results. They in effect dismiss the problem, suggesting that the original findings of the negative impact of NIT on marital stability have been discredited.

While there is indeed disagreement in the research community, that disagreement surrounds apparently strong evidence that the NIT experiments had a significant impact on increasing marital breakups. The strongest evidence comes from studies of the Seattle-Denver NIT experiments. Groeneveld, Tuma, and Hannan (1980) found that, after comparing all experimental subjects (receiving NIT payments) to the controls, NIT programs increased the marital dissolution rate—the rate at which existing marriages were terminated, either by divorce or permanent separation— by 56% for blacks and 72% for whites (p. 665). A later study conducted by Hannan and Tuma (1990) found that the percentage of the sample experiencing a marital dissolution was higher among recipients of income maintenance—36% higher for blacks and 40% higher for whites (p. 1,274). Only negligible effects on marital dissolution were found among Chicano recipients, but for this group the NIT seemed to prevent marriages from forming in the first place. Income maintenance under the experiments "substantially and significantly" lowered the rate of marital formation in the Chicano sample, while having no significant effect on marital formation among blacks and white unmarried women (p. 1,274).

As Hum and Simpson (in this issue) mention, Cain (1986) and Cain and Wissoker (1990a) have challenged these findings. What they overlook, however, is that Cain and Wissoker (1990a) "find only mild or insignificant effects on marital stability" after excluding data from a subsample known to have a strong marriage dissolution response to NIT—childless couples— and also after averaging the responses of black, white, and Chicano samples contrary to standard empirical convention in behavioral response studies. Hum and Simpson also fail to note that Groeneveld, Hannan, and Tuma

(1980) and later Hannan and Tuma (1990) offer strong rebuttals to the Cain and Cain-Wissoker critiques.

Assume, for the sake of argument, that Cain and Wissoker (1990*a*, 1990*b*) offer valid objections (and given the complexity of many of the statistical issues, any simple resolution to the dispute seems almost inconceivable). The issue of the effect of NIT on marriage dissolution still deserves serious attention. Moreover, the Cain and Wissoker (1990*a*, 1990*b*) criticisms may be wrong.

The studies reporting substantial and negative effects are currently the subject of some dispute but have not been discredited. In science, a hypothesis or research finding that is undisputed is also likely to be regarded as trivial; only interesting research generates attempted refutation. Just because the studies reporting significant negative effects of NIT on family stability have critics does not justify ignoring those studies, granting implicit veto power to any kibitzer. Even David Ellwood, a respected liberal researcher into poverty, maintains that the high percentage point estimates of the effect of NIT in increasing family splits are "troubling" (1988, p. 97).

Given the recent controversy surrounding the so-called feminization of poverty, and the likelihood that divorced women and children are at greater risk of falling below the poverty line, any effect of a NIT on marriage dissolution rates constitutes an important policy-relevant economic problem associated with such income assistance plans. In 1987, 45% of U.S. families headed by single mothers had incomes that fell below the poverty line.[4] If NIT schemes tend to increase divorce, this would have the very tangible effect of also increasing the population living in poverty. The findings about marital dissolution from the Seattle-Denver NIT studies, while hotly contested, are by themselves sufficient to contradict Hum and Simpson's (in this issue) optimistic pronouncements concerning the supposed absence of serious ill effects.[5]

[4] Although this was precisely the same percentage as in 1970, over the intervening period the actual number of such families more than doubled (increasing from 1,509,000 to 3,129,000) (U.S. Census Bureau 1990, p. 462, table 750).

[5] Some liberal enthusiasts for a national NIT even responded to the studies that reported higher marital dissolution rates by insisting that this could be a desirable outcome. For instance, according to Rainwater (1986): "Suppose all the excess disruptions involved battered wives, or wives of alcoholic men. Would the policy implication still be that the NIT has an unconstructive effect? Do we believe that the extra money caused people to be self-destructive in the sense that they ended unions in which they would have been better off remaining? What happened to consumer sovereignty? If we want to give couples the best chance to stay together and if more income will make things better, we have only to choose the more generous plan" (p. 200). Rainwater does not address the poverty-increasing effects likely to result from such "excess disruptions."

V. Long-Term Consequences of an NIT

Hum and Simpson (in this issue) claim that a GAI is likely to do no harm to society if it is adopted on a large scale. But this assertion extrapolates on the basis of a set of extremely limited experiments, studies focusing almost exclusively on immediate work disincentive effects within that experimental group. As we have seen, the work disincentive effects they so readily dismiss as negligible are in reality fairly sizeable. However, even if zero labor supply effects had been found, there are other critical factors to consider in evaluating the probable impact of a nationwide NIT program.

The effects that are probably least tangible might actually be the most significant. These concern the impact of a GAI on the "moral constitution" of the community (see Buchanan 1986). This refers to the array of internalized behavioral rules and constraints that help determine the response of individuals to changes in their environment. Murray points out that a strong part of the motivations of people revolves around the social mores prevalent in a peer group (see Murray 1990). If, for example, the customs of the relevant community are such that unwed teenage pregnancy is severely frowned upon, then this will tend to strongly inhibit such behavior. This applies even when such behavior is financially encouraged by the welfare state. In contrast, if such psychic barriers are not in operation, the effect of the subsidy in encouraging illegitimate births will be strengthened.[6]

This example is relevant to the problem, inherent in any NIT, of separating income from the responsibility to work. If it is a matter of shame or discomfort to be on the dole, then the elasticity of hours of work with respect to welfare "benefits" will be low. The presumption in this case is that the sign of this elasticity will be negative, but it is even conceivable that a positive result could be generated, if the embarrassment is intense. Sociologists refer to this phenomenon as "welfare stigma." If, however, accepting GAI payments is seen as a "right," then this elasticity might be negative, and large.

To the extent that a program is confined to a small group of people, while within the larger society the old traditional values (e.g., it is considered shameful to receive guaranteed income from the state) still prevail, it can be expected to have a heavy impact on the small experimental group. That is, if society as a whole frowns heavily upon accepting money coercively taken from the taxpayer, this will tend to encourage the subjects in the experiment to reduce their hours of work, in response to payoffs, to a lesser extent than it would otherwise. Their level and intensity of ambition for self-improvement through work will not be vitiated as heavily as in

---

[6] Walter Williams (1991, p. 10) points out in this regard: "The most important guideline for holding societies together is not the laws written by legislators, it's in the unwritten codes of behavior that evolve over time. Many of these codes are rooted in moral absolutes." See also Murray (1984).

the case where NIT applies to the whole society and work initiative is decreased as a result.

The NIT schemes appeal to many intellectuals who think of poverty as a kind of disease, which swoops down on the unsuspecting through no fault of their own. Poverty is viewed as a natural disaster and not as an individual *choice.*

If poverty is like a natural disaster, then it makes sense to "solve" the "poverty problem" by simply transferring money to the poor, with no strings attached. With the move of the legislative pen, the formerly poor are lifted out of poverty, and the "problem" is solved. Of course, to the extent that poverty reflects personal choice—for example, to choose leisure over labor market income—and hence individual responsibility, subsidizing such behavior might actually make the "poverty problem" worse. The proponents of NIT tend to prefer the "natural disaster" model of poverty.

For example, liberal economist Paul Samuelson testified before Congress in support of NIT that "the curse of the poor is literally their poverty. Give them more money" (quoted in Coyle and Wildavsky 1986, p. 169). Severing the market link between labor effort and income, between providing goods and services to consumers and earning a livelihood, has been an explicit goal of many NIT supporters. This end was taken to its logical extreme by psychologist Eric Fromm, who also testified before Congress on behalf of an NIT and proclaimed that "a psychology of scarcity produces anxiety, envy, egotism, [but] a psychology of abundance"—which a GAI would supposedly help achieve—"produces initiative, faith in life, solidarity" (Coyle and Wildavsky 1986, p. 169). Thus, some proponents of a GAI actually hope for dramatic changes in the moral constitution of society, away from an ethic of achievement and success based on effort, to an "ethic" of income as a basic right. The success of such a moral constitutional sea change might be profound in terms of economic efficiency loss.

The "success" of the existing plethora of expensive welfare programs in reducing the "poverty problem" is less than obvious. Murray (1984) generated a firestorm of controversy by arguing that 20 years of ever-increasing welfare spending had in reality led to an increase in the percentage of the population living in poverty. The welfare establishment responded vigorously and successfully criticized several technical points in Murray's original thesis.[7] But Murray's critics could not explain away the most crucial fact: the massive government spending on welfare programs since the late 1960s seems to have had little impact on the poverty rate. The largest cash aid program for assistance to the poor is Aid to Families with Dependent Children (AFDC). In 1970 only 11% of U.S. families with children lived in poverty; by 1987 this proportion had risen to 15%. Across all American

---

[7] Murray himself (1990) provides one of the best overviews of these various critiques, followed by his own response in each case.

families, the percentage in poverty fell, but only slightly (from 15% to 13%).[8]

This is relevant because, whatever the negative effects of existing welfare programs on the labor effort of the poor, an NIT would be worse. It is rarely noted that the labor supply disincentive effects, and all other adverse consequences, found to be associated with NIT in the limited duration, local experiments were in addition to the disincentive effects resulting from the existing pattern of welfare payments. The NIT effects were all found after a comparison to a control group receiving no NIT. The control group was not "pure" but was receiving all the normal welfare benefits of the 1970s, which were extensive and growing during the period of the experiment. In short, the various NIT effects, regardless of their relative magnitude, were "extra" costs that resulted from replacing normal welfare programs with a GAI (Murray 1984, p. 153).

To reiterate, the NIT experiments were of quite limited duration. The participants in most of the U.S. experiments knew at the outset that the experiment would only last 3 years; only in the Seattle-Denver experiments were a portion (one-quarter) of the participants enrolled in a 5-year experiment. Only a tiny number of participants in the Seattle-Denver experiments were told to expect continued availability of the NIT for 20 years.[9]

The problem is that these experimental durations were too short to allow for full adjustment to the availability of a GAI on all participants. Recipients knew exactly how long they would remain eligible for the NIT and adjusted their behavior accordingly. Those receiving the temporary NIT payments understood that the existing GAI was only short-lived. Also, over the course of the experiments recipients were found to alter their behavior from pre-NIT patterns only gradually, not instantaneously. For example, Robins and West (1980) found that the period required for 90% adjustment of NIT recipients in terms of their labor market behavior was 2.4 years for husbands, 3.6 years for wives, and 4.5 years for female heads of household. Full responses took longer in the case of Chicanos and blacks and took longer for 5-year than for 3-year programs (p. 524). Burtless (1986) argues that on balance the NIT experiments probably underestimated the permanent response given a guarantee equal to the poverty line and a "tax rate" (i.e., a reduction of GAI payments based on outside earnings of the recipient) of 50% or more (p. 31). Since any politically feasible NIT would almost certainly have such a high "tax rate"—otherwise,

[8] Over the same period, the number of families (with children) falling below the poverty line increased from 3,330,000 to 5,229,000 (U.S. Census Bureau 1990, p. 462, table 75).

[9] See Groeneveld, Tuma, and Hannan (1980). The 20-year experiment included only about 170 Denver households in its final sample, considered too few observations for reliable statistical study. See Robins and West (1980, p. 524).

the plan would have to have a low benefit level—this situation is the relevant one for evaluating a permanent NIT. In other words, the limited-duration experiments probably underestimated the probable long-term labor supply disincentive effects associated with a nationwide NIT, even ignoring the possible adverse impact on the work ethic among the poor.

VI. Why NIT Supporters Seem Immune to Evidence

So, contrary to the positive assessment of the NIT experiments by Hum and Simpson (in this issue), the actual experimental results all turned out negatively. The best that can be said is that things could have been worse— a recommendation the captain of the Titanic might have given regarding collisions with icebergs. On the basis of available evidence, NIT seems to offer little attraction to an unbiased observer.

Yet despite these unhappy results, the idea of a NIT continues to exercise broad appeal among social scientists, although this support bears little resemblance to the "consensus" that appears to be "emerging" in the minds of Hum and Simpson and others of that persuasion. This continuing broad appeal, in the face of an avalanche of negative results, deserves some comment.

The NIT appeals to some economists, regardless of ideological persuasion, because in principle it maximizes the opportunity set available to recipients and eliminates unnecessary bureaucracy.[10] This reasoning led Milton Friedman (1962) to propose a version of the NIT, for example. But NIT appeals to liberal social scientists because it elevates welfare payments to the level of a basic right, whereas it is now merely a kind of taxpayer-supported charity. An NIT would sever any link between the minimal income necessary for physically comfortable survival and moral responsibility. Poverty is viewed, not as a personal choice in terms of allocating time and resources between leisure and labor, but instead as an exogenous shock, something bad that just happens to people through no fault of their own.

The alternative proposition, that poverty often reflects individual choice, does not imply that those people "choose" to live in distress or misery. Rational individuals sometimes choose to remain poor because they perceive themselves to be better off as a result. Labor is onerous and produces disutility. Responsible behavior is sometimes not very enjoyable. It is only by lucky happenstance that individuals sometimes find pleasure in pursuing

---

[10] Motivations may or may not be impenetrable, but, even if they are not, it may still be argued that only psychologists, not economists, have any comparative advantage in their discernment. The alternative view is that, at least as regards the holding of alternative perspectives on issues such as the costs and benefits of programs like NIT, economists do indeed have an inside track. The following speculations may perhaps best serve as hypotheses attempting to explain an important aspect of social science reality.

activities that also produce marketable output and, hence, income. Some individuals consequently choose to sacrifice comparatively little of their scarce time and effort to the labor market because they perceive themselves better off by allocating their resources to other activities.

The perspective on the poor that is common among NIT supporters—that paying people not to work will have no impact on the demand by recipients for leisure over labor—may be wrongheaded, and patronizing, but at least it has the saving grace of an (arguably) altruistic basis: a desire to help the poor to help themselves. Give them money, and they will no longer be poor. Self-reliance and self-sufficiency are just a few welfare checks away. One underlying rationalization for an NIT scheme is that a GAI would improve the ability of the poor to work themselves out of poverty. However, there is another, less altruistic motive that is possibly involved in the commitment of many social scientists to justifying the NIT experiments. Those experiments constituted a kind of welfare for the busy bees in the "welfare research industry."

The NIT experiments (including the Canadian Mincome) collectively cost millions to administer, plus millions more in salaries and research grants of NIT researchers. The NIT research has occupied the attentions of a cottage industry on American and Canadian college campuses for almost 20 years. Hum and Simpson (in this issue) are quite defensive about the value of this research to society, for instance, noting (in a classic non sequitur) that the "money involved was small in relation to annual expenditures on social programs in Canada and the United States" and that therefore "it was likely a solid investment" (p. S288). Presumably, then, if the government spent millions digging a huge hole in the ground, and then just filled it up again, that expenditure would "likely be a solid investment," too, because the cost was small in relation to the total amount of government spending!

Many commentators claim that social experimentation such as that involving the NIT somehow plays a vital role in policy formation. But this claim is highly disingenuous. The Nixon administration's proposal for a national NIT died in Congress long before the experiments were completed. The NIT experiments were sponsored by the Office of Equal Opportunity for the blatant purpose of lobbying Congress to sponsor a national NIT plan. Many economists, politicians, and bureaucrats were "sold" on the idea of a NIT long before the results of any of the experiments were available.

In another recent paper on the same theme, Hum and Simpson express even greater enthusiasm for bigger and better future tax-funded social experiments—for instance, employment and training (1991, p. 92). They also insist that the purportedly small estimated labor disincentive effects found in the case of the NIT experiments shows that other income supplementation programs (including current welfare programs and unemployment insurance) cost less than usually supposed. Even income taxation looks better in the light of the NIT experiments (1991, p. 91). In short, Hum and Simpson dream of expanding the NIT experiment cottage industry into a statist Disneyland for professional welfare researchers. Never mind that the NIT labor disincentive effects are "trivial" only in their minds.

The problem with the bulk of the NIT research has not been a lack of sophistication in the technique but rather in the absence of precisely formulated predictions. Lacking any benchmark in the form of testable hypothesis, virtually any experimental result could be marketed by NIT proponents as a "success." There were never any precisely stated hypotheses about predicted disincentive effects, and consequently the "experiments" were from the first lost at sea. Therefore, even fairly major disincentive effects could be dismissed by NIT proponents as less than someone predicted. This is exactly what happened in Congress in the early 1970s, when political supporters of NIT seized on preliminary experiment results to promote their cause. And this is also exactly what Hum and Simpson do in their present article (in this issue).[11]

VII. If Not NIT, Then What?

In the face of the absurdities and inefficiencies of the existing bureaucratic maze of welfare programs, which already reward irresponsibility and penalize self-reliance, it makes no sense to "reform" that system by pouring more money into it. Make no mistake, a nationwide NIT would, even if it completely replaced all existing income transfer programs, transfer substantially more tax money to welfare recipients, as most NIT supporters freely acknowledge. Moreover, as Murray (1986, p. 204) points out, such a wholesale replacement is almost inconceivable—it would require the immense political will, for example, to refuse to offer additional assistance to the poor who mismanage their household income. Thus, a nationwide NIT would bear little resemblance to Milton Friedman's (1962) original

---

[11] They attribute to unnamed politicians and the "general public" the belief that "an NIT would promote idleness among the able-bodied" (Hum and Simpson, in this issue, p. S267), and quote Stanley H. Masters's derisive characterization of the general public's "stereotype" of the poor as "lazy bums" and "profligate boozers" (p. S285). The implication is that the NIT experiments disproved these "stereotypes" because not all recipients of a GAI quit their jobs and became full-time drunks. Unfortunately, significant marginal labor disincentive effects were quite consistent with most able-bodied poor people keeping their jobs. In one sense, the findings were quite remarkable: even when paid a salary by the government for loafing, most poor people continued to work for a living and maintain themselves and their families in a self-responsible manner. This was surely a tribute to the work of the ethic of the poor. Unfortunately (for the Hum-Simpson thesis), the experiments also demonstrated that replacing the current welfare arrangements with an NIT would significantly reduce work effort among the poor.

proposal to replace the existing welfare system. It would more likely be an expensive add-on, grafted to the already bloated set of programs.

Consider the following. In the United States, a family of four below the poverty level can expect to pay an average of 15.3% of their annual earned income to federal, state, and local government in the form of taxes.[12] For an average annual income of $11,700, the average tax bill would come to $1,755. In 1987, a total of $18.44 billion was spent on the largest American welfare program, Aid to Families with Dependent Children, by all levels of government in the United States to assist 11,064,000 recipients.[13] This implies an average annual AFDC payment of $1,667. In other words, AFDC transfers to the poor are more than wiped out by the tax bill imposed on the working poor.

Or consider the net impact of government regulations and restrictions on the welfare of the poor. The federal milk price support program may double the price of milk to poor families. Minimum wage laws vastly reduce the employment opportunities for inner-city youth, leaving many little option other than committing crime to earn a living. Rent control ordinances cause the stock of housing available to poor people to crumble and decay. The public school systems provided to the urban poor are educational disasters. Licensing requirements prevent poor workers from entering occupations that might allow them to lift themselves out of poverty. Government zoning regulations prevent employment-generating land use. Rather than a "war on poverty," government at all levels conducts a massive "war on the poor." Overregulation destroys jobs and produces poverty.[14]

Government welfare programs do not achieve their intended purpose—eliminating poverty. A nationwide NIT would require greater government spending, and would impose additional costs, while doing no better a job of ending poverty. In fact, the research consensus to date clearly indicates that a nationwide NIT would make the problem worse, by reducing the willingness of recipients to support themselves in the labor market.

Perhaps the ultimate absurdity is that proponents of a GAI choose to ignore the mind-boggling array of government activities that directly burden the poor. The same governments that issue AFDC checks impose highly regressive sales taxes that cut poor people's purchasing power; the same government that provides food stamps to poor people lowers the real value of those certificates by paying farmers to reduce their output of nutritious dairy products; the list goes on and on. The reality would be funny if it were not so tragic; government vigorously restrains the ability

[12] Data from Citizens for Tax Justice, cited in the *Los Angeles Times* (July 25, 1991), sec. D, p. 1.

[13] See U.S. Census Bureau (1990, p. 353, table 580).

[14] See Williams (1982) for a brief summary of the effects of government regulations on the lives of the poor.

of poor people to earn an honest living and then "responds" to the "poverty problem" with bloated and ineffective welfare programs—which require higher employment-reducing taxes that further increase poverty!

Creating a new, expensive welfare program would simply mean throwing good money after bad. The only cost-effective government welfare "reform" would be one that truly helped the poor to help themselves, by reducing the net burden of government in their daily lives. A guarantee of economic freedom (Walker 1988; Block 1991) would alleviate poverty more effectively than a thousand Guaranteed Annual Income programs.

References

Block, Walter. *Economic Freedom: Toward a Theory of Measurement.* Vancouver: Fraser Institute, 1991.

Buchanan, James M. *Liberty, Market and the State: Political Economy in the 1980s.* New York: New York University Press, 1986.

Burtless, Gary. "The Work Response to a Guaranteed Income: A Survey of Experimental Evidence." In *Lessons from the Income Maintenance Experiments,* edited by Alicia H. Munnell, pp. 22–52. Boston: Federal Reserve Bank of Boston, 1986.

Cain, Glen G. "The Issues of Marital Instability and Family Composition and the Income Maintenance Experiments." In *Lessons from the Income Maintenance Experiments,* edited by Alicia H. Munnell, pp. 60–93. Boston: Federal Reserve Bank of Boston, 1986.

Cain, Glen G., and Wissoker, Douglas A. "A Reanalysis of Marital Stability in the Seattle-Denver Income Maintenance Experiment." *American Journal of Sociology* 95 (March 1990): 1235–69. (*a*)

———. "A Response to Hannan and Tuma." *American Journal of Sociology* 95 (March 1990): 1299–1314. (*b*)

Coyle, Dennis J., and Wildavsky, Aaron. "Social Experimentation in the Face of Formidable Fables." In *Lessons from the Income Maintenance Experiments,* edited by Alicia H. Munnell, pp. 167–84. Boston: Federal Reserve Bank of Boston, 1986.

Ellwood, David T. *Poor Support: Poverty in the American Family.* New York: Basic, 1988.

Friedman, Milton. *Capitalism and Freedom.* Chicago: University of Chicago Press, 1962.

Groeneveld, Lyle P.; Tuma, Nancy Brandon; and Hannan, Michael T. "The Effects of Negative Income Tax Programs on Marital Dissolution." *Journal of Human Resources* 15 (Fall 1980): 654–74.

Hannan, Michael T., and Tuma, Nancy Brandon. "A Reassessment of the Effect of Income Maintenance on Marital Dissolution in the Seattle-Denver Experiment." *American Journal of Sociology* 95 (March 1990): 1270–98.

Hum, Derek, and Simpson, Wayne. *Income Maintenance, Work Effort, and the Canadian Income Experiment: A Study Prepared for the Economic Council of Canada.* Ottawa: Minister of Supply and Services Canada, 1991.

———. "Economic Response to a Guaranteed Annual Income: Experience from Canada and the United States." In this issue.

Keeley, Michael. *Labor Supply and Public Policy.* New York: Academic Press, 1981.

Murray, Charles. *Losing Ground: American Social Policy from 1950 to 1980.* New York: Basic, 1984.

———. "Discussion of the Policy Lessons." In *Lessons from the Income Maintenance Experiments,* edited by Alicia H. Munnell, pp. 202–5. Boston: Federal Reserve Bank of Boston, 1986.

———. *In Pursuit: Of Happiness and Good Government.* New York: Simon & Schuster, 1990.

———. "Welfare and the Family: The American Experience." In this issue.

Rainwater, Lee. "A Sociologist's View of the Income Maintenance Experiments." In *Lessons from the Income Maintenance Experiments,* edited by Alicia H. Munnell, pp. 194–201. Boston: Federal Reserve Bank of Boston, 1986.

Robins, Philip K. "A Comparison of the Labor Supply Findings from the Four Negative Income Tax Experiments." *Journal of Human Resources* 20 (Fall 1985): 567–82.

Robins, Philip K.; Tuma, Nancy Brandon; and Yeager, K. E. "Effects of SIME/DIME on Changes in Employment Status." *Journal of Human Resources* 15 (Fall 1980): 548–68.

Robins, Philip K., and West, Richard W. "Labor-Supply Response over Time." *Journal of Human Resources* 15 (Fall 1980): 524–44.

U.S. Census Bureau. *Statistical Abstract of the United States: 1990.* Washington, D.C.: U.S. Government Printing Office, 1990.

Walker, Michael A., ed. *Freedom Democracy and Economic Welfare.* Vancouver: Fraser Institute, 1988.

West, Richard W. "The Effects on the Labor Supply of Nonheads." *Journal of Human Resources* 15 (Fall 1980): 574–90.

Williams, Walter E. *The State against Blacks.* New York: McGraw-Hill, 1982.

———. "Moral Codes That Served a Purpose." *Issues and Views* 7 (Winter 1991): 10.

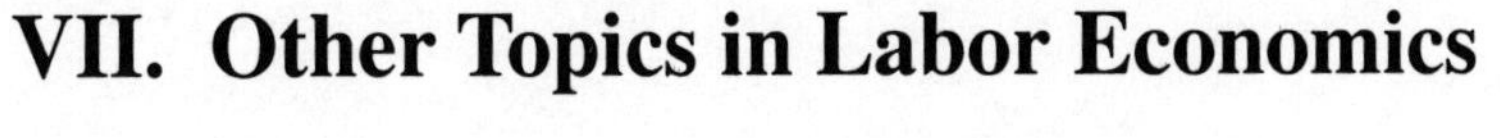

VII. Other Topics in Labor Economics

The Division of Labor Under Homogeneity

A Critique of Mises and Rothbard

By WALTER BLOCK, PER HENRIK HANSEN, and PETER G. KLEIN*

ABSTRACT. Even the most passionate defenders of free trade, such as Mises and Rothbard, claim that trade cannot occur under conditions of strict homogeneity of land, labor, and capital. We show that specialization, trade, and the division of labor can emerge even when resources are initially homogenous, due to "natural heterogeneity," economies of scale, and learning.

Perhaps no proposition in economic theory is better established than the idea of gains from trade. The insight that voluntary exchange is mutually beneficial was articulated as early as the 13th century by Richard of Middleton (de Roover 1963: 338), reaching its full expression in Ricardo's ([1817] 1951) law of association, or the principle of comparative advantage. These insights are not always well understood, of course. The mercantilist view of trade suffers from what Mises (1966: 664) calls the "Montaigne fallacy," the belief that trade benefits one party at the expense of the other.[1]

*Walter Block is Harold E. Wirth Eminent Scholar in Economics at Loyola University, New Orleans, 6363 St. Charles Avenue, New Orleans, LA 70118; e-mail: wblock@ loyno.edu. He is author of a dozen books and over 200 scholarly articles on economics, law, and political philosophy. Per Henrik Hansen is a Ph.D. Fellow in the Department of Economics, Copenhagen Business School, 2000 Frederiksberg, Denmark; e-mail: phh.eco@cbs.dk. He is a former Mises Institute Fellow and a specialist in monetary economics. Peter G. Klein is Assistant Professor in the Division of Applied Social Sciences and Associate Director of the Contracting and Organizations Research Institute, 135 Mumford Hall, University of Missouri, Columbia, MO 65211; e-mail: pklein@ missouri.edu. His research focuses on entrepreneurship, industrial economics, and the economics of organizations and institutions. The authors are grateful to Laurence Moss (the editor) and two referees for insightful comments on a previous draft. The usual disclaimer applies.

American Journal of Economics and Sociology, Vol. 66, No. 2 (April, 2007).

This article attempts to clarify an assumption underlying the principle of gains from trade. Most expositors begin by stating that trade is mutually beneficial only if factors of production vary in their attributes. Resource heterogeneity is thus a prerequisite for the gains from trade. Mises (1966: 157), for example, states that "if the earth's surface were such that the physical conditions of production were the same at every point and if one man were . . . equal to all other men . . . [the] division of labor would not offer any advantages for acting man." Rothbard (1962: 82) adds: "It is a fact that, superimposed on the basic unity of species and objects in nature, there is a great diversity. Particularly is there variety in the . . . factors that would give rise to specialization: in the locations and types of natural resources and in the ability, skills, and tastes of human beings. . . . It is clear that conditions for exchange, and therefore increased productivity for the participants, will occur *where each party has a superiority in productivity in regard to one of the goods exchanged*—a superiority that may be due either to better nature given factors or to the ability of the producer." Further, Rothbard (1982: 35) maintains: "If all people were equally skilled and equally interested in all matters, *and* if all areas of land were homogeneous with all others, there would be no room for exchanges."

We argue here that specialization, trade, and the division of labor can take place even if all human and nonhuman resources were exactly alike. We do not disagree with Mises and Rothbard that resource heterogeneity is sufficient to create gains from trade.[2] However, we deny that resource heterogeneity is a necessary condition in order for gains from trade to be possible. In other words, a division of labor can exist even under resource homogeneity.

Suppose that human beings were all completely interchangeable; they had the same skills, capacities, initiative, intelligence, tastes, and so on as each other. Posit, further, that all land, too, was completely homogeneous: equally fertile, equally watered, equally ground covered, with equal access to whatever amenity anyone cared to name. In addition, we assume that all people and resources are located randomly throughout the world; that is, that there is no grouping of either persons or raw materials based upon any unique or even different characteristics; there cannot be, in any case, because of our other homogeneity assumptions. Under these admittedly unrealistic conditions, would trade be possible, or would all the people depicted in this scenario be forced to rely solely upon self-sufficient production?

Rothbard states in this regard:

> In describing the conditions that must obtain for interpersonal exchange to take place (such as reverse valuations), we implicitly assumed that it must be *two different goods* that are being exchanged. If Crusoe at his end of the island produced only berries, and Jackson at his end produced only the same kind of berries, then no basis for exchange between them would occur. If Jackson produced two hundred berries and Crusoe one hundred fifty, it would be nonsensical to assume that any exchange of berries would be made between them. (1992: 80)

We agree with Rothbard that at least two goods are needed, but maintain that resource heterogeneity does not imply a one-good model. Suppose, for example, that there are two people, Jackson and Crusoe, each with equal abilities, and that there are two goods, fish and berries, such that each of these people can, at the outset, produce an identical number of each product. Let us assume for illustration purposes that both can produce 200 berries, and 20 fish, in a two-day period. We posit a very simple world with no capital goods. Berries are picked by hand, and fish are caught by grabbing them, as is now done by bears. We concede that under these conditions no trade will occur, and certainly productivity could not thereby be enhanced.[3]

However, under reasonably weak assumptions, conditions for mutually beneficial exchange can emerge. First, individuals can never be 100 percent identical, simply because they cannot occupy the same physical space. Even if they are completely identical in every other regard, it is possible (though not necessary) that they have different access to complementary resources. Imagine that the berry bushes are adjacent to the beach. If Crusoe is standing a tiny bit closer to the berry bushes, and Jackson a tiny bit closer to the water, then there are gains from trade by Crusoe specializing in berries and Jackson specializing in fish. That is, the two laborers are identical in each aspect besides location, but they have different costs of production. The one closet to berries has a lower cost of berry production than the one closest to fish, and so on.

More generally, location is a unique attribute, an ontological fact about the real world. One of the characteristics of any resource is its location. That means that, strictly speaking, there is and can be no such thing as completely homogeneous resources. George's (2003) great insight was that no two plots of land are really identical because land always exists at a particular location. Location is nonreplicable, and this gives rise to certain specific patterns of income distribution.[4]

Second, from a subjectivist, Austrian perspective, the value of a resource as used in production is driven not by its objective, physical characteristics, but by its place in the structure of production, as conceived by entrepreneurs.[5] Even resources that are physically identical (subject to the caveat about location just discussed) are not "identical" in the sense relevant to economic analysis. We can imagine that Crusoe's and Jackson's island is also inhabited by entrepreneurs who organize fish and berry production, hiring Crusoe and Jackson as employees. The "true" value of Crusoe's and Jackson's labor in either production process is unknown. One entrepreneur estimates Crusoe's marginal productivity as sufficient to make berry production profitable but not fish production, so she hires Crusoe, placing him into the economy's production process for berries. Another entrepreneur then hires Jackson to produce fish. The labor services of Crusoe and Jackson are now heterogeneous resources, meaning that they are assigned different positions in the economy's capital structure, even though Cruse and Jackson are identical human beings.

Even if we ignored these points about what we might call "natural" or "ontological" heterogeneity, there would be additional sources of potential gains. In both cases, individuals start out identical, but become differentiated by their actions.

Let us develop this point further. Assume now that Crusoe and Jackson are equidistant from berries and fish. Is it still possible for each to "gravitate" to a different resource and, having done so, come to specialize in that one and thereby raise productivity? Such a state of affairs might be considered an accident,[6] but there is no reason to rule it out. Not every Crusoe and Jackson pair might arrange matters in such a way, but it is difficult to see why none of these hypothetical pairs could do so. If one or both production processes are character-ized by increasing returns to scale, then such specialization—even if random—results in net gains. It does not seem unrealistic to expect that over time Jackson and Crusoe—being rational human beings—will realize from their experience in autarchy that the more they produce the better they get at it. They will eventually also realize that the larger the scale of production the better are the opportunities for the use of tools and eventually mechanization of production. When one of them, let us say Crusoe, has so realized, we can imagine that he approaches Jackson and explains this to him. They then agree each to specialize in production of a different good in order to be able to share a larger amount of both at the end of the period.

How would they agree on who would specialize in what? One possibility is that they have different preferences for the attributes of the work environment they consume while working. Producing fish and "fishing" are not separable; one may enjoy fishing while the other enjoys tending berry plants. However, they could also draw lots or simply agree to change production lines every other period.[7] The point is that having realized the prospect of a potential welfare gain, Jackson and Crusoe will be very resourceful. Mises seems to agree with this, as he writes:

> If and as far as labor under the division of labor is more productive than isolated labor, and if and as far as man is able to realize this fact, human action itself tends toward cooperation and association; man becomes a social being not in sacrificing his own concerns for the sake of a mythical Moloch, society, but in aiming at an improvement in his own welfare. (1966: 160)[8]

There is another and perhaps more likely way that specialization could come about. Crusoe—having realized the potential gains of specialization—could start, unilaterally, to change the proportions of his own production of the two goods. Assuming that there are increasing returns to scale, this action by Crusoe would imply that the relationship between the two goods in terms of opportunity costs will be different when seen from the perspective of Crusoe relative to that of Jackson. The different marginal transformation rates between the two goods for Crusoe and Jackson would then provide the basis for beneficial marginal trade between them. This initial change by Crusoe will be reinforced by their new experience.

Even in the absence of scale economies, if there are "learning effects" (i.e., average cost is a decreasing function of accumulated output), then one or both workers can lower his average costs over time.[9] Assume that if the two of them specialized, it matters not one whit in which of these two goods, that the productivity of each would be improved. For example, if Jackson concentrated on picking berries alone, and Crusoe focused on fishing only, let us suppose that the former could pick 500 berries in two days, while the latter could catch 50 fish in that amount of time. Together, they could each potentially[10] have more of both goods than if each worked in isolation, without trade. We do not need heterogeneity to reach this conclusion, only the supposition that specialization enhances productivity. Learning is particularly important in this geographical or "locational" analysis. This is in sharp contradistinction to the typical textbook models of equilibrium economics, which either ignore these crucial elements of economics or give them short shrift.

In short, even under complete homogeneity, specialization, division of labor, and trade can arise. A sufficient condition for such trade to arise is simply that learning and practice bring improvement. This is significant in that it demonstrates the robustness of free trade: it can take place even in the face of homogeneity. Even such notable defenders of free trade as Mises and Rothbard did not make this point explicit in their analysis.[11]

Notes

1. Even today, the Montaigne fallacy plagues discussions of outsourcing, "off-shoring," and increased international capital mobility. Irwin (2005) summarizes recent controversies over trade theory and policy.

2. In making this claim, we are implicitly assuming a whole host of other factors to be in play: sufficient knowledge, a legal system supportive of trade or at least not too disruptive of it, and so forth.

3. If tastes were different, trade might well occur. Let us suppose that Jackson was a vegetarian, a herbivore, and Crusoe a carnivore. Under these conditions, Crusoe would give up his berries for the fish of Jackson; thus, specialization, the division of labor, and greater productivity would ensue. But we are assuming *complete* homogeneity, including tastes, so this result is ruled out.

4. The authors are grateful to Laurence Moss for this point. For more on George, see Moss (2001, 2005).

5. More precisely, it is not physical assets themselves, but their *attributes*—the characteristics, functions, or possible uses of assets, as perceived by entrepreneurs—that are valued and priced on factor markets. On this point, see Foss, Foss, Klein, and Klein (2007).

6. We abstract from all issues related to indifference; for example, if they were both equidistant from berries and fish, would they just stand there like Buridan's Ass, and starve to death? For a debate on indifference, see Nozick (1977), Block (1980, 1999, 2003), Caplan (1999, 2000, 2001), Hülsmann (1999), and Hoppe (2005).

7. If they did this, some of the benefits from practice would be lost, but it still might well be a better outcome than no trade at all.

8. It cannot be denied that heterogeneity is of considerable helpfulness in getting the division of labor underway. And that, really, is an understatement. We readily concede that without heterogeneity, the amount of specialization would be very small. Perhaps it might not even exist at all. But it *might.* There is no theoretical reason why it could not, Mises and Rothbard to the contrary notwithstanding.

9. Both Mises and Rothbard accept the idea of learning by doing. Mises (1966: 164) writes: "Exercise and practice of specific tasks adjust individuals better to the requirements of their performance; men develop some of their inborn faculties and stunt the development of others. Vocational types emerge, people become specialists." Adds Rothbard (1962: 82): "full-time specialization in a line of production is likely to improve each person's productivity in that line and intensify the relative superiority of each."

10. Depending upon the terms of trade that were agreed upon; at the very least, neither would have less.

11. One could respond that neither Mises nor Rothbard intended to generalize their statements quoted above beyond an environment of constant returns and no learning effects. They left no indication that they meant to include such qualifiers, however. We thank a referee for raising this point.

References

Block, Walter. (1980). "On Robert Nozick's 'On Austrian Methodology'." *Inquiry* 23(4): 397–444.

——. (1999). "Austrian Theorizing, Recalling the Foundations: Reply to Caplan." *Quarterly Journal of Austrian Economics* 2(4): 21–39.

——. (2003). "Realism: Austrian vs. Neoclassical Economics: Reply to Caplan." *Quarterly Journal of Austrian Economics* 6(3): 63–76.

Caplan, Bryan. (1999). "The Austrian Search for Realistic Foundations." *Southern Economic Journal* 65(4): 823–838.

——. (2000). "Probability, Common Sense, and Realism: A Reply to Hülsmann and Block." *Quarterly Journal of Austrian Economics* 4(2): 69–86.

——. (2001). "Probability, Common Sense, and Realism: A Reply to Hülsmann and Block." *Quarterly Journal of Austrian Economics* 2(4): 69–86.

Foss, Kirsten, Nicolai J. Foss, Peter G. Klein, and Sandra K. Klein. (2007). "The Entrepreneurial Organization of Heterogeneous Capital." *Journal of Management Studies*, forthcoming.

George, Henry. ([1879] 2003). *Progress and Poverty.* New York: Robert Schalkenbach Foundation.

Hoppe, Hans Hermann. (2005). "A Note on Preference and Indifference in Economic Analysis." *Quarterly Journal of Austrian Economics* 8(4): 87–91.

Hülsmann, Jörg Guido. (1999). "Economic Science and Neoclassicism." *Quarterly Journal of Austrian Economics* 2(4): 1–20.

Irwin, Douglas A. (2005). *Free Trade Under Fire*, 2nd ed. Princeton, NJ: Princeton University Press.

Mises, Ludwig von. (1966). *Human Action*, 3rd ed. Chicago: Regnery.

Moss, Laurence S. (2001). "Why the Preaching Must Never Stop: Henry George's and Paul Krugman's Respective Contributions to the Free Trade Debate." In *The Path to Justice: Following in the Footsteps of Henry George. Supplement to the American Journal of Economics and Sociology* 60(5): 135–161.

——. (2005). "The Henry George Theorem: Turning Henry George on His Head." In *The Economics of Henry George.* Ed. John Laurent. Cheltenham, UK: Elgar Publishers.

Nozick, Robert. (1977). "On Austrian Methodology." *Synthese* 36: 353–392.

Ricardo, David. ([1817] 1951). "The Principles of Political Economy and Taxation." In *The Works and Correspondence of David Ricardo.* Eds. Piero Sraffa and M.H. Dobb. Cambridge: Cambridge University Press.

de Roover, Raymond. ([1963] 1974). "The Scholastic Attitude Toward Trade and Entrepreneurship." In *Business, Banking, and Economic Thought in Late Medieval and Early Modern Europe: Selected Studies of Raymond de Roover.* Chicago: University of Chicago Press.

Rothbard, Murray N. ([1962] 1993). *Man, Economy and State.* Auburn, AL: Mises Institute.

——. ([1982] 1998). *The Ethics of Liberty.* Atlantic Highlands, NJ: Humanities Press.

ACADEMIC TENURE: AN ECONOMIC CRITIQUE

ROBERT W. MCGEE*
WALTER E. BLOCK**

I. INTRODUCTION

A number of arguments have been put forth in favor of academic tenure, the guarantee of lifetime employment in substantially the same position after some initial probationary period. These arguments have centered around cost effectiveness, academic freedom, and pedagogical quality. They invariably conclude that tenure is good and that horrendous things would happen if tenure were abolished or modified. In fact, it is very difficult to find an academic, especially one with tenure, who is not in favor of tenure.[1] "Tenure is to the academic what the closed shop is to a craft union."[2]

In this Article, we take the position of the devil's advocate. We show that many of the arguments usually offered in favor of

---

* Associate Professor of Business Administration, Seton Hall University. B.A., 1969, Gannon University; M.S.T., 1976, DePaul University; J.D., 1980, Cleveland State University; Ph.D., 1986, University of Warwick.

** Associate Professor of Economics, College of the Holy Cross. B.A., 1964, Brooklyn College; Ph.D., 1972, Columbia University.

1. One journalist scoured North America and could find only one academic who would argue against tenure. *See* Walker, *Tenure Debate Falls Short*, THE PROVINCE, Sept. 4, 1983, at 269, 269. He should have polled primary and secondary school administrators. A 1972 poll of school administrators found that 86 percent wanted tenure either reformed or abolished. *See* E. BRIDGES, MANAGING THE INCOMPETENT TEACHER 3 (1984) (citing Cramer, *How Would Your Faucets Work If Plumbers Were Shielded by Tenure Laws?*, AM. SCH. BD. J., Oct. 1976, at 22, 22-24). In four Gallup polls of parents' attitudes toward public education, more than 50 percent said they opposed tenure. *See* A DECADE OF GALLUP POLLS OF ATTITUDES TOWARD EDUCATION, 1969-1978 (S. Elam ed. 1978).

For a strong defense of tenure, see Machlup, *In Defense of Academic Tenure*, 50 AM. A.U. PROF. BULL. 112 (1964), *reprinted in* ACADEMIC FREEDOM AND TENURE: A HANDBOOK OF THE AMERICAN ASSOCIATION OF UNIVERSITY PROFESSORS 306-38 (L. Joughin ed. 1969). Professor Machlup's argument is based on the rather curious premise that although tenure harms academics more than it helps them, the benefits to "society" accruing from tenure more than compensate for the loss by academics because tenure encourages the free flow of debate. There is no evidence, however, that tenure improves the flow of debate beyond what already exists as a result of freedom of the press. Professor Machlup does not recognize that other groups are also harmed by tenure: Consumers of education, for example, must be content to endure a number of mediocre and incompetent professors who would be fired in the absence of the protection that tenure affords. Machlup apparently believes that "society" consists of something more than the sum of all individuals, but society as a whole cannot benefit from tenure—only individuals can benefit.

2. Walker, *supra* note 1, at 269.

tenure do not hold up under close analysis.[3] While tenure can save out-of-pocket expenses, promote independence from outside forces, and increase quality, it also increases overall costs, decreases flexibility, disenfranchises the paying consumer of education, increases dependence on unaccountable insiders, and makes it nearly impossible to remove incompetent and unnecessary professors. While we do not advocate the total abolition of tenure, we would like to see the present tenure practice abolished. We conclude that decisions on the value of tenure should be left to the market. In the absence of the current tenure policy, consumers of education would benefit from lower prices, more choices, and better quality products.

II.   A Critical Analysis of Arguments for Tenure

A.   *Is Tenure Cost-Effective?*

The argument that tenure is cost-effective is a plausible one on the surface. Just ask this question: If two teaching positions paid the same salary, had identical duties, and were identical in every other way—student body, prestige, geographic location, fringe benefits—but one position carried a guarantee of lifetime employment and one did not, which position would one choose? The answer appears obvious. Because most people are risk-averse, they would choose the position that included a guaranteed job for life, other things being equal. Therefore, the possession of tenure has some value.[4] That being the case, professors who have tenure, or who are hired with the possibility of receiving tenure, will work for less money than professors who cannot hope to receive a guarantee of lifetime employ-

---

3. For detailed discussions of tenure and the issues relating to it, see generally The Tenure Debate (B. Smith ed. 1973); and Faculty Tenure (W. Keast & J. Macy eds. 1973).

4. The United States Supreme Court has held that tenure may create a property right that cannot be taken away without due process. *See, e.g.,* Perry v. Sindermann, 408 U.S. 593 (1972) (a written contract with an explicit tenure provision may be evidence supporting a claim of a property right to continued employment); Connell v. Higginbotham, 403 U.S. 207 (1971) (due-process requirements apply when teacher is hired with clearly implied promise of continued employment); Slochower v. Board of Higher Educ., 350 U.S. 551 (1956) (tenured public college professor has interest in continued employment that is safeguarded by due process); Wieman v. Updegraff, 344 U.S. 183 (1952) (college professors have interest in continued employment that is safeguarded by due process). *Cf.* Board of Regents v. Roth, 408 U.S. 564 (1972) (non-tenured faculty member's property interest in employment was created by the terms of his employment and was insufficient to constitute property interest requiring due-process protections before nonrenewal of contract).

                    Academic Tenure   547

ment.[5] Consequently, universities can save money by granting tenure.[6] Once tenure exists at an institution, abolishing it would be equivalent to reducing professors' salaries.[7]

Although this argument is true as far as it goes, it is incomplete. There are other factors to consider in evaluating the tenure system besides initial out-of-pocket costs. If guaranteeing workers lifetime employment were cost-effective, wouldn't it make sense for all employers everywhere to grant lifetime employment to all workers? When viewed from this perspective, the cost-effectiveness argument in favor of tenure seems unpersuasive. If it were valid, all employers could reduce labor costs by guaranteeing lifetime employment. Yet the education industry is practically the only American industry that grants lifetime employment security.[8] One might ask why the practice is not more widespread, especially in the private sector where costs are closely scrutinized.

One reason why the idea has not spread throughout the pri-

---

5. *See* Machlup, *supra* note 1, at 323-26. This salary effect is a double-edged sword. On the one hand, it attracts individuals to academia, which increases the supply of professors and lowers salaries generally (as supply increases, price decreases). On the other hand, consumers of education benefit because they can purchase more professors for their money because of the lower unit cost. This raises an interesting point. All professors' salaries tend to be reduced because of tenure. Yet only a minority of professors realistically benefit from the academic freedom that tenure affords. For example, political science professors tend to be shielded more by academic-freedom protections than do mathematicians and French grammarians. Would it not be more equitable, then, to allow those professors who do not see the need for protecting their academic freedom to bargain it away for higher salaries? Allowing this possibility would increase freedom of choice without jeopardizing the positions of those professors who would be content with lower salaries plus tenure.

6. This argument is a pervasive one, one that is always advanced whenever the institution of tenure is attacked or questioned. Twenty-nine economics department faculty members from the University of British Columbia used this argument when the tenure policy was being challenged in Canada. *See* Block, *Put an End to Academic Tenure*, Econ. Aff., July-Sept. 1984, at 37, 37-38.

7. *See* Milne, *Arthritic Academia: The Problems of Government Universities*, in Occupational Regulation and the Public Interest 193, 198 (R. Albon & G. Lindsay eds. 1984). Of course, if tenure were abolished, professors would tend to demand higher monetary salaries because, as pointed out above, tenure is a form of compensation.

8. The civil service also guarantees lifetime employment, but people rarely argue that providing guaranteed lifetime employment to civil servants reduces labor costs. In fact, there is a widespread perception that the civil service is laden with inefficient or incompetent workers who are underworked and overpaid. Judges often have lifetime tenure, ostensibly so that they will be insulated from political pressure. A problem with lifetime judicial tenure is that it is nearly impossible for the citizenry to remove from the bench an unwanted judge who does not want to resign. The United States Constitution even prevents Congress from reducing judges' pay, although it does not provide that judges must get cost-of-living raises. *See* U.S. Const. art. III, § 1. Therefore, in the absence of a pay raise, inflation will result in a de facto pay reduction. Such a pay cut, however, would apply to all judges, not just those Congress wished to single out for retribution.

vate sector is that lifetime employment guarantees reduce flexibility. If General Motors, Ford, and Chrysler guaranteed their workers lifetime employment, the companies would have gone out of business long ago. To stay solvent, a company must have the flexibility to discard unneeded resources: low-quality or excess machinery, and low-quality or excess employees. The private sector must flexibly comply with the demands of the market in order to survive and prosper.

The education industry need not be so flexible, because it is virtually insulated from competition. Although colleges and universities compete for students, various accreditation agencies and governments dictate much of what they can and cannot do. Colleges and universities cannot compete freely by offering a product that is radically different from what other similar institutions are offering because they must meet the requirements of the accreditation agencies and government regulators, or risk having their accreditation revoked.[9] The insulation from competition in the education industry, and the resulting ability of many educational institutions to survive without being flexible, allows the tenure system to survive. As one academic economist has observed:

> [T]enure is neither necessary nor efficient. Its survival depends upon the absence of private ownership and is also en-

---

9. Elaboration of this point is beyond the scope of this Article. Accreditation agencies are monopolies. The United States Department of Education places its stamp of approval on only one accreditation agency for each geographic location. For example, all schools in Ohio are accredited by the North Central Association. Any agency other than the North Central Association that accredits schools in Ohio is an unrecognized agency. *See* J. BEAR, HOW TO GET THE DEGREE YOU WANT 32-35 (8th ed. 1982).

The Department of Education also approves accreditation agencies for specific disciplines. For example, business schools are accredited by the American Assembly of Collegiate Schools of Business (AACSB). Much of what the AACSB requires before it will grant accreditation, though, actually decreases the quality of the educational institution. Accreditation pressures force schools to offer courses that are irrelevant and unwanted by educational consumers, and taught by faculty who may not be the best qualified. *See* R. McGEE, A MODEL PROGRAM FOR SCHOOLS OF PROFESSIONAL ACCOUNTANCY 103-26 (1987) (providing further elaboration).

At least two of the six regional accreditation agencies in the United States are placing pressure on universities to increase ethnic and racial diversity in their faculties and student bodies. This means that schools are being pressured to hire faculty and accept students on the basis of race, ethnic background, and gender, rather than ability. For example, the Middle States Association delayed reaccrediting Baruch College of the City University of New York for three months because of alleged deficiencies in these areas. Critics of this policy say that the accreditation agencies are going too far, but not much can be done because of their monopoly status. *See 2 of 6 Regional Accreditation Agencies Take Steps to Prod Colleges on Racial, Ethnic Diversity*, Chron. Higher Educ., Aug. 15, 1990, at A1, col. 3. *See also Accrediting Quotas*, Wall St. J., Dec. 14, 1990, at A18, col. 1.

couraged by subsidization of education by non-customer income sources [(taxpayers and philanthropists)]. Without a private profit-seeking system and without full-cost tuition, the demand for tenure increases and the cost of granting it appears to be cheaper because the full costs are not imposed on those granting it. Competition among schools, teachers, and students provides protection to the search for the truth without tenure.[10]

The inflexibility that results from a policy of guaranteeing lifetime employment means that colleges and universities are not able to fire professors who are either incompetent or no longer needed.[11] If educational consumers—students—want more business or engineering courses, a university might try to meet this demand by hiring additional professors to teach these subjects. But it cannot fire tenured professors in fields not favored by consumers. As a result, classes in unpopular disciplines have fewer students, while classes in popular disciplines grow unwieldy. The quality of the classroom experience could be enhanced if class size in the popular disciplines were reduced, but university budgets are tight. Without extra funds, additional professors in the popular disciplines cannot be hired unless unneeded professors who teach other subjects can be dismissed. If tenure exists, students who study popular subjects must often do so in large classes because universities do not have the flexibility, at least in the short term, to allocate their resources according to consumer demand. Tenure thus prevents efficient resource allocation, to the detriment of consumers.[12] Universities must wait for the faculty in overstaffed departments to retire or die before resources can be reallocated.[13]

---

10. A. ALCHIAN, *Private Property and the Relative Cost of Tenure*, in ECONOMIC FORCES AT WORK 177, 201 (1977).

11. It is sometimes possible (but expensive) to encourage professors who are no longer needed to leave through early retirement and attractive severance packages. Universities, in effect, buy out the professors' property rights in their tenure agreement. Britain recently used this approach to rid its universities of unwanted faculty. *See* H. FERNS, HOW MUCH FREEDOM FOR UNIVERSITIES? 43 (1982).

12. For a discussion of flexibility and tenure, see Morris, *Flexibility and the Tenured Academic*, HIGHER EDUC. REV., Spring 1974, at 3.

13. *See id.* at 3. Universities have a number of other options in addition to waiting for unneeded faculty to retire or die. They can be given increased courseloads or unwanted administrative duties, which might induce some faculty members to leave. They could deny salary increases, which would decrease their purchasing power as inflation reduces the value of the dollar. Implementing such policies, however, increases animosity and could lead to a strike, which most university officials would rather avoid. Furthermore, such policies often take years to address the problem.

Although tenure may allow universities to hire faculty at lower salaries, total salary costs may be higher under a tenure system because universities will be forced to hire more faculty to teach popular disciplines and will not be able to fire unneeded faculty who teach unpopular subjects. If professors in disciplines with small consumer demand were given classes of thirty students instead of ten, two-thirds of them could be fired, and the funds saved could be used to hire professors in disciplines that students want to study, such as business.[14] When this inflexibility factor is considered, one can conclude that tenure actually increases salary costs.[15]

B. *Does Tenure Promote Academic Freedom?*

Another popular argument in favor of tenure is that without it, professors would hesitate to speak out on controversial subjects for fear of being fired.[16] This argument seems plausible on the surface. Certainly, if one need not worry about being

---

14. Some might argue that the students do not know what is best for them and that college administrators and accreditation agencies are better able to determine what students should study than the students themselves. This assumes that administrators and accreditors have the right to force their views and values on students, who may have far different values and views. Some schools thus require students to take Western civilization courses that consist of works written exclusively by white males. Other schools force students to take "watered down" Western civilization courses that include authors of questionable value just because they happen to be black, hispanic, or female. Some colleges force students to take sensitivity courses or courses that assert a particular political agenda. Many colleges force students to take mathematics, literature, foreign languages, or other courses that many students consider nonessential or a waste of time and money. For example, schools that are accredited by the American Assembly of Collegiate Schools of Business require students to take background "common body of knowledge" courses for the M.B.A. degree that often have little or nothing to do with what the student wants to learn. These courses often comprise up to 50 percent or more of the total coursework required for the M.B.A., thus doubling the cost and the time needed to complete the degree. The requirements for the M.S. degree in taxation are especially outrageous in this regard. In order to take the 10 or so courses needed for the degree, tax students are forced to take many courses totally outside the field of taxation. *See* R. MCGEE, *supra* note 9, at 18-24. Why would an education bureaucrat or committee necessarily know the needs of students better than the students themselves? In what other industry would the providers of a product be able to tell the buyers what they must buy? If the education monopoly were crushed, students could study the subjects that they value and save years of study and thousands of dollars.

15. This does not take into account the possible detrimental effects of the tenure system on the quality of the professoriate. *See infra* pp. 554-57.

16. Some would describe tenure as "the best guarantee of academic freedom." Lovain, *Grounds for Dismissing Tenured Postsecondary Faculty for Cause*, 10 J.C. & U.L. 419, 419 (quoting COMM'N ON ACADEMIC TENURE IN HIGHER EDUC., FACULTY TENURE 21 (1973)). The United States Supreme Court has recognized that academic freedom is a special concern of the First Amendment. *See, e.g.,* Keyishian v. Board of Regents, 385 U.S. 589 (1967) (teachers cannot be dismissed for refusing to sign a certificate that indicates whether or not they had ever been Communists).

fired for speaking out on controversial issues, one will be more likely to speak out than if one fears for his livelihood. We know that many theories that we now recognize as truth were once considered heretical: The world is round, not flat; the earth revolves around the sun, not the other way around; slavery is evil, even though it was thought of as normal and even necessary in the past. Members of the academic profession have been the proponents of these new ideas that have changed our ideas about truth. The argument runs that, in the absence of tenure, professors who speak out on controversial subjects or espouse unpopular views may jeopardize their careers. The free flow of information will be slowed or stopped.

But academics are not the only ones who challenge authority or existing orthodoxy. Novelists, playwrights, editors, news commentators, journalists, song writers, clergy, film producers, actors, cartoonists, and whistleblowers from various walks of life also espouse unpopular views and challenge orthodoxy. Should each of these groups also be given guaranteed jobs for life so that they can feel free to challenge the establishment? In the case of news journalists, tenure is not needed because journalists are motivated by the market. They seek out the truth and try to present it better and faster than the competition, so that their employer can capture a larger audience. Might not the same argument be made for academics? If universities were entrepreneurial, academics who sought the truth would be valued commodities; they would not need to worry about tenure, because their careers would be secure so long as they sought the discovery and dissemination of truth.[17] On the other hand, if teaching the truth offends or harms some individuals or groups, should not these individuals or groups have the right to combat the dissemination of truth by refusing to pay someone to disseminate it? To argue otherwise would be to conclude that professors deserve some special claim to the pocketbooks of the students or taxpayers who pay their salaries.[18]

The academic-freedom argument also suffers from its weak premise—that academics must be completely "free" and "independent." From whom should academics be independent? Academics are always worried about threats from outsiders.

---

17. *See* H. FERNS, *supra* note 11, at 44.
18. *See* A. ALCHIAN, *supra* note 10, at 199.

But what about the individuals who pay their salaries—education consumers and taxpayers? If a journalist writes something that is unpopular, consumers can exercise their view of the piece by not buying the newspaper that prints the article. Television viewers can turn to another channel if they dislike a commentator's views. Yet taxpayers and consumers of education must often support the careers of academics of whom they disapprove. In effect, those who advocate tenure on the grounds of academic freedom claim that academics have a right to the hard-earned dollars of others even if those who earn the dollars do not support the academics' views. In the view of tenure advocates, academic freedom justifies forcing others to support an academic's views with their cash.

Ironically, a major threat to outspoken professors can be found within the academy itself. Tenured academics decide whether an untenured professor is awarded tenure. Untenured faculty can easily be weeded out if they offend the members of the faculty tenure committee.[19] Unless he can show that racial, ethnic, or gender bias influenced the tenure decision, a professor who is denied tenure has little recourse.[20] Many scholars have had their careers short-circuited by their fellow professors because of their political leanings.[21] If a group of liberals, conservatives, or Marxists wants to exclude those who disagree with their views, all they have to do is deny them tenure. Yale

---

19. The tenure process works in different ways at different schools. At Seton Hall University, for example, there are three tenure committees, one each at the departmental, school, and university level. At other schools, there is only one committee.

20. Finding proof of discrimination is often difficult. *See generally* Note, *Title VII and the Tenure Decision: The Need for a Qualified Academic Freedom Privilege Protecting Confidential Peer Review Materials in University Employment Discrimination Cases*, 21 SUFFOLK U.L. REV. 691 (1987); Palombi, *The Ineffectiveness of Title VII in Tenure Denial Decisions*, 36 DEPAUL L. REV. 259 (1987); Mahoney, *Title VII and Academic Freedom: The Authority of the EEOC to Investigate College Faculty Tenure Decisions*, 28 B.C.L. REV. 559 (1987).

21. *See, e.g., A Teacher Disillusioned With Utopia*, Boston Globe, Dec. 20, 1990, at 97, col. 3 (professor of comparative literature dismissed from Hampshire College for "his refusal, despite his leftish sympathies, to reduce the European literature he teaches to a one dimensional story of imperialism, colonialism and Third World oppression"); Dershowitz, *Quota System Wrong for Harvard Law*, Boston Herald, May 8, 1990, at 29, col. 1 (female assistant professor denied tenure at Harvard Law School because the "women [faculty members] argued that she, though a woman, was not a feminist"). *See generally* ACADEMIC LICENSE: THE WAR ON ACADEMIC FREEDOM (L. Csorba ed. 1988) [hereinafter ACADEMIC LICENSE]; R. KIMBALL, TENURED RADICALS: HOW POLITICS HAS CORRUPTED OUR HIGHER EDUCATION (1990).

A professor's political opinions often enter into the hiring, firing, and tenure decisions. *See, e.g.,* P. LAZARSFELD & W. THIELENS, THE ACADEMIC MIND 392 (1958); W. BUCKLEY, GOD AND MAN AT YALE 136-90 (1951); E. ROOT, COLLECTIVISM ON THE CAMPUS 289-335 (1955); Menn, *The Tenure Process and Its Invisible Kingmaker*, in HOW HARVARD RULES 271, 271-78 (J. Trumpdour ed. 1989).

No. 2]   *Academic Tenure*   553

University's president acknowledged this problem in 1972: "In strong universities, assuring freedom from intellectual conformity coerced within the institution is even more of a concern than is the protection of freedom from external interference."[22] As another academic observed:

> [It is a] tattered secret that faculty members may devise a pattern of departmental appointments that shuts out significant schools of thought, may enter into a Faustian bargain with the grant-givers that trades off mental independence for a larger bank roll, and may make political judgments when reviewing the professional merits of their peers.[23]

Some academics go even further and prevent individuals from speaking at the university if they espouse views with which they disagree.[24] Officials of the South African government, officials from the Reagan administration, Nicaraguan contras, and other non-liberal types often are the victims of such anti-free speech activists these days.[25] During the 1950s, academics who espoused Communist, Marxist, or collectivist views experienced similar difficulties.[26] Even when an unpopular or "politically incorrect" individual is permitted to speak, he will often pay a price in the form of harassment, heckling, and picketing. Academics who promote cultural diversity often use these disruptions as an excuse to exclude certain individuals from the university and foreclose debate on issues of the day.[27] Academic freedom can be threatened from within the university as

---

22. Brewster, *On Tenure*, 58 Am. A.U. Prof. Bull. 381, 382 (1972).

23. Metzger, *Academic Freedom: A Symposium*, 13 N.Y.U. Educ. Q. 4, 5 (1982).

24. The American Association of University Professors (AAUP) takes the position that college and university students have the right to listen to anyone whom they wish to hear, "and affirms its own belief that it is educationally desirable that students be confronted with diverse opinions of all kinds . . . [A]ny person who is presented by a recognized student or faculty organization should be allowed to speak on a college or university campus." *Summary of Forty-Third Annual Meeting*, 43 Am. A.U. Prof. Bull. 359, 363 (1957), *reprinted in* Academic Freedom and Tenure: A Handbook of The American Association of University Professors 112-13 (L. Joughin ed. 1969). This position has sometimes been ignored when the person invited to speak has espoused a conservative position. *See, e.g.*, Silber, *Free Speech and the Academy*, Intercollegiate Rev., Fall 1990, at 33, 34 (Marxist English professor attacked former contra leader Adolfo Calero at Northwestern University).

25. For a listing of campus free speech abuses, see Academic License, *supra* note 21, at 305-12.

26. *See, e.g.*, Keyishian v. Board of Regents, 385 U.S. 589 (1967) (refusal to sign a certificate indicating membership in the Communist Party); Sweezy v. New Hampshire, 354 U.S. 234 (1957) (refusal to answer questions about political beliefs and associations).

27. *See* Chavez, *The Real Aim of the Promoters of Cultural Diversity Is to Exclude Certain People and to Foreclose Debate*, Chron. Higher Educ., July 18, 1990, at B1, col. 2. *See also* Timmons, *Fraudulent Diversity*, Newsweek, Nov. 9, 1990, at 8.

well as from without, and the tenure process does not guarantee that untenured professors will be protected from internal assaults on the right to say or write what is on their minds.

C. *Does Tenure Insure a High-Quality Professoriate?*

According to the tenure theory, only those professors who prove themselves through excellent teaching, research, and service are awarded tenure;[28] their weaker brethren fall by the wayside. At some of the "better" universities, however, receiving an award for good teaching is considered the kiss of death for an untenured professor.[29] Anyone who spends so much time preparing for class must somehow be deficient in research, or so the rationale goes.[30] In one recent period, three out of four recipients of Harvard's teaching award were denied tenure.[31] Stephen Ferruolo, a widely known medieval history scholar, won the Dinkelspiel award for outstanding teaching at Stanford in 1982, and was denied tenure shortly thereafter.[32] Bruce Tiffney won Yale's teaching award two weeks after being denied tenure.[33] Faye Crosby won the Yale teaching award in 1982 and was denied tenure three years later.[34] " 'It's extremely unlikely,' says Douglas Kankel, a tenured associate professor in Yale's Biology Department, 'that if you are a professor with an exceptional teaching background, you will survive the

---

28. A fourth criterion—collegiality—is sometimes added to this list. One court defined collegiality as "the capacity to relate well and constructively to the comparatively small bank of scholars on whom the ultimate fate of the university rests." Mayberry v. Dees, 663 F.2d 502, 514 (4th Cir. 1981), *cert. denied*, 459 U.S. 830 (1982). While personality is not usually an official criterion for tenure, it often plays an indirect role in determining whether a faculty member is awarded tenure. For a discussion of this point, see Zirkel, *Personality as a Criterion for Faculty Tenure: The Enemy It Is Us*, 33 CLEV. ST. L. REV. 223 (1984-85).

A sore spot among those who regard teaching as paramount and everything else as secondary is the rather low esteem in which teaching is held by those in control of promotion and tenure. These persons argue that students are paying tuition to be taught by professors, not graduate assistants, and they are not paying tuition to subsidize faculty research. Yet research is rewarded, and teaching is not. In some universities, it is possible to go through an entire undergraduate program without being taught by a single member of the faculty. The student may be exposed only to a series of graduate assistants. *See* C. SYKES, PROFSCAM: PROFESSORS AND THE DEMISE OF HIGHER EDUCATION 35 (1988) (citing R. DUGGER, OUR INVADED UNIVERSITIES 170 (1974)).

29. *See* C. SYKES, *supra* note 28, at 58 (citing Barol, *The Threat to College Teaching*, NEWSWEEK ON CAMPUS, Oct. 1983).

30. *See id.* at 53.

31. *See id.*

32. *See id.* at 53-54.

33. *See id.* at 54.

34. *See id.*

tenure process.' "[35] One critic has characterized the trend as systemic: "The pattern extends throughout higher education. Virtually every university in the country has a similar story. These cases are dramatic, irrefutable evidence that the academic culture is not merely indifferent to teaching, it is actively hostile to it. In the modern university, no act of good teaching goes unpunished."[36]

Even if the tenure process serves as a gauntlet that only the excellent survive,[37] that does not mean that the excellent untenured professor will continue to be excellent after receiving tenure. A professor may become out-of-date, lazy, incompetent, or senile after receiving tenure, yet need not fear being fired. Furthermore, the tenure process is marred by favoritism, politicking, and the "good old boy" network. Professors who receive strong tenure recommendations from the department chair and dean have a better chance of receiving tenure than do other, perhaps better-qualified nominees who receive lukewarm recommendations.[38] A majority of the professors who sit on university tenure committees are from fields other than that of the tenure nominee, know little about what the nominee has accomplished in his area of research, and depend heavily on the recommendations of others. When examining a nominee's list of publications, some committee members have even been known to ask which journals are refereed and which are not, because they are totally unfamiliar with the literature in the nominee's field.

Another argument against the present tenure rule of "up-or-out"[39] is that the probationary period is too short to evaluate properly a faculty member's potential.[40] This is especially true

---

35. *Id.* (quoting Professor Douglas Kankel).

36. *Id.*

37. Professor Machlup suggests that at least part of the "deadwood" problem is caused not by tenure, but by the inability of university administrations and tenure committees to weed out inefficient and incompetent professors, who also happen to be their friends, before they receive tenure. *See* Machlup, *supra* note 1, at 313-14, 317.

38. One prominent educator has pointed out that this politicking takes valuable time and energy away from teaching. Such politicking, however, is encouraged by the tenure process. *See* G. ROCHE, EDUCATION IN AMERICA 106-07 (1977).

39. In most institutions, professors who do not receive tenure within approximately six years are dismissed.

40. The probationary period varies with the institution. It is generally about six years but may be longer or shorter. Professors who have taught at another institution may have a reduced probationary period at their new institution. The AAUP has recommended a seven-year probationary period. The arbitrariness of this rule has been criticized by a number of commentators, including John R. Silber, now President of

whenever the probationary period is short and the faculty member is engaged in a type of research that takes years to complete. The probationary period might end before the professor has published anything, and thus, a professor with excellent potential may be dismissed.[41] The university and the professor lose in such cases. If the professor can find another academic position, one university's loss is another's gain. But professors who lose their positions must endure the hassle and trauma of finding another position and relocating to another city, often disrupting their family life in the process.

The "up-or-out" policy of tenure also increases turnover rates among untenured faculty. As a result, costs rise as universities engage in constant searches for replacement candidates. The quality of the faculty also generally drops whenever universities fire untenured professors who have a few years experience and replace them with untenured professors who have little or none.[42]

Junior professors who teach such "drill" courses as elementary accounting, beginning French, and freshman English do not need the protection of academic freedom provided by the tenure system as much as professors who teach more theoretical courses.[43] At the same time, many of these young professors face such heavy teaching loads that the burden of a tenure-level research program is at best impracticable. If tenure cannot be completely abolished, at least it should be ended for these junior professors, who stand to lose more from tenure than they can gain. In this way, they can have the opportunity to be employed at the same institution for more than six years and will not have to look for a new job every few years.[44]

As Charles Sykes has noted:

> Tenure corrupts, enervates, and dulls higher education. It is, moreover, the academic culture's ultimate control mechanism to weed out the idiosyncratic, the creative, the nonconformist. The replacement of lifetime tenure with fixed-term

Boston University. *See* Silber, *Tenure in Context*, in THE TENURE DEBATE 34, 48-50 (B. Smith ed. 1973).

41. *See* Machlup, *supra* note 1, at 315.

42. *See id.* at 318-19.

43. We do not concede that anyone needs tenure to protect the right to speak and write. We merely assume it here for the sake of argument.

44. Some academic fields are extremely overcrowded. Professors in these fields who are fortunate enough to find one job may not be so lucky if they must find another job after a few years. Some individuals who are good classroom teachers may be forced to leave teaching altogether.

 Academic Tenure 557

renewable contracts would, at one stroke, restore accountability, while potentially freeing the vast untapped energies of the academy that have been locked in the petrified grip of a tenured professoriate.[45]

The argument that tenure insures a high-quality professoriate cannot withstand scrutiny. If tenure cannot provide special benefits in the form of enhanced academic freedom or lower costs—and it appears likely that this is the case—the arguments for its continuation are correspondingly weakened.

III. ADDITIONAL REASONS FOR ABOLISHING TENURE

The tenure system came under increased attack after World War II,[46] and two national commissions cited it as a possible cause of student unrest during the 1960s.[47] The Linowitz Commission called tenure "a shield for indifference and neglect of scholarly duties."[48] The Scranton Commission said that tenure protects "practices that detract from the institution's primary functions, that are unjust to students, and that grant faculty members a freedom from accountability that would be unacceptable for any other profession."[49] The American Association of University Professors (AAUP) and Association of American Colleges (AAC) were understandably upset at these criticisms, but their own commission concluded that the fault was not with the tenure system itself but with its application and administration.[50] The AAUP's commission recommended that

> "adequate cause" in faculty dismissal proceedings should be restricted to (a) demonstrated incompetence or dishonesty in teaching or research, (b) substantial and manifest neglect of duty, and (c) personal conduct which substantially impairs the individual's fulfillment of his institutional responsibilities. The burden of proof in establishing cause for dismissal rests upon the administration.[51]

---

45. C. SYKES, *supra* note 28, at 258.

46. *See* Lovain, *supra* note 16, at 420 (citing COMM'N ON ACADEMIC TENURE IN HIGHER EDUC., FACULTY TENURE 8-10 (1973)).

47. *See id.*

48. *Id.* (quoting SPECIAL COMM. ON CAMPUS TENSIONS, AM. COUNCIL ON EDUC., CAMPUS TENSIONS: ANALYSIS AND RECOMMENDATIONS 42 (1970)).

49. *Id.* (quoting PRESIDENT'S COMM'N ON CAMPUS UNREST, REPORT OF THE PRESIDENT'S COMMISSION ON CAMPUS UNREST 201 (1970)).

50. *See id.* at 421 (citing COMM'N ON ACADEMIC TENURE IN HIGHER EDUC., FACULTY TENURE 20 (1973)).

51. *Id.* (quoting COMM'N ON ACADEMIC TENURE IN HIGHER EDUC., FACULTY TENURE 75 (1973)).

Other commentators have attacked the tenure system for other reasons. Attorney John Gierak cites Michigan's Teacher Tenure Act[52] as the "single biggest obstacle to correcting the problem of poor teachers in [Michigan] public schools."[53] He gives three compelling reasons for this conclusion:

> [1] The teacher tenure system is wholly unnecessary to protect teachers from arbitrary and capricious actions of school boards, because teachers have more than adequate protection under collective bargaining agreements administered by powerful teachers unions.
>
> [2] Teacher tenure procedures are exorbitantly expensive in terms of both time and money, due in part to the fact that board of education members are required under the system to spend endless hours sitting as a "jury" on charges of teacher incompetency and misconduct.
>
> [3] The enormous amounts of time and money that must be invested in a teacher tenure case are powerful forces built into the tenure system that strongly discourage action against poor teachers. The impact of these cost considerations upon how districts deal with incompetent teachers cannot be overestimated, especially in small districts and those districts experiencing financial difficulties, which include the vast majority of all public school districts in the state.[54]

Times have changed drastically since Michigan passed its teacher tenure law in 1937. Labor law in Michigan and the rest of the nation has evolved to the point where tenure is no longer needed.[55] Teachers are now well-protected against arbitrary dismissal by various teachers' unions, to which the vast majority of teachers belong.[56] Union grievance procedures are much like those afforded by the Teacher Tenure Act. Special legisla-

---

52. MICH. COMP. LAWS § 38.91 (1990) (MICH. STAT. ANN. § 15.1971 (Callaghan 1990)).

53. Gierak, *Abolish the Teacher Tenure Act to Improve the Quality of Public Education*, 68 MICH. B.J. 1098, 1098 (1989). Although Gierak focuses primarily upon primary and secondary schools, his arguments and our subsequent discussion are also applicable to higher education.

Professor Steven Amberg takes a contrary view of the need to abolish the Teacher Tenure Act. *See* Amberg, *Abolish the Teacher Tenure Act? Not While the Need Still Exists*, 68 MICH. B.J. 1104 (1989). However, Amberg ignores Gierak's arguments, does not make any valid arguments of his own, and blames incompetent teachers on administrators who fail to supervise or train them properly.

54. Gierak, *supra* note 53, at 1098.

55. *See id.* at 1099.

56. For a discussion of unions and tenure, see Olswang, *Union Security Provisions, Academic Freedom and Tenure: The Implications of* Chicago Teachers Union v. Hudson, 14 J.C. & U.L. 539 (1988).

tion is thus no longer needed to protect teachers' rights.[57]

Part of the problem with tenure acts like Michigan's is that they make dismissal of an incompetent teacher exorbitantly expensive. In Michigan, for example, legal fees for an action to discharge an incompetent teacher are typically between $50,000 and $70,000 and in some cases exceed $100,000.[58] Dismissing a teacher is time-consuming—a board of education may have to sit through ten or more hearings before it can take action.[59] In Michigan, if a board of education determines that a teacher should be dismissed, the teacher can request another hearing before the state's Teacher Tenure Commission. If the tenure commission upholds the board's dismissal, the teacher may appeal to the circuit court, then to the appellate court, and finally to the Michigan Supreme Court. In order to pay the legal costs of defending its action, the school board may need to lay off an additional teacher for three or four years.[60] In view of the time required to serve as a board member throughout this process, some school districts may find it difficult to obtain volunteers to sit on their boards of education, especially in large school districts, where several teachers may be attempting to have their dismissals overturned.[61]

With such expensive and drawn-out procedures required to dismiss an incompetent teacher, it is always possible that the school administration will decide that retaining an incompetent teacher is the lesser of two evils.[62] If that happens, the incompetent teacher is left free to damage hundreds or even thousands of students over the course of his career.[63] Allowing incompetent teachers to continue teaching also harms the morale of good teachers and administrators. This problem can be solved by abolishing the Teacher Tenure Act and instead rely-

---

57. Almost every state legislature has passed some type of statewide tenure legislation. *See* L. STELZER & J. BANTHIN, TEACHERS HAVE RIGHTS TOO 1 (1980).

58. *See* Gierak, *supra* note 53, at 1100.

59. *See id.* Gierak points out that some hearings take much longer. In Cooper v. Oak Park Public Schools, 624 F. Supp. 515 (E.D. Mich. 1986), board hearings took almost 10 months. *See id.* at 516.

60. *See* Gierak, *supra* note 53, at 1100.

61. *See id.* at 1101.

62. Gierak cites this fact as a major contributing cause to the mediocrity of the public schools. *See id.*

63. Several studies by the American Association of School Administrators found teacher incompetence to be a major problem in public schools. In a 1977 study, school administrators estimated that between 5 and 15 percent of their teachers were unsatisfactory performers. *See* E. BRIDGES, MANAGING THE INCOMPETENT TEACHER 1 (1984) (citing S. NEILL & J. CUSTIS, STAFF DISMISSAL: PROBLEMS AND SOLUTIONS 5 (1978)).

 Harvard Journal of Law & Public Policy [Vol. 14

ing upon binding arbitration,[64] which, according to Gierak, offers a number of advantages over the present system:

> 1. Arbitration is much quicker, which results in less disruption to educational programs, less potential back-pay liability, and increased fairness. . . .
> 2. Arbitration is much less expensive, for it reduces attorneys fees and costs. . . .
> 3. Arbitration does not necessarily involve board members. . . .
> 4. Because arbitrators are mutually selected by the parties, their decisions are generally better received and perceived to be more fair. . . .
> 5. Arbitration decisions are seldom appealed, which lends greater finality to the process. . . .[65]

Gierak goes on to observe that eliminating the Teacher Tenure Commission and its accompanying overhead would save taxpayers millions of dollars.[66]

While Gierak's arguments do not necessarily apply directly to situations in other states, many states have processes similar to that of Michigan.[67] And, although Gierak aims his attacks at the public primary and secondary school systems, his analysis largely applies to the collegiate level. Unionism is not as rampant at the college level as it is in the primary and secondary public schools, and university trustees do not necessarily become as involved in tenure disputes as do board of education members, but abolishing tenure at the college level can improve the quality of education by ridding the school of incompetent teachers.

IV. Conclusions and Recommendations

Academic tenure has advantages and drawbacks. It may save out-of-pocket expenses, but it also decreases flexibility, and the

---

64. *See* Gierak, *supra* note 53, at 1101-03.

65. *Id.* at 1102. Gierak uses as a case-in-point Beebee v. Haslett Public Schools, 406 Mich. 224, 278 N.W.2d 37 (1979), in which "the tenure charges filed with and upheld by the board against a teacher in 1968 were not finally resolved until 1979, a period of litigation spanning 11 years." Gierak, *supra* note 53, at 1103 n.13. Gierak does not mention how much money was wasted in the course of this litigation or how many students were permanently damaged during the 11 years that this incompetent teacher was in the classroom.

66. *See* Gierak, *supra* note 53, at 1103.

67. As noted above, most other states also have teacher tenure acts. For discussions of Louisiana's Teacher Tenure Act, see Comment, *Louisiana's Teacher Tenure Act*, 34 Loy. L. Rev. 517 (1988); and Resetar, *The Louisiana Teachers' Tenure Act—Protection from Dismissal for Striking Teachers?*, 47 La. L. Rev. 1333 (1987).

resulting inflexibility may cost more than the out-of-pocket expenses saved. Tenure enhances independence from outside forces but reduces the influence of those who pay and consume—taxpayers and students—while increasing the influence that tenured professors have on their untenured colleagues. The tenure process might increase the quality of the professoriate if it can truly separate high-quality teachers from those of low quality, but once a professor is tenured, it is difficult for the university to fire him unless he becomes incompetent or unneeded. Indeed, courts have ruled that universities may not fire a tenured professor even for ethical lapses.[68] Moreover, dismissing a tenured professor or teacher who does not want to go can be time-consuming and expensive.

So what is the solution? Is tenure good or bad? Nothing in the discipline of economics allows an unambiguous answer. The question is similar to "What is the optimal allocation of labor and capital in a given plant or industry?" Economics, however, can provide a methodology for answering this question. Let the market determine the worth of tenure. Give buyers and sellers of educational services freedom of choice. Allow colleges and universities the freedom to determine whether they want to offer tenure and how they want their tenure policy to be structured, without fear of losing accreditation.[69] Then, educational consumers would have the option of going to a school that has a tenure policy or one that does not. If tenure is a good management policy, it will tend to show in the consumer's perception of the university's reputation,[70] and that will have an effect on the university's enrollment and the amount of tuition it can charge.

Education consumers' dollar votes will be most effective, however, in rewarding good universities and punishing poor universities only if government is completely removed from the picture.[71] Government tends to reward universities with fund-

---

68. *See* Burton v. Cascade School Dist. Union High School No. 5, 353 F. Supp. 254, 255 (1973), *aff'd*, 512 F.2d 850 (9th Cir.), *cert. denied*, 423 U.S. 839 (1975); Texton v. Hancock, 359 So. 2d 895 (Fla. Dist. Ct. App. 1978). *See also* Magner, *Can't Fire Professor for Ethical Lapses, Rutgers Told*, Chron. Higher Educ., Aug. 15, 1990, at A2, col. 2.

69. For a discussion of how tenure policies may be changed or abolished, and how schools that do not have tenure policies have fared, see R. CHAIT & A. FORD, BEYOND TRADITIONAL TENURE: A GUIDE TO SOUND POLICIES AND PRACTICES (1982).

70. We do not mean to limit our recommendations to universities. The economics of tenure also apply to primary and secondary schools, although the problems and opportunities are somewhat different.

71. Readers may be shocked by the suggestion that education be completely priva-

ing regardless of quality, whereas consumers reward universities (by paying tuition) based on their perceptions of quality. Consumers will have little influence if a university's budget is provided entirely or mostly by government and accredited by a government-approved accreditation agency. Universities can best serve consumers only if they are free to offer the best product they can without artificially-imposed government obstacles and prohibitions. If universities were allowed to become entrepreneurial, they would be forced to serve consumers or suffer a loss of enrollment.[72] Under the present system, universities can thrive while ignoring consumers so long as they continue to receive generous government funding and, at the same time, can prevent other universities from competing effectively

---

tized. We have been taught to believe that free public education is desirable and even a right. Discussing this issue in depth is beyond this Article's scope. Fortunately, several books and articles have been written on the benefits of totally private education, which can be much cheaper and qualitatively better than tax-supported education. For discussions of this point, see S. BLUMENFELD, IS PUBLIC EDUCATION NECESSARY? (1985); J. COLEMAN & T. HOFFER, PUBLIC AND PRIVATE HIGH SCHOOLS (1987); J. COONS & S. SUGARMAN, EDUCATION BY CHOICE (1978); S. DENNISON, CHOICE IN EDUCATION (1984); R. FITZGERALD, WHEN GOVERNMENT GOES PRIVATE: SUCCESSFUL ALTERNATIVES TO PUBLIC SERVICES 139-48 (1988); F. FORTKAMP, THE CASE AGAINST GOVERNMENT SCHOOLS (1979); M. PIRIE, PRIVATIZATION: THEORY, PRACTICE AND CHOICE 194-97, 231-35, 283-87 (1988); R. POOLE, CUTTING BACK CITY HALL 172-88 (1980); THE PUBLIC SCHOOL MONOPOLY: A CRITICAL ANALYSIS OF EDUCATION AND THE STATE IN AMERICAN SOCIETY (R. Everhart ed. 1982); E. SAVAS, PRIVATIZING THE PUBLIC SECTOR: HOW TO SHRINK THE GOVERNMENT 102-03 (1982); Spring, *The Public School Movement vs. the Libertarian Tradition*, J. LIBERTARIAN STUD., Spring 1983, at 61; E. WEST, EDUCATION AND THE STATE: A STUDY IN POLITICAL ECONOMY (2d ed. 1970).

72. Professor Armen Alchian suggests that the incidence of tenure is directly correlated to the percentage of university income that comes from endowments and taxpayers. That is, the higher percentage of funding that comes from sources other than consumers (students), the more likely there is to be a high incidence of tenure, because the system is insulated from consumer demand. Such universities can continue to grant tenure, despite its inefficient features, because consumers do not have much clout if the vast majority of university funding comes from other sources. If consumers paid 100 percent of the cost of their education, tenure would be far less likely to exist as a management policy, as evidenced by the fact that at proprietary schools (which are run like businesses), tenure is practically nonexistent. Thus, tenure can best be eliminated by getting government out of education. *See* A. ALCHIAN, *supra* note 10, at 194.

Alchian also points out that teaching will grow in importance as the percentage of total university funding that comes from tuition increases. *See* A. ALCHIAN, *The Economic and Social Impact of Free Tuition*, in ECONOMIC FORCES AT WORK 203, 220 (1977). Students pay tuition to be taught rather than to subsidize faculty research and service, and if they are paying the full cost, they will be able to exercise tremendous influence by threatening to withhold their tuition or take their money to a competing institution. As it stands, tuition covers a small part of total funding, especially at tax-supported institutions. The inescapable conclusion is that teaching will improve if government is removed from the business of funding education. Even if this does not occur, emphasis on teaching could improve if government provided support in the form of vouchers instead of direct funding. A voucher policy still raises a moral question, because under such a policy, the government is still using the force of taxation to make some persons (taxpayers) pay for a benefit that goes to others (students).

by blocking such competitors' innovative courses and delivery systems.[73]

If a consumer-oriented educational system were available, a number of tenure options would likely result. Some universities would probably retain the traditional tenure system. Others would abolish it entirely, thus becoming like every other industry that refuses to guarantee its workers lifetime employment. A number of combinations and permutations would also likely develop. Perhaps five-year renewable contracts would replace tenure in some cases. Through hybrid instruments of this type, professors could be somewhat insulated from external and internal influences but could still be dismissed if their services became unneeded or unwanted. Other universities might decide to grant tenure to some professors and give term contracts to everyone else.[74] The market would discover other options as well, if allowed to operate. Tenure is a management policy. As is the case with all management policies, companies with good policies tend to thrive and companies with poor ones tend to lag behind the competition. The market determines what a good management policy is, and what is not, far better than theoreticians gathered around the fireplace in the faculty lounge.

---

73. The accreditation system prevents competition. A university cannot just offer an innovative course whenever and wherever it wants. It must first look over its shoulder to see whether the relevant accreditation agency will permit the new course. The American Assembly of Collegiate Schools of Business (AACSB), for example, routinely prohibits business schools from starting or expanding a new program without its approval—and approval is often denied in advance by telling the business school at the time of its accreditation visit that it may not expand a certain program or start any new programs unless it first complies with specified conditions and receives AACSB approval. Thus, a business school may be prevented from offering M.B.A. courses at a company's facilities and may be forced by the accreditation agency to phase out such programs already in existence in order to maintain accreditation—even if consumers want the programs to be expanded.

74. Present university policy is either to grant tenure after some period of time, generally about six years, or to give a one-year notice of termination. Thus, professors who do not receive tenure are fired with one year's advance warning. If a termination letter is not received after a certain cut-off date, the professor is often considered tenured. The tenure or term contract option, by contrast, would allow untenured professors to continue teaching without forcing the university to terminate after some arbitrary period of time. Until a few years ago, academics in British schools were generally granted tenure automatically after a one- to five-year probationary period. *See* H. FERNS, *supra* note 11, at 42 n.2.

COMMENT ON CANICE PRENDERGAST'S "A THEORY OF 'YES MEN'"

WALTER BLOCK

Prendergast (1993) has discovered a new "market failure": In the market, "subordinates have an incentive to conform to the opinions of their superiors" (p. 757).

She bases this on three assumptions. First, "the profitability of most organizations depends upon how cheaply they can collect information." Therefore, for any given level of quality, the cheaper the better. Second, "the worker [must] be provided with incentives to collect information" (p. 757) so as to overcome his reluctance to engage in such distasteful activity. Third, there is no objective information against which employee efforts can be measured.

If these three conditions hold, they give rise, in Prendergast's view, to the "yes-man" phenomenon. Here, the worker, instead of ascertaining the facts of the case—that is, the profitability of a new project, the desirability of new machinery or new hires, or the evaluation of a subordinate's productivity—tells the manager what he thinks the manager wants to hear.

This thesis is a more-than-passingly curious one, as the uninitiated would have supposed that any "yes-man" tendency would be summarily dealt with by the market through the imposition of bankruptcy, stock sell-offs, or "hostile" takeovers.[1] After all, profits are earned in accordance with consumer satisfaction, and this can hardly be accomplished if employees, instead of ferreting out accurate information so as to produce high-quality goods and services at low prices, are busily engaged in "yessing" their managers.[2] Management

---

WALTER BLOCK is the Harold E. Wirth eminent scholar endowed chair in economics at Loyola University. Unless otherwise identified, all page notations refer to Prendergast (1993).

[1]Where is Michael Milken? Our need for him has never been more desperate. See Manne (1966a,b); see also Block and McGee (1989).

[2]This applies, for example, to Bill Gates, president of Microsoft, one of the most profitable corporations in U.S. history. Karen Fries, a top executive, describes personal meetings with Gates: "'It's like going to a haunted house. Scary, but fun. Bill loves people who

cannot succeed unless it "makes the customer an offer he cannot refuse." If instead of helping the managers do this, workers tell them what they think they want to hear, this will, at the very, least function as a competitive disadvantage. It therefore behooves us to delve into the reasoning of Prendergast, to determine just how she is led to such a counterintuitive supposition.

There is no problem with the first of her three assumptions. The importance of collecting accurate information in an expeditious manner has a strong pedigree within the profession.[3] Nor can there be any serious objection to the second assumption. Work has disutility, and the principle-agent problem must somehow be overcome. But the third presents difficulties. Why should there be no objective information against which workers' efforts can be evaluated? Or, to express this in terms more amenable to economic analysis, where is the evidence that the costs of unearthing such information would be more costly than the losses of "yes-manning"?

Prendergast offers us three reasons why, in effect, it would be too costly for the manager to "wait and see if the project is profitable before deciding how well to reward the worker." More explicitly, she specifies what the problem is with trying "to obtain more objective information on the true parameter before rewarding the worker" (p. 757).

First, it may be "difficult" to "obtain an objective measure of the worker's input" (ibid.). But this is just like estimating the marginal-revenue product of actual and potential employees. Of course it is difficult to do: there are joint products; productivity depends upon how the worker "fits in" with others; it is impossible to keep one's eye on a given person all day long; etc. But this challenge occurs in every nook and cranny of the economy. Why this obstacle should be elevated to the status of a market failure is not at all clear. Nor is the solution theoretically very formidable: those entrepreneurs who can carry out such tasks prosper; those who cannot, do not.

Second, some amount of time may pass between the worker's task being completed and the reward (or penalty) being imposed. This calls for "commitment" and "institutional memory" (ibid.). Again, this does not seem to be an insurmountable barrier. There can be no denying that we live in an imperfect world; it simply does not pay to memorize trivial details. But we are vouchsafed no reason to believe that most firms do not set their rewards and penalties at least roughly in accord with the actual accomplishments of their employees—that they do not "remember" who was naughty and who was nice when it comes time, much later, to hand out bonuses, raises, and promotions.

---

stand up to him. He hates yes people.' At Microsoft, she adds, everything depends on how well you 'defend your point of view'" (*Newsweek*, 11 July 1994, p. 41).

[3]See Hayek (1948), Stigler (1961), Kirzner (1973), Klein (1996, 1999), Foss (1994), Lewin (1999).

When firms do not set wages and perks at least roughly comparable to productivity, labor mobility tends to push the market toward equilibrium in this regard. The worker who feels unduly belittled can leave; the firm that feels cheated by the employee can fire him. There is always a passage of time between work and reward, moreover. What makes this case special?

Third, Prendergast posits some sort of radical unpredictability[4]: at the time the worker made his decision, he had no way of knowing whether he was right or wrong, because there was "noise." Project profitability, for example, depends upon so many other people and phenomena (e.g., the weather, technological change, etc.) "in a way that cannot be predicted when the initial inference is made" (pp. 757-58). But Knightian uncertainty is a concomitant of all entrepreneurial activity. We must again ask, why is this case so special? No one can ever predict for sure; if they could, the most extreme versions of rational expectations would be true, and there would be no need for a market at all.

But even these extreme assumptions will not yield the conclusion desired by this author. If there are one hundred workers, each making predictions with no ostensible justification at the time they were made, still, some will do better than others. These are the ones who will tend to be better rewarded by the marketplace.

Further, there is always the expedient of manufactured prediction. A firm interested in the ability of its labor force can very cheaply administer a test[5] to its workers, where the correct answers are known in advance only to the manager.[6] Under the conditions posited by Prendergast, the management would have every incentive to hide[7] its true opinions. In this way, it could nip "yessing" in the bud.

In any case, it is imperative for the Prendergast thesis that there be "no obvious metric by which to measure whether a worker's report is correct" (p. 758).

---

[4]See Lachmann (1977, 1986) for an elaboration of the thesis that the future is inherently unknowable.

[5]Certainly, corporations go to great lengths to attempt to measure performance, and reward it accordingly, lest they suffer an excessively high quit rate precisely for those employees they wish to keep on the basis of these exams. If "yesmanning" was so pervasive, firms would hardly engage in such activities. For more on this see Brickley (2001). I owe this point to an anonymous referee.

[6]It may be difficult for the orchestra conductor to know which instrumentalists are carrying their full weight and which are shirking. After all, during the cacophony of a symphony, it may be difficult to tell one from the other. But if there are problems here, there are remedies, too. First, conductors are hired particularly on the basis of being able to separate the sheep from the goats even when there is a lot of "noise" occurring. Second, he can always ask each member of the orchestra to play by himself, during practice, in an attempt to "objectively measure" the various contributions of the musicians. The various artificial tests of predictive power would be the business analog to the musical case.

[7]Or perhaps even better, falsify them (so as to trap the unwary).

She needs this conclusion in order to explain the rise of "yesmanship." For if there are objective criteria, telling the manager something false in order to butter him up is unlikely in the extreme to be a successful strategy.[8]

She states:

> In the absence of reliable objective measures of performance, firms generally resort to subjective evaluation procedures, where managers compare their own findings to those of their subordinates. When the point of reference for adequate performance is the manager's opinion, an endogenous desire arises for the worker to conform to the opinion of the manager. This arises because the only way to induce the worker to exert effort is by comparing the findings to the manager with those of the worker. (p. 758)

There are problems here. Why is it that firms only "generally" resort to subjective evaluation procedures? Under the assumptions imposed by Prendergast, there is no other alternative but to act in this way. Certainly, she has not left room for any objective criteria. The fact that she falls short of writing consistently with her own theory may perhaps indicate that even she does not fully take it seriously.

Notice, too, the almost imperceptible change from "opinion" of the manager in the second sentence quoted above to her "findings" in the third. Only one of these can be correct. If we are truly in a kaleidic (Lachmannite) world, where the future is absolutely unknowable, then there can only be "opinions" about what is to come, not "findings." To the extent that there are indeed "findings," we have not moved entirely from the real world to one of Prendergast's construction.

Most importantly, there is nothing wrong with yesmanship under Prendergast's unrealistic assumptions! Let us assume with her that there are no objective criteria for correctness in prediction. But let us also take it for granted that the manager is somehow best able to discern things in this fog that are reality[9]—after all, he wouldn't be manager if he didn't have some sort of comparative advantage over his workers. Under these artificial conditions, it is a sign of productivity that the workers be able to anticipate the manager. Under these extreme assumptions, yessing, which is ordinarily counterproductive, becomes, paradoxically, productive.[10]

---

[8]With the exception of cases where the manager is using compliance of his underlings as a consumption good. But in this case, it is the function of the CEO to root out such behavior. And if this somehow fails to occur, there is still the fail-safe market mechanism of lower stock prices, "hostile" takeovers, and bankruptcy.

[9]"In the land of the blind, the one-eyed man is King."

[10]In the case of cooperative music playing, "yessing"—doing exactly what the conductor wants even if this means repressing one's own individual interpretation—is precisely the way to maximize production. The conductor is to the orchestra as the pianist is to the

Prendergast wants to have her cake and eat it too. She gives us such unrealistic assumptions that "yessing" actually becomes productive. Under these assumptions there is nothing left for the worker to do except agree with his boss. It is impossible for "the worker to act honestly" (p. 769) if we interpret "honest action" in its usual meaning so as to obviate the possibility of agreeing for the sake of agreeing. For honesty means telling the truth without fear or favor, whether or not the boss agrees with the truth. But in the artificial world constructed by Prendergast, there is no truth separate and apart from the manager's opinion. On the contrary, it is his views which define truth. That is why it is impossible to "be honest" without "yessing."

There are three main responses to the Prendergast thesis. One, we do live in a rational world where it is possible to predict (even though there are no guarantees of success). Here, the decisions of workers can be objectively evaluated, and rewarded or punished accordingly, even though this cannot be done costlessly. In this model, there is simply no room for the yes man. The market will continually root out this behavior, as it is incompatible with ascertaining the facts on the basis of which consumers can be satisfied.

Two, Prendergast is right: the world (or at least a significant part of it) is essentially causeless; nothing can be predicted; "there is no obvious (or even nonobvious) metric by which to measure whether a worker's report is correct" (p. 758). Here, the only thing to go by is the opinion (not "finding") of the manager. But in this case, it is perfectly "reasonable" to engage in yes-man behavior. This is because of an implicit premise that the boss knows more than the worker and that, therefore, the way to increase productivity is to somehow train the latter into being able to mimic the former. That is to say, "yessing" now, paradoxically, becomes productive.[11]

It is rather disingenuous of Prendergast, therefore, to claim that "the incentive to conform implies inefficiencies . . . and is likely to lead to more centralized decision making than in the absence of the desire to conform" (p. 757). The incentive to conform implies inefficiencies, but only in the real world–the one in which market forces continually grind down all such tendencies. In her artificial world, conformity with the boss does not at all imply inefficiency; rather, the opposite. Nor is it true in the model she has constructed that conformity "will lead" to more centralized decision-making (e.g., the boss will do things his way, and ignore the advice of the workers). Instead, this is logically implied by her model. It follows from the fact that there are no independent

---

keys; and, just as the last thing the pianist wants is keys with "individuality" that "think for themselves," so too does the conductor strive to promote "yesmanship" amongst the orchestra members. (This, of course, does not apply directly to soloists, an entirely different and more complex matter.)

[11]It becomes productive solely due to the force of Prendergast's unrealistic assumptions.

criteria upon which an opinion different from the managers could possibly be justified.

From this statement, also, it would appear that the desire to conform is exogenous. This is indeed true in the real world, where the market is busily engaged in stamping it out, exactly as it treats all other forms of antiprofitable behavior (shirking, cheating, stealing, etc.) But in the Prendergast world, "yessing" is not exogenous. On the contrary, it again follows from the "fact" that there are no independent standards with which to judge the worker's decision-making. All that can be done in this case is to rely on the manager's opinion— and on the ability of the worker to ferret it out of him and replicate it.

There is a third reply to Prendergast as well.[12] The not-too-implicit implication of her theory is that government intervention is justified in order to eradicate, or at least reduce, the inefficiency of the market failure of yesmanship. After all, given the market-failure literature, to characterize something as a member of that category is to call for the state to come in and fix it.[13] But it is within the bowels of government where the real yes-men problem lies. Here, there is no automatic feedback mechanism of the market to rely upon, to quell any incipient tendencies in the direction of yes-manning. At best they are extraordinarily weak; at worst they are nonexistent.[14] Thus, far from Prendergast being able to make her point that the yes man phenomenon is a market failure, the truth of the matter is that it is a government failure. If market firms rely upon or are involved in yesmanship, they lose money and tend to go bankrupt; if and to the extent that governments become embroiled in this economic virus, the forces for their dissolution are either very much attenuated, or totally absent.

REFERENCES

Block, Walter, and Robert W. McGee. 1989. "Information, Privilege, Opportunity and Insider Trading." *Northern Illinois University Law Review* 10(1): 1–35.

Brickley, James A., Clifford W. Smith, Jr., and Jerold L. Zimmerman. 2001. *Managerial Economics and Organizational Architecture*. 2nd Ed. New York: McGraw Hill Irwin.

---

[12]This was suggested to me by an anonymous referee. I thank him for it.

[13]See Cowen (1988).

[14]True, the Public Choice School posits that the political process constitutes a feedback mechanism that would be able to ameliorate yes-man tendencies; see on this Buchanan and Tullock (1971), Dauterive and Barnett (1984), Grier and McGarrity (1998), Holcombe (1986), McGarrity and Lloyd (1995), O'Brien and Logan (1989), Sutter and McGarrity (2000), DiLorenzo (1988). The ballot-box vote, however, is a very imperfect substitute for the dollar vote. The former occurs, typically, once every four years and is in effect limited to a package deal offered by two not dissimilar political parties; the latter occurs every day and has no such limits. Further ruining the analogy, the political process is based on compulsion, the market on voluntary interaction.

Buchanan, James M., and Gordon Tullock. 1971. *The Calculus of Consent: Logical Foundations of Constitutional Democracy.* Ann Arbor: University of Michigan Press.

Cowen, Tyler, ed. 1988. *The Theory of Market Failure: A Critical Examination.* Fairfax, Va.: George Mason University Press.

Dauterive, Jerry W., and William Barnett. 1984. "The Economics of Special Interest Groups: An Alternative Framework." Paper presented at the Annual Meeting of the Southwestern Society of Economists, San Antonio.

DiLorenzo, Thomas J. 1988. "Competition and Political Entrepreneurship: Austrian Insights into Public Choice Theory." *Review of Austrian Economics* 2:59-71.

Foss, Nicolai Juul. 1994. "The Theory of the Firm: The Austrians as Precursors and Critics of Contemporary Theory." *Review of Austrian Economics* 7(1): 31-65.

Grier, Kevin, and Joseph McGarrity. 1998. "The Effect of Macroeconomic Fluctuations on the Electoral Fortunes of House Incumbents." *Journal of Law and Economics* 41(1): 143-61.

Hayek, F.A. 1948. "Socialist Calculation I, II, & III." In *Individualism and Economic Order.* Chicago: University of Chicago Press.

Holcombe, Randall, G. 1986. "Non-optimal Unanimous Agreement." *Public Choice* 48:229-244.

Kirzner, Israel M. 1973. *Competition and Entrepreneurship.* Chicago: University of Chicago Press.

Klein, Peter G. 1996. "Economic Calculation and the Limits of Organization." *Review of Austrian Economics* 9(2): 3-28.

———. 1999. "Entrepreneurship and Corporate Governance." *Quarterly Journal of Austrian Economics* 2(2): 19-49.

Lachmann, Ludwig. 1977. "The Role of Expectations in Economics." In *Capital, Expectations, and the Market Process.* Kansas City: Sheed Andrews and McMeel.

———. 1986. *The Market as an Economic Process.* New York: Basil Blackwell.

Lewin, Peter, and Steven E. Phelan. 1999. Firms, Strategies, and Resources: Contributions from Austrian Economics. *Quarterly Journal of Austrian Economics* 2(2): 3-18.

Manne, Henry A. 1966a. *Insider Trading and the Stock Market.* New York: Free Press.

———. 1966b. "In Defense of Insider Trading." *Harvard Business Review* 113 (Nov/Dec).

Manne, Henry A., Joseph McGarrity, and Robert Lloyd. 1995. "A Profit Analysis of the Senate Vote on Gramm-Rudman." *Public Choice* 85:81-90.

O'Brien, J. Patrick, and Robert R. Logan. 1989. "Fiscal Illusion, Budget Maximizers, and Dynamic Equilibrium." *Public Choice* 63:221-35.

Prendergast, Canice. 1993. "A Theory of 'Yes Men.'" *American Economic Review* 83(4): 757-70.

Stigler, George. 1961. "The Economics of Information." *Journal of Political Economy* 69.

Sutter, Dan, and Joseph McGarrity. 2000. "A Test of the Structure of PAC Contracts: An Analysis of House Gun Control Votes in the 1980s." *Southern Economic Journal* 67(1): 41–63.

Cyberslacking, Business Ethics and Managerial Economics

Walter Block

ABSTRACT. Often, new technology brings in its train unprecedented problems. As far as computers, e-mail and the internet are concerned, this certainly holds true in many arenas. But there is one aspect of this new technology which does not present additional difficulties: cyber-slacking. The managerial challenges posed by employees using these amenities for job search, shopping sprees, personal relationships, in a word, general goofing off, have long ago already been overcome by employers. There is "nothing new under the sun" in at least this one dimension of the computer age.

Nothing new under the sun

According to some pundits, cyberslacking presents a new and unique managerial problem. They wax on eloquently to the effect that this is a new and never before encountered challenge for managers. It is the contention of the present article, in sharp contrast, that "there is nothing new under the sun" in this regard; that each and every so called new obstruction presented by our new ways of communicating with each other electronically has been successfully faced by businessmen in the past.

According to Friedman (2000, p. 1562), "Cycberslacking involves visiting pornographic sites and news sites, shopping, stock trading, vacation planning, gaming, chatting, in other words, engaging in general non-business Internet activities on company time and using company resources." In addition, this practice includes looking for a new job on the internet, comparing present salaries and working conditions with that available elsewhere purely as a matter of curiosity, doing homework on company time, exchanging e-mail with friends, etc., etc.

But there is nothing on this list of activities which was unknown before the advent of personal computers, e-mail, the web and all the rest. Pornography, surely, was a staple of pre "modern" days, with calendars of naked women adorning the walls of many offices and factories. How many comedic sketches on TV made use of a man reading, ostensibly, a business related journal, while inside was tucked something very different? As for "news sites," has no one ever seen an employee reading old fashioned "news sites," e.g., newspapers and magazines, while on the job? Yes, in days of yore one could not shop while still at the desk or workbench. But an employee could peruse advertisements in the aforementioned newspapers and magazines, clip them out, and plan on a shopping spree. Stock trading was accomplished in the days of our working grandparents by telephoning a stock broker while on company time. Vacation planning, too, was not unknown in the days of typewriters. Chatting, forsooth, was a staple around the water cooler (this was before the appearance of designer bottled water for those readers unfamiliar with this appliance). As for gaming, the computers have no monopoly over chess, checkers or solitaire. Similarly, there were always want adds for those looking for a new position, and job comparison shopping; homework could be done with pen and ink

Walter Block is Full Professor in, and Chairman of, the Department of Economics and Finance at the University of Central Arkansas, Conway AR 72035; wblock@mail.uca.edu, 501 450 5355. He has published numerous articles in business ethics, economics, environmentalism, and is the author of the book Defending the Undefendable.

 *Journal of Business Ethics* **33**: 225–231, 2001.
© 2001 *Kluwer Academic Publishers. Printed in the Netherlands.*

while at work, and exchanging e-mail with friends had its parallels with snail mail, courtesy of the post office.

Friedman (2000, pp. 1562–1563) claims to find a disanalogy in the fact that "there are indirect costs as well: increased Internet use, beyond what is necessary for the business, requires (sic) purchase of additional bandwidth and consumption of unneeded resources. Increased Internet usage brings security problems as well. The more web sites visited unnecessarily... the greater the exposure to viruses . . . Corporate intelligence is also at risk . . ."

But all of these phenomenon have their counterparts in the pre computer era. Writing letters to friends on typewriters or by hand used up stationary and typewriter ribbons, or pens, and added to the wear and tear of these writing implements. If bygone era employees were not "taking care of business," security problems could also arise. A company with excessive goofing off on the part of its workforce would be subject to loss of business intelligence and exposed not to viruses, but to theft, pilferage, etc. Conceivably, the modern problem may be more costly than before, but this is surely a difference of degree, not kind.

Morality

Is cyber (or any other kind of) slacking on the job immoral? Is it akin to theft?

Friedman appears to be of two minds on this. On the one hand, he (2000, p. 1562) characterizes such acts as "indirect theft," and states (2000, p. 1563), "A fair extension of the notion of stealing to embrace more than property, but in fact anything whatsoever that is of value to an individual or organization, is in order. The time employees misuse, but for which they are compensated, as well as the monetary value of Internet access privileges via company equipment are properly considered objects of theft. Something of economic value has obviously been stolen."

On the other hand, Friedman (2000, p. 1563) gives it as his opinion "If, however, the apparent Internet addict somehow still renders value to the company, perhaps even as a result of the cyber-activity, which might foster subconscious problem solving or provide a necessary break from drudgery or intense creative endeavor; there is obviously then no swindle." He even (2000, p. 1564) goes on to positively characterize "the widespread tendency to use the web for healthy relaxation or occasional business, yes, even for moderate cyberslacking. Productivity might actually increase with such a corporate culture."

So, which is it? Theft, or increased productivity? The problem with putting the matter in this way is that the answer could well be "both." This is because the two of them are not contrary to each other. That is, it is like asking of an object, which is it?, round or blue, as if the answer had to be one or the other. Rather, there are two entirely separate questions involved in cyber slacking. In order to make sense of this phenomenon, we must consider them separately.

Whether it is theft or not depends, entirely, upon the contact in force, whether explicit or implicit. If the employer posts signs on the walls to the effect that time wasting (whether of the cyber or old fashioned variety) will be considered theft, and offenders will be fired, and stipulates this in all its employment contracts, there is an easy answer to the question. Then, yes, cyber slacking is explicit theft in such contexts. If not, then it is not explicit theft.

What about implicit theft? This, too, depends upon the contract in force. By implicit contract I mean, paradigmatically, a situation where we go into a restaurant, order and eat a meal, and then, when presented with the bill protest on the ground that we were never given a meal contract to sign beforehand. The menu, I claim, constitutes an implicit contract to pay the price stated. Similarly, in the other direction, if a man goes into a fast food chain, and orders a burger, there is an implicit contract which can only be overridden explicitly, as to price. For example, if, after he has consumed the burger he is presented with a bill for $1 million (no menu has been presented to him, there are no prices posted on the wall), this should not be upheld in any rational court of law, on the ground that there was no explicit contract which superceded the implicit one for a "reasonable" price.

Consider as an early slacker Smith's (1776, p. 9) famous (at least within the realm of economics) boy who, bored with his job, placed a string in such a manner so that it would do the job assigned to him, so that he could go out and play:

"In the first fire-engines, a boy was constantly employed to open and shut alternately the communication between the boiler and the cylinder, according as the piston either ascended or descended. One of those boys, who loved to play with his companions, observed that, by tying a string from the handle of the valve which opened this communication to another part of the machine, the valve would open and shut without his assistance, and leave him at liberty to divert himself with his play-fellows."

Is this boy a hero, or is he to be condemned by his employer for job neglect and theft? It all depends. If there is an explicit contract in force prohibiting such "slacking," then the latter, if the employer wishes to pursue it. If not, then it is not theft (or services). Management may of course fire an employee[1] who acts in this manner; however, he should not be counted as guilty of theft since, for one thing, the job he was hired to do was being done, albeit not by him, but by the string he set up. There is another consideration as well. Whether theft of services occurred in the case of this inventive boy depends upon expectations, and there are no clear ones which apply to this situation. Here, as a matter of law, we must resort to the concept of innocent until proven guilty (Rothbard, 1990).

If a complete stranger goes to the premises of a business, and walks out with a typewriter under his arms without paying for it, this is as clear a case of robbery as we are ever likely to be presented with. However, if he is an employee of that firm and undertakes the exact same physical act, then, in the absence of any explicit contract between employer and employee, things are much less clear. Again, possibly, this should also be considered theft; after all, the typewriter, say, will have the owner's seal affixed to it. On the other hand there is always the possibility that the firm has inaugurated a policy whereby such implements can be borrowed by its workers either to engage in company business on its behalf while at home, or, even, purely for the personal enjoyment of the employee, as part of a fringe benefit. Context, expectations, implicit contracts – in the absence of explicit rules or contracts – all figure, heavily, in the legal analysis of such an act.

Hidden employer monitoring?

What of the legitimacy of an employer spying[2] on employees to reduce or eliminate their cyber (or other) slacking? Again, the same analysis holds. If there is in force an explicit contract or agreement between the two allowing or preventing this behavior, then our answer is clear: whatever is specified in this document determines the propriety or not of surreptitious surveillance.

Suppose, now, that there is no contract in force specifying such activity. Is it then legitimate? My claim, here, is that something of the same order concerning expectations now applies to employer spying: that there will be none, unless specifically allowed. That is to say, it is my reading of employer – employee relationships that in the absence of any specific agreement to the contrary, eavesdropping, hidden cameras, and other accoutrements of the cloak and dagger set would properly be interpreted as invasive. This applies to the employer secretly monitoring the modern day practice of web surfing as well as to the more traditional peeping, such as overseeing the lavatory use of unsuspecting employees.

Productivity

An entirely separate question from the legality or morality of cyber (or other) shirking is whether it improves productivity. Consider, again, Adam Smith's boy with the string. On the one hand, he has clearly increased production. For with a cheap piece of string, he has totally replaced himself, thus allowing the employer to replace him, and the economy to grow.[3] On the other hand, if the other boys witness such unpunished behavior, they will be inclined to disregard their duties; if they are unable to match the creative genius of the first boy (as will all too often be the case) productivity may well decline. There is no axiomatic way to determine which of these effects will be the stronger.

Take another case. Suppose there are two employees, both on an eight hour shift. One works steadily, all throughout the day. The other engages in either cyber or old fashioned slacking for four hours, and then works twice as hard, or efficiently, for the other half day. At the end of their labors, they have each produced exactly the same amount for their employer. Is this a plausible scenario? Certainly. We have all slacked off in some way or other, and then worked "twice" as hard, to make up for lost ground. Whether the goofing off or the inspired (or perspired) bout outweighs the other is impossible to say.

The bottom line, at least as a first approximation, is that inputs are largely irrelevant to success in business; outputs are pretty much all that count. That is, for all intents and purposes it doesn't matter much how much so called neglect of duty takes place at the work bench or assembly line; what is of vital importance, in contrast, is the quantity and quality of the goods and services which are forthcoming. Of course, this is not to deny that the two are hardly unrelated to one another; other things equal, the harder people work, the less they slack off, the more they produce. But the point is, other things are not necessarily equal. Being given a modicum of independence may or may not help increase the number of goods and services which come tumbling off the assembly line. It is a matter of managerial skill to be able to adopt the correct policy to any given business situation.

It is of course more than conceivable that additional final product can be had out of workers at the lower end of the skill distribution by severely penalizing dereliction of duty, while at the upper end, where more initiative and brainwork is required, the best policy on the part of the administration may be one of benign neglect; that is, to entirely ignore the issue of shirking for parts of the day, as long as the job gets done. But again, there is no guarantee that this should be the case, nor, even, that it would not be reversed (e.g., the brainier workers need more control) in some cases.

Yes, it seems obvious to some that cyber slacking would lead to lower productivity. Even the very term "slacking" seems to indicate this. One difficulty, however, with this way of looking at the matter is that it is mired in the Marxian labor theory of value. In this view, goods and services have value in accordance with how much labor has been inputted into them. But this is entirely erroneous, as can be seen by the fact that a mud pie, and a cherry pie, may take an identical amount of labor to create, but one is worthless, the other valuable.[4] Or, alternatively, that a good idea may come about through dint of hard work, or, while in the midst of doing what to outsiders might be considered a waste of time.

Another way of seeing the fallacy in supposing that "slacking" necessarily leads to poorer results is to borrow a leaf from the labor economist's analysis of fringe benefits or working conditions. Consider equation (1).

$$\$wage + wc = total\ wage \qquad (1)$$

It indicates that money wages (\$ wage) plus the amount spent on working conditions (typically, air conditioning, rugs, drapes, canteen, etc.) together comprise the total wage. For example, equation (2)

$$\$500 + \$100 = \$600 \qquad (2)$$

specifies a total wage of \$600, a money wage of \$500, and expenditures on working conditions of \$100.

One point to keep in mind is that the employer cares not one whit how the total wage is divided up into money wages and expenditures on working conditions; he only has eyes for the bottom line, the total wages he must spend on his work force. In order to see this, consider the following cases. First, assume the employees are a bunch of immigrants, working in the domestic country, but sending the lion's share of their earnings to their families abroad. To them, money spent on fringe benefits are fripperies, worth virtually nothing to them, while the size of the pay packet means all. To them, in effect, equation (2) implies a total wage of only \$500, since they ignore the $100 ostensibly spent for their creature "comfort." They have a marginal revenue productivity, in equilibrium, of $600, otherwise they would not be earning that amount in total. This means they would desert their present employer for a money wage of as little as $501 (or more). Under these conditions, a different firm employing them will be able to earn a pure profit of $600–$501, or $99 in the time period under consideration. This will act as a magnet, drawing other employers to reduce expenditures on working conditions, and increase take home pay. If there are enough of these workers to staff an entire operation, their present employer, if he does not mend his ways and allocate money more in line with the desires of his workers, is in dire danger of losing them all to a "raiding" corporation.

Second, assume that the employees are a bunch of rich yuppies, "beautiful" people, whose parents subsidize their earnings. They do not really need the $500 take home pay, and would be willing to accept only, say, $400, if the employer would spiff up the factory (personal trainers, happy hours, latte coffee machines, etc.) to the tune of $200. Any employer not catering to their wishes, again, risks losing them to a more accommodating competitor.

Another point to keep in mind is that cyber "slacking," for all intents and purposes, may be analyzed as if it were but one more aspect of working conditions. It matters little whether the working conditions money is spent on piped in music or painted walls or comfortable chairs for the employees, or on the encouragement of their "illicit" use of the internet while on company time.

When we put these two considerations together, we arrive at the insight that it is to the employer's best interest to tailor the level of cyber slacking, just as in the case of the other aspects of working conditions or fringe benefits, to the wishes of the employees. On the assumption that recreational use of the internet increases productivity, this conclusion follows necessarily,

since both worker satisfaction and the amount of goods and services forthcoming from the employees will increase. But this holds true even in the case where cyber slacking actually reduces goods and services produced, provided only that the workers value this more than they do the lost income to themselves.

Public policy?

How many free market economists does it take to change a light bulb? None. They leave it to market forces.

At the end of the day there is no right or wrong answer to the question of whether the profit maximizing strategy leads in the direction of quelling slacking, actually promoting it,[5] or being indifferent to it. As we have seen, it is entirely an empirical matter, with no principles at all unambiguously pointing in any direction or other. No broad managerial advice, then, will suffice for all business. Each must learn, through a trial and error process, which style works best. Happily, to the extent that there is a free enterprise system undergirding the efforts of all entrepreneurs, we can be assured that the appropriate techniques – whatever they are – will emerge and survive.[6]

A similar analysis applies to the issue of whether universities ought to adopt tenure as a managerial policy. Much heat but little light has emanated from both sides of this debate, each one intent on showing that tenure either provides flexibility and "academic freedom," on the one side, or intellectual arteriosclerosis on the other. The answer arising from this quarter is that there is no one solution to this problem. All that can be said is that if institutions of higher learning are all privatized, and if some adopt tenure and others not, then a market test will determine whether one policy is unambiguously better than the other, or whether profits can be earned, and the enterprise survive and prosper under both systems.[7]

Notes

[1] For the argument that management may fire workers "at will" provided there are no labor contracts in force preventing this, see Block (1991, 1996a, 1996b), Kauffman (1992), Petro (1957), Poulson (1982), Reynolds (1982, 1984, 1987), Hutt (1973, 1989).

[2] For an analysis of the legal environment in this regard, see Rosenberg (1999).

[3] On the benefits of labor saving innovations which make jobs redundant, see Hazlitt (1979).

[4] Marxists typically reply to such refutations of their theory with the claim that they mean "socially useful" labor, not any old labor such as that incorporated into a mud pie. But this renders their claim sterile and tautologous: the amount of socially necessary labor now becomes the result of value, not its cause. For a refutation of this Marxist doctrine, see Bohm-Bawerk (1959); see particularly his Part I, Chapter XII, "Exploitation Theory of Socialism-Communism."

[5] Some companies actually institutionalize slacking off, with happy hours, free or subsidized lunches, weight and exercize rooms, trainers, basketball courts, etc.

[6] All during the writing of this piece I have been playing solitaire with myself (physical cards, not electronic) and chess (electronic, against a computer) whenever I ran into an awkward spot, or had to rephrase my presentation, or just plain old needed a break. An outside observer might well have characterized my efforts as half work, half "slacking." Yet, for all that, I did complete the research and writing necessary for publication. I have no doubt that I could have done this more quickly had I stuck strictly to "business," but I am convinced that at least some of the insights informing the piece came to me while I was otherwise engaged in these ways, and allowed my subconscious to grapple with the difficulty of explaining myself in typed words.

[7] For further elaboration of this point see McGee and Block (1991).

References

Block, Walter: 1996, 'Labor Market Disputes: A Comment on Albert Rees' 'Fairness in Wage Distribution'", *Journal of Interdisciplinary Economics* **7**(3), 217–230.

Block, Walter: 1996, 'Comment on Richard B. Freeman's 'Labor markets and institutions in economic development'", *International Journal of Social Economics* **23**(1), 6–16

Block, Walter: 1991, 'Labor Relations, Unions and Collective Bargaining: A Political Economic Analysis', *Journal of Social Political and Economic Studies* **16**(4) (Winter), 477–507.

Bohm-Bawerk, Eugen: 1959 (1884), *Capital and Interest* (Libertarian Press, South Holland, IL). George D. Hunke and Hans F. Sennholz, trans.

Friedman, William H.: 2000, August, 'Is the Answer to Internet Addiction Internet Interdiction?', in H. Michael Chung (ed.), *AMCIS 2000: Proceedings of the 2000 Americas Conference on Information Systems*.

Hazlitt, Henry: 1979, *Economics in One Lesson* (Arlington House Publishers, New York).

Hutt, W. H.: 1989, 'Trade Unions: The Private Use of Coercive Power', *The Review of Austrian Economics* **III**, 109–120.

Hutt, W. H.: 1989, 'Trade Unions: The Private Use of Coercive Power', *The Review of Austrian Economics* **III**, 109–120.

Hutt, W. H.: 1973, *The Strike Threat System: The Economic Consequences of Collective Bargaining* (Arlington House, New Rochelle, NY).

Hutt, W. H.: 1973, *The Strike Threat System: The Economic Consequences of Collective Bargaining* (Arlington House, New Rochelle, NY).

Kauffman, Bill: 1992, 'The Child Labor Amendment Debate of the 1920's; or, Catholics and Mugwumps and Farmers', *The Journal of Libertarian Studies* **10**(2) (Fall), 139–170.

McGee, Robert W. and Walter Block: 1991, 'Academic Tenure: A Law and Economics Analysis', *Harvard Journal of Law and Public Policy* **14**(2) (Spring), 545–563.

Rothbard, Murray N.: 1990, 'Law, Property Rights, and Air Pollution', in Walter Block (ed.), *Economics and the Environment: A Reconciliation* (The Fraser Institute, Vancouver).

Petro, Sylvester: 1957, *The Labor Policy of the Free Society* (Ronald Press, New York).

Poulson, Barry W.: 1982, 'Substantive Due Process and Labor Law', *The Journal of Libertarian Studies* **VI**(3-4) (Summer/Fall), 267–276.

Reynolds, Morgan: 1982, 'An Economic Analysis of the Norris-LaGuardia Act, the Wagner Act and the Labor Representation Industry', *The Journal of Libertarian Studies* **VI**(3-4) (Summer/Fall), 227–266.

Reynolds, Morgan O.: 1984, *Power and Privilege: Labor Unions in America* (Manhattan Institute for Policy Research, New York).

Reynolds, Morgan O.: 1987, *Making America Poorer: The Cost of Labor Law* (Cato Institute, Washington DC).

Rosenberg, R. S.: 1999, 'The Workplace on the Verge of the 21st Century', *Journal of Business Ethics* (October), 3–14.

Smith, Adam: 1776/1965, *An Inquiry into the Nature and Causes of the Wealth of Nations* (Modern Library Canaan, New York).

Economics and Finance Department,
Harold E. Wirth Eminent Scholar Chair in
Economics,
College of Business Administration,
Loyola University New Orleans,
New Orleans, LA 70118,
U.S.A.
E-mail: wblock@mail.uca.edu

PATERNALISM IN AGRICULTURAL LABOR CONTRACTS IN THE U.S. SOUTH: IMPLICATIONS FOR THE GROWTH OF THE WELFARE STATE

Walter Block*

Abstract

Paternalism is a relationship where a person usually in a powerful position helps someone below him in the pecking order. Alston and Ferrie disparage this practice, and the author criticizes them for this unwarranted stance. The latter also maintains, vis a vis the former, that the mechanization of southern U.S. agriculture was not responsible for the rise of the welfare state in that country.

Key-word: Paternalism.

Resumo

O paternalismo é um relacionamento em que uma pessoa, geralmente em uma posição poderosa, ajude a alguém abaixo. Alston e Ferrie depreciam essa prática, e o autor os critica por causa dessa infundada posição. O autor também afirma que a mecanização da agricultura do sul dos EUA não foi responsável pela ascensão do estado de bem-estar nesse país.

Palavra-chave: Paternalismo.

* Harold E. Wirth Eminent Scholar Endowed Chair in Economics
6363 St. Charles Avenue, Box 15, Miller 321
College of Business Administration
Loyola University New Orleans
New Orleans, LA 70118
www.cba.loyno.edu
work: (504) 864-7934
secy: (504) 864-7944
fax: (504) 864-7970
email: wblock@loyno.edu

Perspectiva Econômica	Vol. 35	Nº 112	Outubro/Dezembro	2000	p. 81-93

82

1 – Introduction

Under the above title, Alston and Ferrie (1993, hereafter AF) set themselves two main tasks. First, explicitly, as the title of their paper indicates, they attempt to explain the growth of the welfare state. Their reasoning is that the rise of mechanization in southern agriculture in the mid 20th century supplanted the paternalistic labor market that had been in place there since the end of slavery in the mid 19th century. This, in turn, they maintain, reduced previous southern political opposition to the new welfarist dispensation, since the representatives of this region acted as agents of the rural elites, and they no longer feared substitutes for paternalism such as the welfare state. The line of causation, then, runs from exogenously introduced mechanization of southern farms to the end of the demand for paternalistic labor markets on the part of rural elites, to a cessation of opposition to national welfare programs which had been successfully opposed before mechanization, but not afterward.

The second task these authors set themselves is a far more implicit one: to disparage, criticize, demean, and otherwise call into question in every way they can the ancient and honorable practice of paternalism.

They offer a very interesting set of theses. The first is addressed to a very pivotal epoch in our history, the second to a very important philosophical and moral issue. Both thus deserve critical examination. This comment will put forth alternative explanations and criticize several aspects of their analysis, roughly in the order as presented in their paper.

Let us consider the second, minor claim first, before dealing with the major historical exegesis. AF are very bitter about paternalism. They never come out and explicitly castigate it per se on moral grounds (such sentiment would of course be wildly misplaced in a scholarly journal dedicated to value free economic analysis) but they do everything short of that to place this institution in an unfavorable light. For example, they wax eloquent about the evils of the social control imposed upon the black tenant farmers of the south, which is an element of paternalism. This is in sharp contrast to the outrages of the physically coercive abuses heaped upon these people which were not paternalistic, in that there was no claim, implicit or explicit, that they were for the blacks' own good.

83

2 – Paternalism

AF start off, properly enough, by defining "paternalism." In their view, it is "an implicit contract whereby workers exchange dependable labor services for a variety of goods and services" (852).[1] These latter consisted of "credit, housing, medical and old-age assistance,... protections from acts of violence" (852), "education, old age security, and welfare" (853), "improved housing, garden plots, firewood, and plantation schools and churches" (855), "payment of doctor's bills, the establishment and maintenance of schools and churches, ... various unspecified forms of entertainment, ... pa(yment of) legal fines ... parole sponsors" (856), "interceding in commercial transactions, obtaining medical care, providing influence or money to bail a son out of jail or settling family disputes" (857).

This rather narrow conception of paternalism concentrates solely on employment. In contrast, the dictionary (Webter's, 1957) defines the term more inclusively as "the principle or system of governing or controlling a country, group of employees, etc., in a manner suggesting a father's relationship with his children." Here, a strong element of benevolence is implied; at the very least, an act which is patently *male*volent could never be counted as paternalistic.

What are we to make of the following quotations, then, in this regard:

> *The planters increasingly offered protection to their faithful black workers as the social and legal environment became more hostile to blacks – a hostility which, over several decades, the white rural elite was instrumental in creating ... planters needed to ensure that no other party stepped forward to act as the workers' protector in commercial and legal dealings. In short, planters had an interest in maintaining a racist state and preventing federal interference in race and labor issues. (856)*
> *For much of the late 19th and early 20th centuries, individual Southern plantation owners had the local political influence to ensure the delivery of protection and, by the turn of the century, the collective political influence at the state level to create a discriminatory socio-legal environment from which they then offered dispensation. (858)*

It is patently obvious that whatever system is being described here it is *not* paternalism. Would a "father" treat his "children" in such a manner? Hardly; or only if he were an abusive parent, which does not at all capture the richness of the word "paternalism."

Nor is this a mere quibble about semantics. Rather, this constitutes the employment of the "straw man" argument. AF set out to castigate an institution, but instead of adhering faithfully to its actual characteristics, they manufacture an en-

---

1 – Unless otherwise noted, all page citations are to this one article, AF (1993). The author wishes to thank Mark Thornton for helpful comments.

84

tirely new set out of the whole cloth; since these are highly negative, and rather inflammatory, they attempt to color the institution of paternalism in their negative glow.

A far better description of what AF call "paternalism"[2] in the above citations would be "thralldom," "exploitation," or better yet, "protection racket." This is *precisely* what the Mafia is accused of doing: demanding "protection money" to guarantee safety from their own depredations. Alternatively, if AF insist on investing the word with this totally new pejorative meaning, all men of good will can readily agree with them that paternalism is a completely nefarious system. But this would be a phyric victory, indeed, on their part.

Let us now consider several of the other arguments put forth by AF in their critique of paternalism. These authors seem particularly exercised[3] by the verbal and bodily contortions (Scheflen, 1972) typically undergone by the reconstruction blacks forced to exhibit "racial etiquette": the tipping of the hat, the bowing and scraping, the avoidance of looking white women in the eye, etc. (865, fn 34). But this was enforced by threat of violence: blacks who failed to act in such a manner ran the risk of being lynched (Brundage, 1993; Cutler, 1969; Wright, 1990; Finkelman, 1992). It had nothing to do with what we may call "pure" paternalism: the paternalism of AF minus the violent aspects. Further, AF tells us that "Blacks were expected to show deference to whites in general under the system of social control, but in particular to employers who provided paternalistic benefits" (861). The difficulty, here, is that they vouchsafe us no evidence for the latter part of this claim. On the contrary, there was no great differential in the amount of "respect" a black was expected to show his employer, vis a vis other powerful whites. What *all* whites had in common in this sorry time of our history was the threat of violence for "disrespect." Paternalism *alone* did not garner any significant *additional* kow towing over and above that accorded other (especially powerful) whites. If the non violent part of what AF call paternalism were solely responsible, there would be no need for *any* excessive "respect" for other whites. That it was given nonetheless is surely strong evidence that what motivated blacks was the ever present threat of violence – not fulfilling their part of the so called paternalistic bargain.

This can easily be seen by considering other cases of paternalism which operated in a context totally free of threats of violence. For example, private char-

---

2 – Kleinig (1983, p. 5) does not conflate paternalism and the initiation of violence or threats thereof. He states: "... there are ... ways in which paternalistic limitations of liberty of action can be achieved, but where coercion need not be involved. If Ruritania restructures its aid program so that needy citizens are given food ... instead of cash, their liberty to spend is limited though no coercion takes place." Similarly, McLaurin (1971, p. 16) criticizes the view put forth by AF: "... there was an essential difference between the plantation and the factory, between the slave or tenant farmer and the mill operative. Physical force, not paternalism, kept the Negro on the plantation."

3 – And properly so, from my own moral perspective.

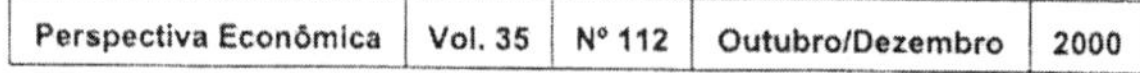

Perspectiva Econômica	Vol. 35	Nº 112	Outubro/Dezembro	2000

85

ity. The Salvation Army and the Shriners are world renowned for providing for needy people; this is "pure" paternalism, utterly divorced from even a hint of threatened violence. The same applies to I.B.M. and many Japanese firms. They act paternalistically (Bennett, 19xx), but engender no excessive deference, because they offer no threats of physical violence.[4]

Another problem with the AF analysis stems from their overly narrow conception of paternalism; they see it as intrinsically bound up with the employee – employer relationship. But there is more to paternalism than can be found in their philosophy. To wit, the governmental welfare system they see as a substitute for paternalism is itself an embodiment of that very institution. In AF's view, such government programs as Social Security, the Farm Security Administration (860), Great Society programs (865) and the "welfare state" (866) are not themselves paternalistic, but are instead substitutes for paternalism.[5] This is false. Welfare recipients may not have to demean themselves in the exact same manner as perpetrated on reconstruction blacks, but they do have to go through several embarrassing hoops (Murray, 1984; Piven and Cloward, 1971) in order to collect the funds allotted to them. Sometimes this reality is subjected to an ill disguised veneer. For example, calling welfare recipients "clients" is a transparent way to try to hide the facts of the case. But the subservient status of those on the dole is nonetheless a reality.

It is interesting to compare the murder rate under the modern paternalism (the welfare state) much preferred by AF, to the number of lynchings under the paternalism of reconstructionism they much excoriate. Contrary to the expectations engendered by their view, the later period compares very unfavorably with the earlier.[6] Say what you will about the evils of paternalism of the south at the turn of the century, but the average person – to say nothing of the average black male – had a better chance of living to a ripe old age – at least as far as being victimized

---

4 – Other cases of deference occur under different sorts of "threats." We defer to the postman otherwise he will "lose" our mail; we defer to the clerk at the Motor Vehicle Bureau or we will be forced to wait, interminably, at yet another queue; we defer (tip) the cab driver out of fear that he will make a fuss if we don't; the waitress has the additional threat of poor service; similarly, the baggage handler at the airport can send our luggage to the wrong city, the garbage man can spread refuse in front of our homes, and the full professor can play havoc with the tenure decision of the assistant professor. All of these cases are prime candidates for "excessive deference."

5 – They state: "... paternalism disappeared in the late 1950's and early 1960s" (860).

6 – According to Wright (1990, Appendix A, pp. 307-323) there were only 353 lynchings in Kentucky between 1866 and 1934, most of the victims, but not all, being blacks. In contrast, there were 251 murders in that state in 1991 alone (Statistical Abstract of the U.S., 1993, tables 31, 302). Brundage (1993, Appendix A, pp. 262-263) reports that in 1880-1930 19 whites and 461 blacks were lynched in Georgia and 15 of the former and 71 of the latter in Virginia for these years. In the single year of 1991, almost 900 people were killed in Georgia, and more than 550 in Virginia (Statistical Abstract of the U.S., 1993, tables 31, 302).

86

by murder is concerned – under planter paternalism than under the variety practiced by the welfare state.[7]

But we have not yet mentioned all sorts of other instances of paternalism which are not objectionable in any way, manner shape or form. Professional boxers, football and basketball players, very successful singers, dancers, musicians and movie stars – and innumerable others who earn vast wealth at a very young age – are often put on an allowance by their agents, accountants, business managers (Kleinig, 1983, p. 55-58). This is done totally volitionally,[8] but it is nevertheless paternalistic, in that these high income earners form a child-adult relationship with their financial advisors. In this regard, too, we must count that legion of other professionals who treat us almost as children: psychologists, psychiatrists, social workers, marriage counselors, clergy, spiritual advisors – all of them minister to our (child-like) psyches. They take on a quasi-parental role with us, their charges. They have their counterparts with regard to our bodies: dieticians, weight trainers, golf coaches, etc. These cases all depict a paternalistic relationship – one that is entirely voluntary. We hire our "parents"; we can fire them at will. They all tell us what to do – for our own good!

The problematic nature of the system described by AF as paternalistic is accounted for by the "thralldom" "exploitation," "protection racket" aspects. When these are removed, what still remains of paternalism is unobjectionable.

AF also take great umbrage at "plantation stores or designated stores in the county or town" which supplied "in-kind goods" (865). They imply that there was something intrinsically problematic about such commercial arrangements, but this is clearly not the case. Many department stores, super markets, restaurants and other such commercial establishments have established fringe benefits for their employees in the form of discounts, gifts of merchandise sold in the store, etc. (Licht, 1991, p. 63). Should this practice be interpreted as exploitative, and intrinsically so? Hardly.

They wax eloquent about the indignities heaped upon black shoppers: "merchants did not permit blacks to try on clothing ..." (865), but many stores fol-

---

7 – It is interesting to compare the two paternalisms, state and private, with regard to competitiveness and accountability. Clearly, the latter is preferable to the former in this regard. In each case, if satisfaction is not given, people can "vote with their feet." That is, they can move from state to state or from plantation to plantation. However, the latter is far larger than the former, and thus harder to negotiate. Further, if a state suffers out migration, there is no one in a position to do anything about it - governor, mayor, etc., – who loses money thereby. If this occurs to a plantation, the owner suffers almost immediately, as he cannot function without a work force.

8 – This applies to *adults*; child movie actors are of course treated paternalistically – by their parents – but this is a separate issue.

87

low such policies.[9] They bemoan the lack of "discretion over where .. the black tenants and croppers ... shopped" (865), and single out the "commissary" (866) for condemnation, but numerous institutions follow similar practices, with no obvious attendant problems. For example, guests on cruise ship lines and hotels (at least in the Catskill Mountains of New York State) are totally dependent on the relevant "commissary" during their stays. In the former case, there may be no source of food for hundreds, thousands of miles. Yes, violence and the threat thereof can indeed be used to exploit people, and blacks in the south for 100 years after the civil war were indeed victimized. But, contrary to AF, this had nothing to do with institutional arrangements concerning "commissaries," "plantation stores" or "designated stores."

There is no doubt that blacks were subjected to numerous indignities in those days, eloquently attested to by AF. But this was logically unrelated to paternalism. To see this, reflect on the fact that Jews and orientals were also treated in harsh, unfeeling and intolerable ways (Sowell, 1983, 1994), but all in the total absence of paternalism. Yes, blacks were treated badly, and also engaged in paternalistic contracts. But neither implies or is implied by the other.[10]

Then there is the issue of share cropping, "share contracts and paternalism" (864). Contrary to AF, there is nothing intrinsically exploitative about such commercial arrangements either Cash, 1954, p. 174). At the very least, many people who have been "victimized" by them would be very surprised to learn that they had been at all exploited. For example, many large legal and accounting firms award shares to each of the (senior) partners. The support staff, and the junior professionals, are ofttimes on fixed salaries. Many fishing boats follow similar procedures. For example, the captain receives 10 shares, the mate 5, the skilled hands 3 and the cabin boys 1. Sometimes professional athletic teams share the proceeds of post season games, sometimes awarding injured players partial shares. Some large industrial firms engage in profit sharing with their workers (Raff, 1991). This is all part and parcel of residual income claimancy; only instead of a clear delineation between the risk taking entrepreneur and the risk avoiding employees, as in the usual case, sharing, or share cropping blurs the line.

To summarize the points made in this section let us by all means join AF in looking askance at the protection rackets perpetrated on the blacks by the rural

---

9 – True, they do not usually disciminate on the basis of race in these policies, at least in the modern era, but rather prohibit *all* customers from trying on clothes. In this case, the blacks of whom AF speak were no worse off than all customers in the present day. For an economic analysis of racial discrimination, see Block, 1992) and Epstein (1992).

10 – Friedman (1994) wrote scathingly of the paternalistic practices of the Herschey Chocolate company in Herschey, Pennsylvania – the company store, payment in kind, lack of shopping alternatives, and other practices of single industry towns – along much the same lines as AF. But as long as these locational choices were voluntary on the part of the employees of this firm, it is hard to see how any such policies can be coherently objected to.

88

white southern elites. But AF throw out the paternalist baby with the coercive statist bathwater. A clearer distinction between these two very different phenomena is needed.

3 – Causal explanation

We begin this section on a philosophical note. AF propose a causal chain. First, mechanization leads to a reduction in paternalism; this, in turn, leads to a rise of the welfare state.

How are we to interpret this explanation? Is it a general law of economics,[11] similar to the claim that rent control leads to a reduction in the quantity and quality of rental housing, or that the minimum wage law causes teen and other unskilled unemployment? Or does their exegesis amount to a one time once and for all accidental concatenation of events, such as the assertion that in Cleveland, in 1982, the demand elasticity for roast beef was 10%? If it is the latter, it is well nigh worthless as an *explanation* of events; at best it can be a description, or the uncovering of a coincidence. But if the former, it must pass the test of generalizability; it must exhibit a modicum of lawfulness. That is, we must be able to verify the proposition by ascertaining whether other instances of the rise and fall of paternalism lead to similar results. If so, this is evidence in behalf of their thesis; if not, then against. We must also compare their interpretation with others which have been put forth to account for the rise of the welfare state.

For example, let us contrast the situation of blacks in the south with native Americans on Indian reservations (Prucha, 1985). This is a reasonably good analogy insofar as paternalism is concerned, for if there is a paradigm case of this phenomenon, it is probably the reservation system. As much as paternalism may have characterized the relationship between black freedmen and plantation owner south of the Mason Dixon line, this applies even the more to that between the Bureau of Indian Affairs (BIA) and those people unfortunate enough to be located on reservations.[12] If the phenomenon focussed on by AF is more than a mere coincidence, then, we could confidently expect similar occurrences in this case. To wit, we would look forward to an initial resistance to the welfare state on the part of the BIA, followed up by acceptance, or at least by a reduction in the op-

---

11 – See Hoppe (1991, 1992), Mises (1957).

12 – With regard to accountability, the reservation Indian is in a worse position than the (freed) plantation black, because there is more effectiveness in the competition in the private sector than in the public.

position. In the event, of course, no such situation has ensued, which tends to disparage the AF thesis.[13]

Naturally, in historical analysis, one can not hope for definitive breakthrough explanations which must, by the force of their logic, be accepted by all and sundry forthwith. Instead, we are forced to resort to what AF reasonably call "circumstantial evidence" (861).[14]

In this regard we have already posited some half dozen incidents of paternalism which were not linked with anything of the sort posited by AF; this certainly cannot be counted as evidence in their behalf. Moreover, their historical claim is certainly a counterintuitive one, since most explanations of the rise of the welfare state are very far removed indeed from southern agricultural concerns.

Most commentators, for example, give responsibility for the birth of the modern welfare state to the efforts of Karl Marx, or to those of the Fabians. Piven and Cloward (1971), cited favorably several times by AF (865, 869), offer a revisionist perspective very much inconsistent with their own. They account for this system as an attempt to quell the incipient tendency of the poor to riot by throwing money at them when they were restive, and withdrawing it when they were not. Then there is the left wing revisionist story offered by Kolko (1963) which maintains that big business interests in the North and East (not the South) saw welfarism as a means of solidifying their control over the economy. There are also those who work in the tradition of what might be called right wing or Chicago revisionism (Krueger, 1974; Posner, 1975; Tullock, 1967; Buchanan, Tollison and Tullock, 1980) which would explain the advent of this system in terms of "rent seeking," the rise in the power of social workers and other such professionals. But we have barely begun to explore the wealth of theories of the welfare state. Any reasonably complete listing must also include the thoughts of Kristol (1978), Raimondo, (1993), Schaffer (1991). It must be counted as a weakness of the AF presentation that they not only eschew criticizing these alternatives to their own theory but fail to even mention them. Nor do AF apply their theory to other countries. Yet, if it is a new general rule they have discovered, as opposed to a coincidence, it would apply there too.

In addition to the failure to criticize alternative hypotheses, there are several weaknesses in the account of the welfare state offered by AF itself. For one thing, it is predicated upon a very well functioning political process, or "political marketplace." That is to say, AF claim that the Southern contingent to the Congress were the agents of the rural elites, accurately reflecting their wishes and doing their bid-

---

13 – Another analogy concerns victimless crimes such as gambling, drugs, tobacco. U.S. history is replete with greater and lesser amounts of paternalism being imposed on those who engage in these activities. Yet these waves of changes have not given rise to phenomena analogous to the ones posited by AF.

14 – This is akin to Mises' (1957) distinction between theory and history.

90

ding:[15] "After the Civil War and especially after the disfranchisement of blacks and poor whites, Southern Democratic congressmen viewed the rural elite as their constituents" (858). It is hard to reconcile this claim with the fact that they were only subject to recall every two or six years, respectively, for House and Senate members, and that in any event they were judged on so much more than reduction in opposition to the onrushing welfare state. It is difficult to see why Congressmen from the Southern states would have paid a political penalty if they had maintained their resistance to the Great Society. Many did, with no apparent loss in power or prestige.[16]

Even the interpretation placed by AF on the voting behavior of these politicians is problematic. They construe opposition to labor unionism, to increasing socialism, and their defense of states' rights and free enterprise, as paternalism and racist imperialism. For example, "That role (as paternalists) ensured the opposition of planters to federal interference in Southern labor and race relations in the first half of the 20th century" (853, material in brackets added by present author). It does not seem to have occurred to AF that there could be other, more responsible reasons, even ideological ones, for opposing the leftward march of the Great Society.

Another difficulty arises with regard to their explanation on the basis of increasing mechanization. If increasingly sophisticated machinery lead to a reduction of paternalism and hence lowered the previous barriers to the welfare state during the mid 20th century, it is undeniable that automation has increased at an even faster pace between that time and the present. Given the causal explanation offered by AF, the demand for the welfare state must be even stronger in the modern day. How, then, can we account for *opposition* to the statist welfare system throughout the U.S., and in the south – perhaps the most conservative area in the nation (Cash, 1954; Reed, 1992) – in particular?

A further point in this vein: mechanization does not seem to be a necessary part of the process. The proportion of tenants to all householders in New York City, for example, is very high compared to other cities. Many landlords act paternalistically toward them, helping them out with jobs, loans, obtaining bail for their children, etc., in much the same manner as in the time period discussed by AF. Yet increasing mechanization plays no role in this process; nor has it lead to any appreciable increase in the demand for government welfare; if anything, the reverse is true on the part of landlords, a very conservative group of people. This further indicates that the phenomenon discussed by AF may be more of a coincidence than part of an economic law.

---

15 – Recent economic literature is replete with numerous cases of "market failure." So popular is this refrain that it is unnecessary even to cite it. In contrast, the "political marketplace," curiously, seems unaffected by any sort of "failure" at all. We live in curious times.

16 – One party government was the rule of the day in Southern states in this epoch, and that this would have attenuated the competitiveness of the "political marketplace" even more.

91

In addition to the foregoing, there are several other analytic problems which mar their presentation:

State AF: "... the South's judicial system displayed a clear bias, meting out sentences to blacks in the South far more severe than those given for corresponding crimes in the North" (855). There may be bias against blacks involved here, but if so, AF fail to show it. Suppose for example that for a given crime, the punishment for blacks in the South is 10 years and that it is 5 years for blacks and whites in the North. This case would appear to fit AF's charge, but if the punishment for this crime for southern whites is also 10 years, there is no discernable bias against blacks. Rather, the South merely metes out more punishment than the North,[17] but both could do it in a completely color blind manner.

AF see the existence of "farm-specific knowledge" entirely too much in a one sided direction. They make the point that this "gave landlords an incentive to curb the migration of tenants with such knowledge" (857). To be sure, they are correct in this contention. But the obverse holds as well: the existence of "farm-specific knowledge" also gives tenants an incentive to curb their own outmigration, since their marginal revenue product is by definition higher on their original farm than on any other to which they might move. Similarly, if it is true that "The advances in science that accompanied mechanization increased and stabilized yields, making the farm-specific knowledge of tenants less valuable" (861), then not only did this "reduce... the economic incentive to provide paternalism" (860) it *also* gave impetus for tenants to leave such places. This is because the value of the human capital specific to the Southern plantation is shared by both employer and employee (see Becker, 1964).

4 – Conclusion

AF offer us a very intriguing set of hypotheses: the welfare state is due to the fall in the supply of paternalism on the part of rural Southern elites due to mechanization, and a good thing too, since the welfare state is a vast improvement over the old system.

In contrast, I have tried to show several weaknesses in their analysis; that southern plantation paternalism, albeit harmful to the blacks who suffered under it, has certain distinct advantages over its governmental counterpart; that the objectionable part of this paternalism was the admixture of violence and threats; that in their absence, "pure" paternalism (like voluntary trade) must necessarily benefit both parties to it, otherwise one or the other would not agree to take part.

---

17 – The encarceration rate in prison for the South is 267.0, per 10,000 population 18 years and older. For the Midwest it is 145.5, for the Northeast 119.2 and for the West 152.6. Statistical Abstract of the U.S., table 347.

92

Bibliography

Alston, Lee J. and Ferrie, Joseph P., "Paternalism in Agricultural Labor Contracts in the U.S. South: Implications for the Growth of the Welfare State", *American Economic Review*, Vol. 83, No 4, September 1993, pp. 582-76.

Becker, Gary, *Human Capital*, New York: The National Bureau of Economic Research, 1964.

Bennett, John, *Paternalism in the Japanese Economy*, 19xx.

Block, Walter, "The Economics of Discrimination", *The Journal of Business Ethics*, Vol. 11, 1992, pp. 241-254.

Brundage, W. Fitzhugh, *Lynching in the New South: Georgia and Virginia, 1880-1930*, Chicago: University of Illinois Press, 1993.

J. M. Buchanan, R. D. Tollison and G. Tullock, eds, *Towards a Theory of the Rent Seeking Society*, College Station: Texas A&M University Press, 1980, pp. 51-70.

Cash, W. J. *The Mind of the South*, Garden City, N.Y.: Doubleday, 1941.

Cutler, James Elbert, *Lynch Law: An Investigation into the History of Lynching in the U.S.*, New York: Negro Universities Press, Greenwood, 1905, 1969.

Epstein, Richard A., *Forbidden Grounds: The Case Against Employment Discrimination Laws*, Cambridge: Harvard University Press, 1992.

Finkelman, Paul, *Lynching, Racial Violence, and Law*, 1992.

Friedman, Milton, "Paternalism in Herschey, PA," *Wall Street Journal*.

Hoppe, Hans-Hermann, "Austrian Rationalism in the Age of the Decline of Positivism," Ebeling, R., ed., *Austrian Economics: Perspectives on the Past and Prospects for the Future*, Hillsdale, MI.: Hillsdale College Press, 1991.

Hoppe, Hans-Hermann, "On Praxeology and the Praxeological Foundation of Epistemology and Ethics," Herbener, J., ed., *The Meaning of ludwig von mises*, Boston: Dordrecht, 1992.

Kleinig, John, *Paternalism*, Totowa, N.J.: Rowman and Allanheld, 1983.

Kolko, Gabriel. *Triumph of Conservatism*. Chicago: Quadrangle Books, 1963.

Kristol, Irving, *Two Cheers for Capitalism*. New York: Basic Books, 1978.

Krueger, Anne, "The Political Economy of the Rent-Seeking Society," *American economic review*, June, 1974, Vol. 64, pp. 291-303

Licht, Walter, "Studying Work," in Jacoby, Sanford M., ed., *Masters To Managers: Historical and Comparative Perspectives on American Employers*, New York: Columbia University Press, 1991.

McLaurin, Melton Alonza, *Paternalism and Protest: Southern Cotton mill Workers and Organized Labor, 1875-1905*, Westport, CT: Greenwood, 1971.

Mises, Ludwig von, *Theory and History*, New Haven: Yale University Press, 1957.

Murray, Charles, *Losing Ground: American Social Policy from 1950 to 1980*, New York: Basic Books, 1984.

Piven, Frances (sic) Fox, and Richard A. Cloward, *Regulating the Poor: The Functions of Public Welfare*, New York: Random House, 1971.

Posner, Richard, "The Social Cost of Monopoly and Regulation". *Journal of political economy*, august 1975, Vol. 83, pp. 807-827.

Prucha, Francis Paul, *The Indians in American Society*, Berkeley: University of California Press, 1985.

Raff, Daniel M.G, "Ford Welfare Capitalism," in Jacoby, Sanford M., *Masters to Managers: Historical and Comparative Perspectives on American Employers*, New York: Columbia University Press, 1991.

Raimondo, Justin, *Reclaiming the American Right*. Burlingame CA: Center for Libertarian Studies, 1993.

Reed, John Shelton, "The Mind of the South and Southern Distinctiveness," in Eagles, Charles W., ed., *The Mind of the South*, Jackson, Miss: University Press of Mississippi, 1992.

93

Schaffer, Ronald, *America in the Great War: The Rise of the Welfare State*, Oxford: Oxford University Press, 1991.

Scheflen, Albert E, *Body Language and the Social Order*. Englewood Cliffs, N.J., 1972.

Smith, John David, *Anti-Black Thought, 1863-1925*, vol. 2, New York: Garland, 1993.

Sowell, Thomas, *The Economics and Politics of Race: An International Perspective*, New York, Morrow, 1983.

Sowell, Thomas, *Race and Culture: A World View.* New York: Basic Books, 1994.

Tullock, Gordon, "The Welfare Cost of Tariffs, Monopolies and Theft," *Western Economic Journal* (now *Economic Inquiry*), June 1967, Vol. 5, pp. 224-232.

Webster's New World Dictionary of the American Language, Encyclopedia Edition, New York: World Publishing, 1957.

Wright, George C., *Racial Violence in Kentucky, 1865-1940: Lynchings, Mob Rule and "Legal Lynchings,"* Baton Rouge: Louisiana State University Press, 1990.